# THE SCIENCE OF FLIGHT

## A Gateway to New Horizons

"Custom Edition for the US Air Force
Junior Reserve Officer Training Corps (AFJROTC)"

## JONES & BARTLETT
## LEARNING

*World Headquarters*
Jones & Bartlett Learning
5 Wall Street
Burlington, MA 01803
978-443-5000
info@jblearning.com
www.jblearning.com

Jones & Bartlett Learning books and products are available through most bookstores and online booksellers. To contact Jones & Bartlett Learning directly, call 800-832-0034, fax 978-443-8000, or visit our website, www.jblearning.com.

Substantial discounts on bulk quantities of Jones & Bartlett Learning publications are available to corporations, professional associations, and other qualified organizations. For details and specific discount information, contact the special sales department at Jones & Bartlett Learning via the above contact information or send an email to specialsales@jblearning.com.

**Editorial Credits**
High Stakes Writing, LLC, Editor and Publisher: Lawrence J. Goodrich
Department of the Air Force Editor: Linda F. Sackie
Editor, HSW: Karen Annett
Primary Writer, HSW: Katherine Dillin

**Production Credits**
Chief Executive Officer: Ty Field
President: James Homer
SVP, Editor-in-Chief: Michael Johnson
SVP, Chief Technology Officer: Dean Fossella
SVP, Chief Marketing Officer: Alison M. Pendergast
SVP, Curriculum Solutions: Christopher Will
Associate Production Editor: Tina Chen
Manufacturing and Inventory Control Supervisor: Amy Bacus
Associate Photo Researcher: Lauren Miller
Text Design: Anne Spencer
Cover Design: Kristin E. Parker
Composition: Mia Saunders Design
Illustrations: Morales Studios, Norton Graphics
Cover Image: **Navigation tools:** © Anthony DiChello/ShutterStock, Inc.; **Weather map:** Courtesy of NOAA; **Parachuters:** Courtesy of US Marine Corps photo/Lance Cpl. Reece E. Lodder; **Thunderbirds:** Courtesy of USAF/Tech. Sgt. Justin D. Pyle
Chapter Opener Background Credits: **Pages 2–3,** © yuyangc/ShutterStock, Inc.; **Pages 110–111,** © AbleStock; **Pages 214–215,** Courtesy of US Marine Corps/Lance Cpl Reece E. Lodder; **Pages 260–261,** © haveseen/ShutterStock, Inc.
Lesson Opener Background Credits: **Pages 4–5,** Courtesy of NASA; **Pages 18–19,** Courtesy of Library of Congress, Prints & Photographs Division [LC-DIG-ppprs-00574]; **Pages 34–35,** Courtesy of NASA Langley Research Center (NASA-LARC); **Pages 54–55,** © Christopher Parypa/ShutterStock, Inc.; **Pages 68–69,** © Laser143/Dreamstime.com; **Pages 92–93,** Courtesy of NASA; **Pages 112–113,** Courtesy of NOAA; **Pages 132–133,** © Zaporozhchenko Yury/ShutterStock, Inc.; **Pages 150–151,** Courtesy of the National Museum of the USAF; **Pages 168–169,** Courtesy of USAF; **Pages 192–193,** Courtesy of NASA; **Pages 216–217,** Courtesy of USAF/MSgt Kevin J. Gruenwald; **Pages 240–241,** Courtesy of NASA/JPL; **Pages 262–263,** © Photodisc; **Page 280–281,** Copyright © NAS Fort Lauderdale Museum; **Pages 302–303,** Courtesy of USAMHI; **Pages 320–321,** © Daboost/ShutterStock, Inc.; **Pages 336–337,** © Guy Shapira/ShutterStock, Inc.
Printing and Binding: Courier Companies
Cover Printing: Courier Companies

Some images in this book feature models. These models do not necessarily endorse, represent, or participate in the activities represented in the images.

ISBN: 978-1-4496-3065-2

Library of Congress Cataloging-in-Publication Data
The science of flight : a gateway to new horizons. -- 1st ed.
       p. cm.
   Includes index.
   ISBN 978-1-4496-3065-2
1.  Aeronautics. 2.  Meteorology in aeronautics. 3.  Flight--Physiological aspects. 4.  Aids to air navigation.
   TL570.S36 2012
   629.13--dc23
                                      2011046049
6048

Printed in the United States of America
16 15 14 13 12    10 9 8 7 6 5 4 3 2

# Contents

# Airplanes FLY

## Chapter Outline

"It is possible to fly without motors,
but not without knowledge and skill."

*Wilbur Wright*

# LESSON **1**

## ✈ Principles of Flight

### ✎ Quick Write

_____

_____

What kinds of things do you think Chuck Yeager had to learn to accomplish all that he did?

### ✦ Learn About

- the theory of flight
- airfoils and flight
- Newton's laws of motion and aircraft design
- Bernoulli's principle, airfoils, and flight
- the effect of relative wind on flight
- the effect of angle of attack on flight

**S**ometimes you set out to do one thing in life—but life has other ideas. That's what happened to a young West Virginian named Chuck Yeager.

The son of a poor oil driller, Yeager loved tinkering with things. He could take apart an engine and put it back together with no problem. He joined the Army rather than the Navy because he thought the Army recruiter was more persuasive. He ended up as a mechanic in the US Army Air Forces.

The budding mechanic had no real interest in learning to fly. But in 1942 the Air Forces started a "Flying Sergeants Program" to train enlisted men as pilots. Seeing an opportunity to gain a promotion and a pay raise, Yeager enrolled.

After receiving his wings Yeager shot down an enemy plane in 1943 on his seventh mission. On his next sortie, however, his plane was shot down. He bailed out over France, where the French Resistance helped him escape to Spain. He made it back to England and back to his unit. Soon after, he shot down five enemy planes in one day, becoming an "ace in a day." He even shot down a Messerschmitt Me-262 jet with his propeller-driven P-51 Mustang.

After the war Yeager was stationed at Wright Field in Ohio as the maintenance officer. This involved flight-testing all the field's planes. His growing experience and obvious pilot skills caught the attention of superiors, who invited him to become a test pilot.

Yeager not only became a test pilot—he became the most famous American test pilot ever. Flying the X-1 experimental aircraft on 14 October 1947, he became the first human to fly faster than the speed of sound. In 1953 he flew 2.4 times the speed of sound in the X-1A. Near the end of the flight, the aircraft began spinning out of control and rotating on all three axes of flight—straight lines from nose to tail, from wingtip to wingtip, and from top to bottom. After spinning wildly for 50 seconds, Yeager regained control and landed safely. His story was told in the 1983 movie, *The Right Stuff*.

Yeager went on to have several further adventures in flying, including 127 combat missions in the Vietnam War. In 1969 the Air Force promoted him to brigadier general, making Yeager one of only a handful of men who had started as an enlisted Airman and risen all the way through the ranks to general. In 1976 President Gerald Ford presented him with the Congressional Medal of Honor for that first supersonic flight. President Ronald Reagan presented him with the Presidential Medal of Freedom in 1985—giving Yeager the highest military and civilian awards for individual service and achievement.

Yeager served on the Rogers Commission that investigated the crash of the space shuttle *Challenger* in 1986. On 14 October 1997, the 50th anniversary of his breaking the sound barrier, Yeager did so again in an Air Force F-15 fighter.

Chuck Yeager's story stands out as a lesson that people from simple beginnings can achieve great things. Who would have guessed that a humble young West Virginia mechanic who was unimpressed with airplanes would become one of the most famous flyers in the world?

Vocabulary

- theory
- aerodynamics
- airfoil
- camber
- chord line
- wind tunnels
- jet propulsion
- air pressure
- relative wind
- knot
- angle of attack
- pitch
- critical angle of attack
- stall

# The Theory of Flight

Humans flew in their imaginations long before doing so in reality. Ancient legends tell of winged horses and huge birds lifting people up into the skies. In these stories living men sprouted wings, and dying ones flew up to heaven in reward for leading heroic lives. Such stories stretch back thousands of years to the earliest civilizations.

## Getting Off the Ground

But the reality of human flight came long after the idea. What is perhaps the earliest recorded case of a human being successfully flying with the aid of a mechanism is found in the Chinese *Book of Sui* (AD 636). It relates that in the previous century, in AD 559, the tyrannical Emperor Gao Yang forced prisoners to attach themselves to bird-shaped kites and leap from a high tower. The record says that all died except one, a certain Prince Yuan Huangtou, who managed to stay aloft and land safely 1.5 miles (2.5 km) away. By the time of Marco Polo (thirteenth century), man-lifting kites were common in China.

Flight by balloon, or lighter-than-air flight, did not occur until more than 1,000 years after the *Book of Sui*. In 1783, the French brothers Jacques-Étienne and Joseph-Michel Montgolfier built the first hot-air balloon. They carried out initial experiments without passengers outside the French city of Annonay. On 19 September of that same year, the first balloon flight to lift animals took place at the Palace of Versailles. As the king, queen, and a huge crowd looked on, Joseph-Michel used a balloon to lift a sheep, a rooster, and a duck 1,500 feet (457 meters) into the air. All three landed without injury two miles (3.2 km) away. That same year, on 21 November in Paris, the first manned flight of a Montgolfier-designed balloon took place. The test pilots rose over the city on a flight that lasted 20 minutes.

As often happens with new technology, new achievements multiplied relatively quickly. Some 100 years later the first manned gliders were in use. In the first decade of the twentieth century, human beings were flying in propeller-driven airplanes. Less than 50 years later the first jets were carrying crews. In another 20 years the first astronauts reached the moon.

It took centuries of experimenting, of trying and failing, before people understood how things fly. They slowly had to develop a theory of flight. A theory is *a set of principles scientists use to explain something they have observed*, in this case the process of flight. In this lesson you will read about the basics of modern flight theory.

## Aerodynamics

**Aerodynamics** is *the way objects move through air*. Whether the object is a large airliner, a model rocket, a beach ball, a kite—or for that matter, a car or even a skyscraper—its aerodynamic characteristics affect its motion.

*Wing* TIPS

Aerodynamics comes from two Greek words: *aerios*, which means *air*, and *dynamis*, which means *force*.

## Airfoils and Flight

For an object to fly, you need a force to lift it. One object that creates this lifting force is an airfoil. An airfoil is *a structure—such as a wing or propeller blade—that when exposed to a flow of air generates a force.*

It's important to know an airfoil's parts because their relationship to each other affects the way an aircraft flies. The force an airfoil exerts can be either a lifting force, as in the case of a wing, or a force that pushes or pulls the object—as in a propeller blade. It can also be a combination of the two—as in a helicopter propeller—that both lifts and pulls.

The front edge of an airfoil is called the *leading edge*. This edge cuts through the air ahead. The rear edge is the *trailing edge*. The typical airfoil has a rounded leading edge and a sharp trailing edge (Figure 1.1).

In general, an airfoil's upper and lower surfaces curve. *The curve of an airfoil is its* camber. In most airfoils the camber is greater on the upper surface than on the lower.

*The straight line between the foremost and hindmost points of an airfoil viewed from the side is called the* chord line. The distance between these two points is the *chord length*.

When all these elements of an airfoil come together in just the right way, the airfoil will produce a lifting force when it's propelled through the air.

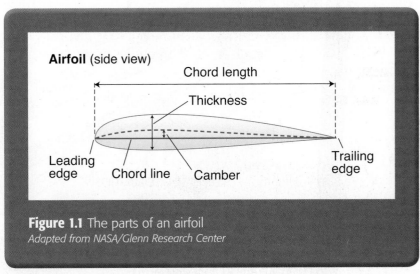

**Figure 1.1** The parts of an airfoil
*Adapted from NASA/Glenn Research Center*

## Airfoils and Wind Tunnels

Aircraft designers—including the Wright brothers—use wind tunnels to test models of new aircraft. Wind tunnels are *controlled spaces for testing airflow over a wing, aircraft, or other object.* The Wrights built one to test airfoils after their 1901 glider failed to produce enough lifting force. They tested more than 200 wing shapes in designing their successful 1902 glider. That glider led directly to the 1903 powered *Wright Flyer.*

In a wind tunnel, a researcher can carefully control the airflow conditions, which affect the forces on an aircraft. By carefully measuring these forces on the model, the researcher can predict how these forces will affect the full-size aircraft. Using a wind tunnel, the Wrights discovered which airfoil shape produced the most lifting force.

Modern wind tunnels are usually designed for a specific purpose and speed. The tunnel may be open and draw in air from outside, or it may be closed and recirculate air inside the tunnel. Figure 1.2 shows a closed tunnel viewed from above. In this tunnel the fan on the far right side continuously moves the air counterclockwise around the circuit.

But airfoils and lifting force are only part of the theory of flight.

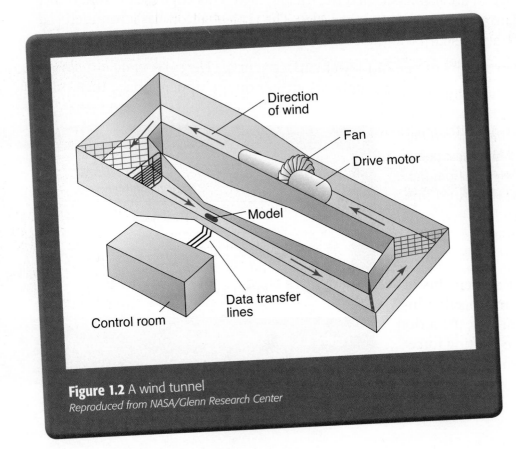

**Figure 1.2** A wind tunnel
*Reproduced from NASA/Glenn Research Center*

**CHAPTER 1**   How Airplanes Fly

# The Wright Way

The Wright brothers didn't just walk out onto the dunes at Kitty Hawk, North Carolina, and launch the first controllable, self-propelled, heavier-than-air machine. They spent years in orderly research and development before that December 1903 flight.

- First, *they researched how things fly*. They checked all the available literature to learn everything known about flight at the time. (And they didn't have the Internet to help them!) They wrote to the Smithsonian Institution in Washington for technical papers on flight. They contacted other experts. They observed large gliding birds. They decided that controlling the flying machine would be the hardest problem to solve.

- Second, *they developed flight control systems*. They tested their theories on a series of gliders (unpowered aircraft). They realized that a flying object had to be controlled along three axes. They developed movable surfaces to control these movements.

- Third, *they tested their ideas*. When their ideas about aircraft control didn't match their calculations, they began to doubt the data they were using. So they developed a wind tunnel, along with techniques to test their airplane models, to more accurately measure the forces affecting a flying object. This led to very successful experiments on their 1902 glider.

- Fourth, *they developed test-pilot skills*. As the first pilots of powered, heavier-than-air flight, they had no teachers or flight schools where they could polish their techniques. They learned their skills in more than 1,000 test flights at Kitty Hawk between 1900 and 1902. From 1903 to 1905 they continued learning on a series of powered aircraft.

- Fifth, *they developed supporting technology*. They built a lightweight engine and propellers to push their aircraft through the air. They developed a unique propeller design using their wind tunnel. After their first flight, they continued to improve the engine, making it more and more powerful.

- Sixth, *they kept on experimenting*. Their longest flight that first day lasted less than a minute and included no maneuvers. But they didn't stop there. They continued to perfect their machine with a series of new aircraft built between 1903 and 1905. This allowed them to stay up for half an hour, to fly figure eights, and even to take passengers for a ride.

## Sir Isaac Newton (1643–1727)

Sir Isaac Newton changed the world. Born in England in 1643, Newton was only an above-average student. But he went home from Cambridge University one summer in 1665, thought a lot about the physical nature of the world, and came back two years later with a revolutionary understanding of mathematics, gravitation, and optics. One of his professors was so impressed by what Newton had done that he resigned his own position at Cambridge so Newton could have it. Newton's calculus provided a new mathematical framework for the rapid solution of whole classes of physical problems. His law of gravitation explained in one simple formula how apples fall and planets move. His laws of motion allowed scientists to predict the movements of physical bodies with unprecedented precision. Newton's insights proved to be so overwhelmingly powerful that he was the first scientist ever knighted for his work.

Sir Isaac Newton
© iStockphoto/Thinkstock

## Newton's Laws of Motion and Aircraft Design

Sir Isaac Newton's three laws of motion will help you understand more about how planes move through the air. They explain forces that result in movement, lack of movement, and acceleration. While discovered back in the seventeenth century, these laws are fundamental to our understanding of flight today.

### Newton's First Law

Newton's *first law* of motion states that *a body in motion tends to stay in motion in a straight line, and a body at rest tends to stay at rest*, unless some outside force causes the moving object to stop, to change direction, or to change speed—or causes a stopped object to move. This is sometimes called the *law of inertia*.

You experience Newton's first law when you're riding in a car and the driver slams on the brakes. Your body is in motion and would keep going right into the dashboard if you weren't wearing your seat belt. The seat belt exerts a force that stops your body from moving. (That's why cars have seat belts and why you should wear them!)

Getting back to airplanes, an aircraft would simply stand still on the ground without some outside force—provided by propellers or jet engines—moving it. When the outside force moves the aircraft fast enough to provide enough lifting force, the airplane takes off and flies.

## Newton's Second Law

Newton's *second law* explains how much force is needed to cause an object to move faster—*to accelerate*. According to this law, *an object's acceleration (a) is proportional to the sum of all forces (F) exerted on it*. Mathematically, you can write this law in two ways:

*Force = mass × acceleration, or F = ma*

*Or*

*Acceleration = Force / mass, or a = F/m*

A newton is a common unit of force. (One newton equals 0.102 kilograms [0.225 lbs.] of force.) Assume that your bicycle has a mass of 9.1 kilograms. You accelerate at a rate of 1.79 meters per second. To find out the net force accelerating the bike, you multiply the mass (9.1) by the acceleration (1.79) and get a net force of 16.289 newtons.

One way to think about this is that the same mass will accelerate twice as fast if you apply twice as much force to it. The other way is to think about how the same amount of force will affect a light object and a heavy one. Suppose you first attached a bottle rocket to a toy boat and lit the fuse. When the rocket fired, the small boat would zip away from you, accelerating rapidly. But a heavy boat powered by the same rocket would accelerate more slowly. In other words, for a given amount of applied force, there is more acceleration when there is less mass and less acceleration when there is more mass. The second law expresses this tradeoff mathematically.

If you think about aircraft, this formula implies that a heavy plane with a small engine would need a long runway because it would take a long time for it to reach takeoff speed. In fact if you know the weight of a plane and how fast you would have to accelerate to reach takeoff speed on a runway of a particular length, you could use Newton's formula to calculate exactly how powerful the engine would have to be to supply enough pushing force to accelerate for takeoff.

## Newton's Third Law

Newton's *third law* is often stated like this: *"For every action there is an equal and opposite reaction."* This means that if one body applies a force to a second body, then the second body applies an equal force in the opposite direction to the first body. An obvious example in aviation is jet propulsion—*a driving or propelling force in which a gas turbine engine provides the thrusting power*. A jet engine applies a force to gas particles it expels in its exhaust. According to Newton's third law, the ejected gas particles apply an equal force in the opposite direction on the jet engine. Because the engine is connected to the aircraft, this force propels the aircraft in a direction opposite to that of the gas particles.

## Bernoulli's Principle, Airfoils, and Flight

Another very important principle for understanding flight is Bernoulli's principle, named after mathematician Daniel Bernoulli. According to this principle, an increase in the rate of airflow causes a decrease in air pressure. Air pressure is *the force exerted by the air on a unit area of surface*. In other words when air moves faster, its pressure drops. For example when air flows through a pipe, the pressure is lower in narrow sections of the pipe because the airflow is faster (Figure 1.3).

This principle allows you to grasp how an airfoil creates a lifting force simply by its shape. When air flows over the curved upper surface of a wing, it speeds up. This increase in speed reduces the pressure above the wing and produces the upward lifting force. The difference in air pressure above and below the wing depends on the wing's shape (Figure 1.4).

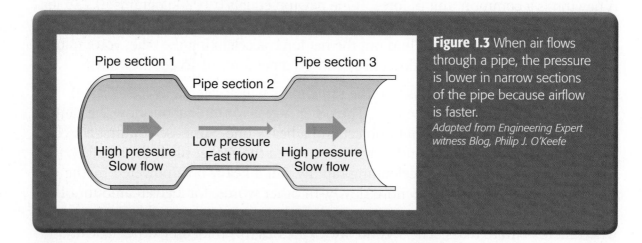

Pipe section 1    Pipe section 3

Pipe section 2

High pressure    Low pressure    High pressure
Slow flow    Fast flow    Slow flow

**Figure 1.3** When air flows through a pipe, the pressure is lower in narrow sections of the pipe because airflow is faster.
*Adapted from Engineering Expert witness Blog, Philip J. O'Keefe*

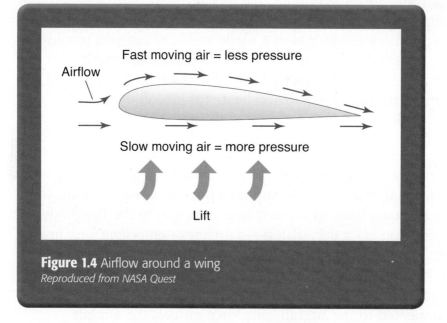

Fast moving air = less pressure

Airflow

Slow moving air = more pressure

Lift

**Figure 1.4** Airflow around a wing
*Reproduced from NASA Quest*

## Daniel Bernoulli (1700–1782)

Daniel Bernoulli was a Dutch-born Swiss scientist. Although early on he developed a fascination for mathematics and wanted to make it his career, his father insisted on medicine as a better way to make a living. Bernoulli did become a medical doctor, but never gave up the study of his first love, mathematics. In fact, after Isaac Newton's death in 1727, he became the foremost mathematician in Europe, and made many important contributions to science. In particular his formulas describing the forces acting on fluids remain fundamental to aeronautics even today.

Daniel Bernoulli
© INTERFOTO/Alamy

## The Effect of Relative Wind on Flight

The concept of relative wind, often abbreviated RW, is also key to understanding flight. Relative wind is *the motion of air as it relates to the aircraft within it*. Relative wind has both a speed and a direction. To determine relative wind, you ignore air disturbances at the aircraft's surface and consider only the movement of air at a distance from the plane.

To better understand the nature of relative wind, suppose that when you get into a parked car the wind is blowing from the side. Even though this is the case, if you stick your hand out of the window as the car begins to move, it will feel as if the wind is coming from in front of the car. And the faster the car goes, the more you will think so. This is because the movement of the air relative to the car is now much greater from front to back than from the side.

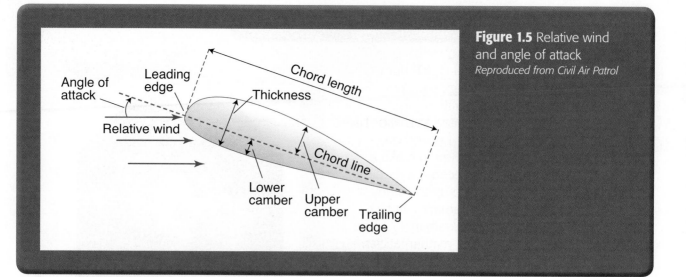

**Figure 1.5** Relative wind and angle of attack
*Reproduced from Civil Air Patrol*

The speed of the relative wind depends not only on an aircraft's speed, but also on the direction it is flying. For example suppose a plane takes off directly into a 20-knot headwind. (A knot is *a nautical mile, or 1.15 statute miles [1.85 km] per hour.*) If the plane's speed relative to the ground is 100 knots, then the relative wind will be 120 knots. In contrast, if a wind of the same speed were blowing from behind the plane, then the relative wind would have a speed of only 80 knots, if the plane's ground speed were unchanged at 100 knots. This explains why planes take off into the wind. The higher speed of the relative wind allows the plane to take off on a shorter runway because the greater relative wind speed decreases the air pressure above the wing and so provides more lifting force (Figure 1.5).

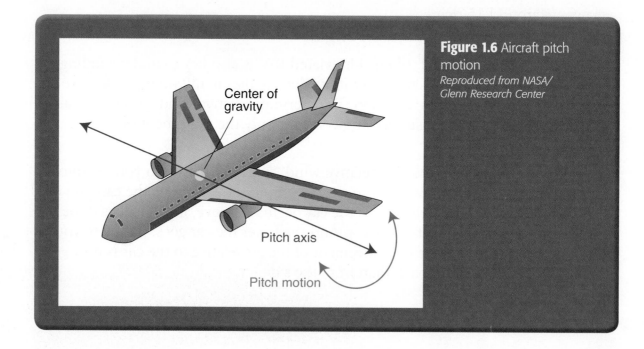

**Figure 1.6** Aircraft pitch motion
*Reproduced from NASA/ Glenn Research Center*

In general, when an airplane is in flight, the relative wind blows in nearly the exact opposite direction to the plane's direction. However, there is often some deviation, and sometimes the deviation can be large, as when the plane is moving slowly and there is a strong crosswind.

An aircraft doesn't have to be pointed in precisely the same direction that it is moving through the air. A plane may continue on the same straight line even when the pilot adjusts the controls to point the nose to the left or right of that line.

So by manipulating the controls, you can vary the angle at which the relative wind hits your plane. For a pilot, this is a critical piece of information.

## The Effect of Angle of Attack on Flight

*The angle between the direction of the relative wind and the chord line of an airfoil is the* angle of attack.

If a plane continues along the same line of motion, but alters its pitch— *the up-and-down movement of the plane's nose*—the angle of attack on its wings will change. Look at Figure 1.6. You will see that if the plane lowers its nose, the angle of attack will decrease. If the nose rises, the angle of attack will increase. This is important because changing the angle of attack alters the amount of lift a wing produces.

As the angle of attack increases, the wings generate more lifting force, but only up to a certain point—the critical angle of attack, somewhere around 15 degrees in most aircraft. The critical angle of attack is *the point at which a plane stalls.* A stall is *a rapid reduction in lifting force caused by exceeding the critical angle of attack.* This reduction is caused by the separation of airflow from the wing's surface (see Figure 1.7).

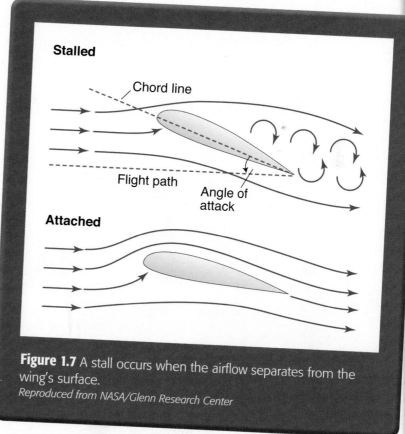

**Figure 1.7** A stall occurs when the airflow separates from the wing's surface.
*Reproduced from NASA/Glenn Research Center*

A stall can occur at any airspeed or pitch attitude. In a stall, controlling the plane becomes much more difficult, and in some cases, impossible. This is especially dangerous when an aircraft is close to the ground. Many airplanes have crashed because the pilot allowed the aircraft to exceed its critical angle of attack and stall.

These laws and principles you've read about in this lesson make up the basic theory of flight. You need to understand them in order to fly an airplane safely, to decide whether it is safe to fly in certain conditions, or to create a successful aircraft design.

But you'll need to know much more. It's important also to understand the various forces surrounding your aircraft and how you use them to get airborne and stay there. The next lesson will look more closely at the four forces of flight.

# ✔ CHECK POINTS

## Lesson 1 Review

Using complete sentences, answer the following questions on a sheet of paper.

1. In what year was the first manned balloon flight?

2. What is aerodynamics?

3. Which edge of an airfoil is rounded? Which is sharp?

4. What did the Wright brothers discover using a wind tunnel?

5. What is another name for Newton's first law?

6. What does $F = ma$ mean?

7. What is an example in aviation of Newton's third law in action?

8. Why does an airfoil's shape create a lifting force?

9. The speed of relative wind depends on the aircraft's speed and what else?

10. At about how many degrees does a plane reach its critical angle of attack?

## APPLYING YOUR LEARNING

11. Why does changing the angle of attack increase or decrease the lifting force?

# LESSON 2

## ✈ The Physics of Flight

### Learn About

- how lift is generated
- how weight affects flight
- how thrust affects an aircraft's movement
- how drag slows an aircraft
- how the four forces of flight interact with each other

**Y**ou don't often hear a lot about Octave Chanute (1832–1910), but he mentored many big names in early aviation—including the Wright brothers. This, although he never flew in a powered airplane. Chanute's great interest in flight led him to lecture on the subject, write books on its history, tinker with and design aircraft, and correspond with aviators. He served as a one-man clearinghouse for information on aviation discoveries.

The French-born American spent his career as a civil engineer working on railroads and bridges. Aviation was a hobby. He devoted himself to it wholeheartedly after retiring from the railroad business at age 57. Among his career achievements was designing the country's first railroad bridge over the Missouri River. He also invented a way to preserve wood for rail ties, specialized in iron bridges, and had a great knowledge of how to build a truss—a support structure—which he applied to bridges and later to his gliders.

Chanute's first book on aviation—*Progress in Flying Machines* (1894)—was invaluable to the Wrights and other aviators in their experiments. It was like an instruction manual; it examined the technical findings already made. The book also analyzed the errors of would-be aviators and so helped others correct them.

Chanute never patented any of his work. In fact he didn't approve of the Wright brothers' efforts to sell their powered aircraft. He argued that knowledge should be freely shared.

Even so the three men kept in close contact over the years, and Chanute saw their test flights numerous times in 1902, 1904, and 1905. He had conducted his own glider experiments in the 1890s from the Indiana dunes on Lake Michigan. He applied much of what he had learned building bridges to his glider construction, especially to wing structure. The biplane—an aircraft with two main supporting surfaces, usually placed one over the other—still uses his strut-wire-braced wing framework. He also laid the groundwork for aircraft control and stability.

The former Chanute Air Force Base in eastern Illinois (closed in 1993) was named after him. The Chanute Air Museum is still located there.

Octave Chanute
*Courtesy of USAF*

## How Lift Is Generated

Aircraft are subject to four forces: lift, weight, thrust, and drag. Newton referred to such forces in his first law (Figure 2.1): "Objects at rest remain at rest and objects in motion remain in motion in a straight line unless acted upon by an unbalanced force." These forces push and pull airplanes in numerous directions—up and down, back and forth, and side to side.

You can measure all four of these forces. They all have a magnitude— *a measurable amount, such as a size, speed, or degree.* For example, you can measure how fast (the magnitude) an aircraft is traveling.

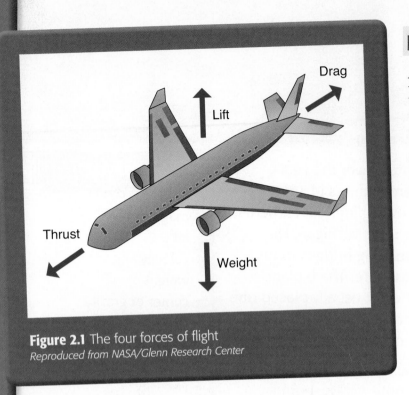

**Figure 2.1** The four forces of flight
*Reproduced from NASA/Glenn Research Center*

In order to safely fly, you must understand these four forces and their effect on your aircraft. This lesson will explain these forces in greater detail and how they interact with one another during flight.

## What Lift Is

In the previous lesson you read that to fly, you need a lifting force. Lift is *the upward force on an aircraft against weight.* At less than the speed of sound, you create lift by increasing the aircraft's forward motion and the airfoil's angle of attack—up to the stall speed.

Remember that an aircraft's motion through the air generates lift as the air passes over an airfoil, or wing. Therefore, lift is an aerodynamic force—one that deals with an object moving through air or air moving around an object.

Airfoils generate most of an airplane's lift, although the amount of lift an aircraft achieves also depends on the plane's shape, size, and velocity—*speed and direction.* The major role played in lift by airfoils explains why the Wright brothers spent so much time researching wing shape and size. As you can see from Table 2.1, the Ohio inventors continued to increase the wingspan's length and the wings' surface area to produce greater lift and control.

## How to Control Lift

When you pilot an airplane, you adjust lift with the controls that allow you to direct the angle of attack. If you need more lift, you can increase the angle of attack (only up to the critical angle of attack, though, because you don't want to stall). If you need less, you can decrease the angle of attack.

# The Wrights and Wings

The Wright brothers focused a lot of attention on developing the wings for their gliders. Over time they increased the wingspan (the distance from wingtip to wingtip) and the chord, increasing the wing area on their gliders. This gave their aircraft, including the powered 1903 *Wright Flyer*, more lift, stability, and control, as well as the ability to fly with more weight. Table 2.1 shows their steady progress.

| Gliders | Wingspan | Wing Area |
|---|---|---|
| 1900 glider | 17 feet | 165 square feet |
| 1901 glider | 22 feet | 290 square feet |
| 1902 glider | 32 feet | 305 square feet |
| 1903 *Wright Flyer* | 40 feet 4 inches | 510 square feet |

**Table 2.1** The Wrights' increasing wingspan

Orville Wright with the 1901 glider at Kitty Hawk
*Courtesy of Library of Congress, Prints & Photographs Division [LC-DIG-ppprs-00574]*

As mentioned earlier, you also need velocity for lift. Stated mathematically, lift is proportional to velocity squared. At low speeds, every time you double your velocity, you get four times as much lift. For example, an aircraft traveling at 200 knots has four times as much lift as an aircraft traveling at 100 knots, all other factors remaining constant.

Besides adjusting your angle of attack, you can also increase or decrease lift by changing your velocity. Your aircraft will climb if you keep a steady angle of attack and increase your velocity. It will also climb if you hold your velocity steady and increase your angle of attack. Therefore, to keep to a level, straight flight path, you must balance angle of attack and velocity.

Air density also affects lift. This density is determined by the air's pressure, temperature, and humidity. The higher you go, the less dense the air is. So at higher altitudes you must fly at a different airspeed for any given angle of attack to maintain lift.

Further, warm air is less dense than cool air, and moist air is less dense than dry air. So on a humid day in the middle of summer, you'll need to fly at a greater velocity to maintain lift than you will on a dry day in the middle of winter.

## How Weight Affects Flight

*The downward force that directly opposes lift* is weight. Weight is always directed toward the center of Earth because of the force of gravity. Weight acts through a single point called the center of gravity (Figure 2.2). The center of gravity is *the average location of an object's weight.*

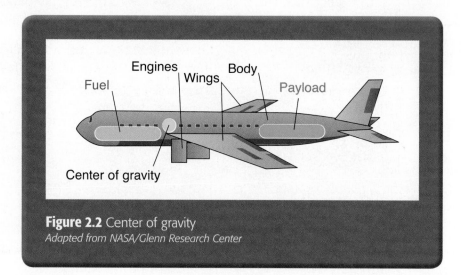

**Figure 2.2** Center of gravity
*Adapted from NASA/Glenn Research Center*

## Newton's Law of Gravity

When Sir Isaac Newton published his three laws of motion in 1686, he also introduced another of his important findings, the law of gravity.

The law of gravity explains the nature of the gravitational force and how it acts on objects on Earth, flying above Earth, and on objects in space. Newton stated his law as follows:

> Between every two objects there is an attractive force, the magnitude of which is directly proportional to the mass of each object and inversely proportional to the square of the distance between the objects' centers of mass.

The more mass an object has, the greater its attractive force. Earth has a tremendous attractive force compared with people and aircraft, or even the moon. It exerts a downward pull called *weight*. Distance also makes a difference. The closer an object is, the more pull it can exert on another object.

Unlike lift, and drag, weight is *not* an aerodynamic force. Weight is the gravitational force on an object due to its mass. The nature of these forces differs: aerodynamic forces are mechanical forces, meaning an aircraft must be in physical contact with air to generate them. On the other hand, an aircraft doesn't have to be in physical contact with Earth (the source of the gravitational force) for gravity to pull down on it.

The magnitude, or size, of an aircraft's weight depends on the aircraft's mass, made up of its parts, fuel, passengers, and freight. Remember that weight and mass are not the same thing: weight relates to gravity, but mass is a measure of inertia. This means the more mass an object has, the more difficult it is to overcome its inertia or resistance to moving.

Think of a cannonball versus a child's beach ball—they may be approximately the same size, but the cannonball is much more difficult to roll forward than the beach ball. This is because the cannonball has more mass than the beach ball. An object's weight varies depending on where it is in the universe. Its mass is always the same.

## Weight's Major Challenges

Weight poses two major challenges in flying: overcoming an aircraft's weight and controlling the aircraft in flight. You overcome weight with lift. As you increase the angle of attack and speed, you get more lift than weight and the aircraft climbs. Just the opposite would also be true: a lower angle of attack and speed decreases the lift. When the weight is greater than the lift, the aircraft descends.

Newton's second law, also explained in the last lesson, addresses the question of overcoming weight. A heavy aircraft requires more acceleration to take off (net force = mass × acceleration, or F = ma), while a lighter plane requires less acceleration.

Controlling the object's flight also requires handling the aircraft's weight and the location of its center of gravity. An aircraft's weight changes during flight because it consumes fuel. (You'll read more about the fuel cells in the next lesson.) The distribution of weight and its center of gravity also shift during flight, so pilots have to constantly trim—or *balance*—the aircraft. (Pilots have to trim the aircraft because of changes in speed, as well.)

## How Thrust Affects an Aircraft's Movement

Another force that affects flight is thrust. Thrust is *a forward force that moves an aircraft through the air*. Its opposite force is drag, *the pull or slowing effect of air on an aircraft*.

An AV-8B Harrier lands onboard the USS *Nassau* following a strike mission into Kosovo in 1999. Harrier jet engines can change the direction of thrust.
*Courtesy of USN/PH1 Richard Rosser*

## Where Thrust Comes From

Engines provide a powered aircraft's thrust. It is important to note that the engine's only job is to provide thrust; it does not lift an aircraft. The wings (airfoils) provide the aircraft's lift; thrust provides forward motion to overcome drag and allows the airfoils to produce lift.

The placement of engines, their number, and their type affect thrust. Some aircraft have the ability to change the direction of the thrust rather than changing the angle of attack. For instance, a Harrier jet can rotate its nozzles to direct thrust downward.

Thrust demonstrates Newton's third law, as illustrated particularly by jet engines. The hot gas pouring out the engine nozzle thrusts the jet forward. The engine pushes on the gas, and the gas pushes on the engine. This is the action and equal and opposite reaction Newton laid out in this law.

**Wing TIPS**

Engines aren't necessary for flight. Gliders don't have engines but can soar through the air. This is because their airfoils provide lift. Gliders do require an initial thrust, however, to get them into the air. This thrust comes from some other power source such as a tow to the skies from a powered airplane.

## Adjusting Thrust

Just as you can increase or decrease the amount of lift on your plane, you can also adjust the amount of thrust your engine produces. If you reduce your engine power while in level flight, your thrust drops and the aircraft slows down. As long as thrust is less than drag, your aircraft decelerates until its airspeed will no longer provide enough lift to support the plane in the air.

On the other hand, if you increase your engine power while in level flight, your thrust grows greater than drag and your airspeed increases. As long as this particular imbalance continues, you will accelerate. When the two forces are equal, you will fly at a constant airspeed.

You can maintain straight and level flight at a wide range of airspeeds. To maintain the balance between lift and weight and so maintain straight and level flight, you must properly coordinate thrust and angle of attack at any speed. Experts group flight into three main categories: low-speed flight, cruising flight, and high-speed flight (Figure 2.3).

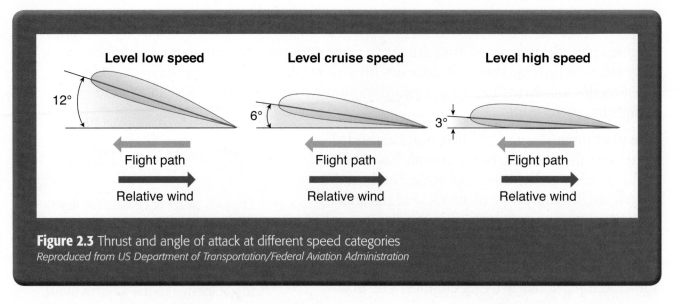

**Figure 2.3** Thrust and angle of attack at different speed categories
*Reproduced from US Department of Transportation/Federal Aviation Administration*

- *Low-speed flight*—When airspeed is low, you must keep the angle of attack fairly high. This greater angle of attack generates a lift force equal to the aircraft's weight. You'll be flying slowly but you can maintain level flight this way.

- *Cruising flight*—In this situation, straight and level flight occurs at a higher speed than during low-speed flight. During this steady flight, you must decrease your angle of attack as you increase thrust and airspeed.

- *High-speed flight*—When airspeed is high (high thrust), you decrease your angle of attack even more. Otherwise, you will continue climbing. You can even fly at a slightly negative angle of attack if your thrust is great enough. By decreasing the angle of attack, you bring lift and weight back in balance and your aircraft remains in level flight.

## How Drag Slows an Aircraft

Drag results from the air resisting an aircraft's forward motion. Drag pulls in the opposite direction of the flight path. An aircraft's shape, its speed, and the air's viscosity—*stickiness, or the extent to which air molecules resist flowing past a solid or other object*—determine the magnitude of the drag.

As an aircraft moves through the air, molecules stick to its surface. This *layer of air near a plane's surface* is the boundary layer. It actually changes the shape of the airflow over the aircraft. So now as the plane flies, the air flows not only around the aircraft body, but it flows around this new shape created by the boundary layer. This flow creates vortices—a vortex is *a kind of whirlpool of wind or water with a vacuum at its center*.

Sometimes when the angle of attack nears the stalling point, the boundary layer separates from the aircraft to create an even more unusual shape. In this way the boundary layer can produce unsteady flying conditions. The boundary layer also plays a large role in determining an object's drag (Figure 2.4).

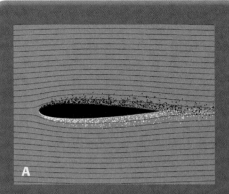

**A**

In the first simulation, the wing is moving at a small angle of attack (here taken as zero). Note that the boundary layer vortices remain close to the wing until they are washed downstream. At zero angle of attack, there is no lift, and there is little drag.

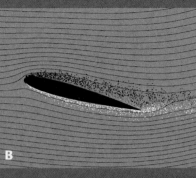

**B**

The wing has started pitching up, but the boundary layer vortices stay close to the wing. The wing is now producing significant lift force and still little drag.

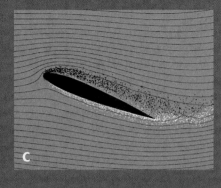

**C**

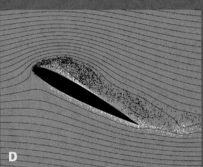

**D**

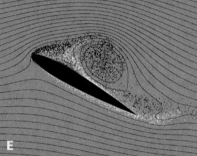

**E**

The angle of attack has become too large. The boundary layer vortices have separated from the top surface of the wing, and the incoming flow no longer bends completely around the leading edge. The wing is stalled, causing a significant drag. However, much of the lift remains because the separated vortices are still above the wing.

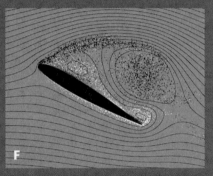

**F**

When the separated vortices are being blown past the trailing edge, the lift starts to drop off.

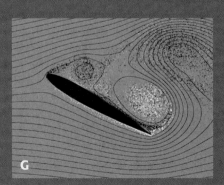

**G**

The wing is now producing little lift and lots of drag.

**Figure 2.4** This series of simulations shows how the boundary layer (the black-and-white dots) can separate from the wing and cause a stall.
*Courtesy of Leon van Dommelen, FAMU-FSU College of Engineering*

**Flat plate**

**Sphere**

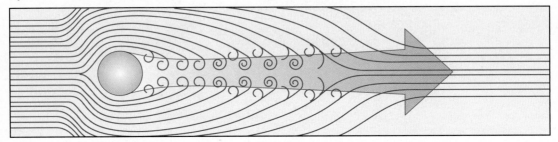

**Sphere with a fairing**

**Sphere inside a housing**

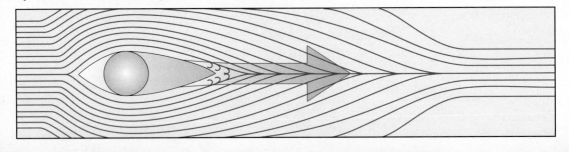

**Figure 2.5** Form drag
*Reproduced from US Department of Transportation/Federal Aviation Administration*

**CHAPTER 1** How Airplanes Fly

## Types of Drag

Drag comes in two main forms: *parasite drag* and *induced drag*. Parasite drag does nothing to help an aircraft fly. It does just what it sounds like—drains flight of needed energy. Induced drag, on the other hand, results from an airfoil developing lift.

### Parasite Drag

There are three types of parasite drag: form drag, interference drag, and skin friction drag. Antennas, engine covers (cowlings), and other items that determine an aircraft's shape create *form drag*. When air passes over an aircraft, it parts when it hits one of these objects, then rejoins when it gets to the other side. Think of a rock in the middle of a stream and how the water flows around the rock. How smoothly the air flows around objects on an aircraft determines how much form drag impedes flight. Engineers try to make aircraft and their parts as streamlined as possible to reduce form drag (Figure 2.5).

*Interference drag* is a bit more complex. When airplane surfaces meet at perpendicular angles (such as the body and wings) or other sharp angles, the air flows over each in different currents. When these currents collide, they create interference drag. Designers try to overcome this drag with such things as fairings (a fairing is a cover or structure used to reduce drag) and by putting some distance between airfoils and not-so-streamlined parts such as radar antennas.

*Skin friction drag* concerns air stickiness and the boundary layer. No matter how smooth an aircraft's "skin," when you get down to the molecular level there is always something for air molecules to stick to and create a boundary layer. Designers reduce skin drag by using flush mount rivets and removing anything that sticks up above the plane's surface. Polishing the surface of the plane also reduces drag. Pilots of smaller aircraft can fight skin drag by keeping their aircraft cleaned and waxed.

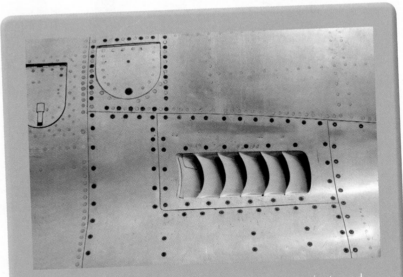

Designers reduce skin drag by using flush mount rivets and removing anything that sticks up above the plane's surface.
© Aaron Kohr/ShutterStock, Inc.

## Induced Drag

Because no work is 100 percent efficient, balanced flight always includes *induced drag*. The presence of induced drag means more energy must be expended to counteract it.

As you've read, airfoils produce lift by taking advantage of the energy of the airflow. Whenever you have lift, the area above the wing has less pressure than the area below the wing. The high-pressure area then flows up toward the low-pressure area on the wing's upper surface. When these two pressures meet at the wingtips, they equalize and shoot air in a lateral (sideways) stream toward the wings' upper surface. The resulting vortices are sometimes visible as you can see in the accompanying photo. If you ever spot wingtip vortices and you are viewing these swirling currents from the aircraft's tail, you will see that they spin clockwise at the left wingtip and counterclockwise at the right wingtip.

Wingtip vortices from a C-17 Globemaster III flying over the Atlantic Ocean near Charleston, South Carolina. The aircraft released flares and the smoke got caught up in the vortices.
*Courtesy of USAF/TSgt Russell E. Cooley IV*

**CHAPTER 1** How Airplanes Fly

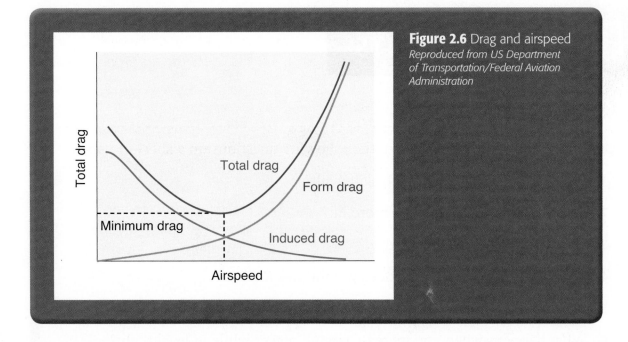

**Figure 2.6** Drag and airspeed
*Reproduced from US Department of Transportation/Federal Aviation Administration*

These vortices produce induced drag (Figure 2.6). The greater the vortices, the greater the induced drag. In a vortex the air flows above the tip and then comes back in a downwash behind the wing. This downwash pulls down and back on the aircraft. This rearward pull is induced drag.

Wingtip vortices can be very dangerous. They have caused or contributed to several aviation accidents.

## How the Four Forces of Flight Interact With Each Other

The motion of the airplane through the air depends on the relative strength and direction of the four forces. If the forces are balanced, the aircraft cruises at a constant velocity. If the forces are unbalanced, the aircraft accelerates in the direction of the largest force.

During steady, level flight the opposing forces are in balance. Simply put, the sum of the opposing forces of lift and weight are zero, and the sum of the opposing forces of thrust and drag are zero. Put yet another way, lift has an equal value to weight and thrust has an equal value to drag so they cancel each other out.

To maintain a constant airspeed, your thrust and drag must remain equal; for a constant altitude, your lift and weight must be equal.

In this lesson you've read about the relationship between the four forces of flight and flight stability. You must control these forces to get an aircraft aloft, remain there, and land safely. The next lesson will study the parts of an airplane and how they help you do that.

## Lesson 2 Review

Using complete sentences, answer the following questions on a sheet of paper.

1. What generates most of an airplane's lift?

2. What do you do if you need more lift?

3. Where is weight always directed?

4. What two major challenges does weight pose?

5. What is an engine's only job?

6. What happens when you increase engine power while in level flight?

7. Which three things determine the magnitude of drag?

8. Which two forms does drag come in?

9. What happens when the four forces of flight are *unbalanced*?

10. To maintain constant airspeed and constant altitude, what must happen?

## APPLYING YOUR LEARNING

11. If an aircraft traveling at 200 knots has four times as much lift as an aircraft traveling at 100 knots, how much more lift does an aircraft traveling at 400 knots have as an aircraft traveling at 100 knots?

# The Purpose and Function of Airplane Parts

## Quick Write

_____

_____

Write about another time in history when a major event spurred technological progress, and list three inventions that resulted.

## Learn About

- how the fuselage and wing shape correspond to an aircraft's mission
- the types, purpose, and function of airfoil design
- the role of stabilizers and rudders
- the positions of flaps, spoilers, and slats on an aircraft
- how the airflow and airfoil affect flight movement
- the purpose and function of propellers, turbines, ramjets, and rocket propulsion systems

**W**orld War II was a scramble among the Allied and Axis powers to build faster, more-lethal aircraft. Each side needed an edge. Aviation technology was still very new. Engine types taken for granted today were just being developed.

The Germans and British applied their attention to the jet engine during the war years because engineers calculated that such engines could go very fast and so would be useful in combat. In fact the jet engine, which uses hot exhaust to generate thrust, works best at high speeds, hundreds of miles per hour at a minimum.

After Hans von Ohain developed Germany's first jet engine, the Germans produced the world's first jet aircraft, the experimental Heinkel He 178. It flew on 27 August 1939. But it didn't see action because what Germany and every country in the war needed was a jet fighter. A fighter offers speed and the ability to maneuver easily. So the Germans continued working on developing jet aircraft.

Meanwhile, a British engineer named Frank Whittle also spent time in the 1930s before the war working with jet engines. The British government was so impressed with his idea (patented in 1930) that they used it to develop an engine for a fighter jet called the Gloster Meteor. British pilots flew it in the final year of the war.

The American company General Electric (GE) also got into the act when the US Army Air Forces decided that the United States should pursue jet engine technology. The British gave GE Whittle's plans, which the company used to produce engines for the Bell XP-59. The US jet first flew on 1 October 1942 but like the Heinkel He 178, it didn't see combat.

Back in Germany an engineer named Anselm Franz completed his country's first mass-produced engine for a jet fighter, the Messerschmitt 262 Schwalbe. The Messerschmitt 262, or Me 262, had a range of 650 miles and could reach 540 miles per hour. While the Me 262 saw combat, it arrived too late in the war to affect the outcome. It also burned up too much fuel, which the Germans had in short supply by the war's end. Germany also looked at using jet engines on other aircraft types.

## Vocabulary

- fuselage
- monocoque
- torque
- aspect ratio
- glide angle
- dihedral angle
- anhedral
- winglet
- mean camber line
- symmetrical airfoil
- concave
- downwash
- flap
- aft
- spoiler
- slat
- propulsion
- propulsion system
- excess thrust
- combustion
- propellant
- oxidizer

NACA
LMAL 37931

## How the Fuselage and Wing Shape Correspond to an Aircraft's Mission

Engineers design planes in many shapes and sizes to carry out different commercial or military missions. Two major airplane parts that vary greatly, depending on the mission, are the fuselage and the wings.

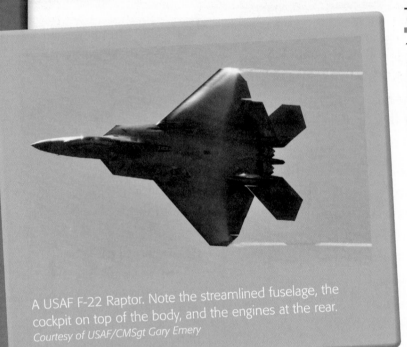

A USAF F-22 Raptor. Note the streamlined fuselage, the cockpit on top of the body, and the engines at the rear.
*Courtesy of USAF/CMSgt Gary Emery*

### The Fuselage

The fuselage, *the aircraft body*, is a long, hollow tube that not only holds passengers and freight, but also holds the pieces of an airplane together. It's hollow to reduce weight and create room for its payload.

While most aircraft bodies are long tubes, they vary in shape to fit the mission. For instance the airliner you board at Thanksgiving to travel to your relatives will have a wide fuselage to accommodate a couple hundred travelers and their luggage (Figure 3.1). If you were a pilot flying a supersonic fighter, however, the fuselage would be slender and streamlined to reduce drag at high speeds.

In an airliner the pilots sit at the front of the fuselage in the cockpit, the passengers and cargo are in the rear, and the fuel is carried in the wings. In a fighter jet the cockpit is part of the fuselage, the engines and fuel are toward the rear, and the weapons are on the wings.

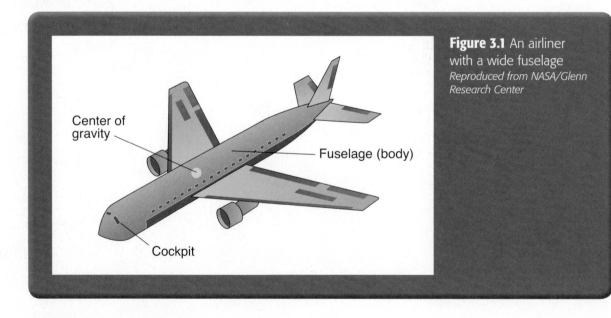

**Figure 3.1** An airliner with a wide fuselage
*Reproduced from NASA/Glenn Research Center*

Center of gravity

Fuselage (body)

Cockpit

## Aircraft Construction

Engineers originally built aircraft with open truss structures, which gave aircraft their strength and rigidity. This was how the Wright brothers and other early enthusiasts constructed their gliders and powered aircraft. The main drawback to this design was that it wasn't streamlined. The wooden structure created a lot of drag. As speeds increased, this drag became a problem.

Most aircraft today are monocoque (pronounced "MAWN-uh-coke" and French for "single shell")—*a one-piece body structure that offers support and rigidity*—or semimonocoque (Figure 3.2). A monocoque shell serves the same function as an open truss structure— it provides strength and support—but it is streamlined. A semimonocoque shell, such as many commercial airliners use, is a combination of shell and framework. The shell and the internal framework share the workload and offer strength and support. The structure inside the shell includes such parts as the *former* (a perpendicular structural support), *bulkhead* (a perpendicular partition in an aircraft's fuselage), and *stringer* (a horizontal support along a fuselage's length). The aircraft's skin is attached to the stringers.

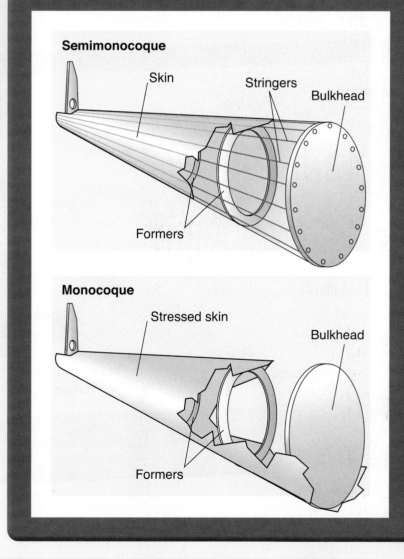

**Figure 3.2** Monocoque and semimonocoque shells and their parts
*Reproduced from US Department of Transportation/Federal Aviation Administration*

Semimonocoque
- Skin
- Stringers
- Bulkhead
- Formers

Monocoque
- Stressed skin
- Bulkhead
- Formers

In the previous lesson you read about the force called weight. The fuselage and the payload carried in the body make up the bulk of an aircraft's weight. The center of gravity is inside the fuselage. During flight the aircraft rotates around the center of gravity because of torque generated by a number of different aircraft parts. Torque is *a twisting or rotating force*. The fuselage must be strong enough to withstand torque.

## Wing Shapes

Wings are one of the main aircraft parts attached to the fuselage. Like the airplane body, wings come in all different shapes and sizes, depending on the plane's purpose and how much lift it requires.

### Wing Positions and Parts

Manufacturers generally place wings in three locations on the fuselage: at the top, toward the middle, and at the lower portion of the body. Engineers refer to these designs as high-, mid-, and low-wing. The position of the wings depends on

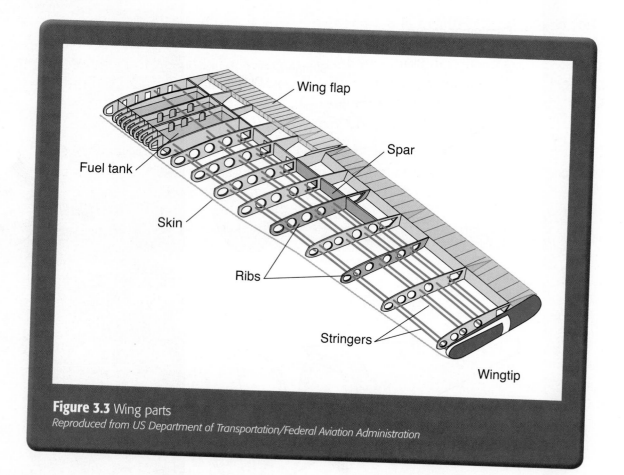

**Figure 3.3** Wing parts
*Reproduced from US Department of Transportation/Federal Aviation Administration*

the aircraft's mission. A monoplane is a plane with a single set of wings, while a biplane is an aircraft with two sets of wings. Monoplane wings can be low, mid, or high. A biplane will have some combination, although most have one high and one low set of wings.

A wing's main internal parts are spars, ribs, and stringers. Spars and stringers run lengthwise. And ribs run a wing's width and determine how thick it will be. Fuel tanks are usually also a part of the wing. They either are mounted inside the wing in containers or are part of the wing's structure (Figure 3.3).

## Wing Size

A wing's aspect ratio—*a measure of how long and wide a wing is from tip to tip*—depends on the plane's function. For example, a glider, which travels at slow speeds, has high-aspect-ratio wings with long wingspans. By contrast, a fighter like the F-16, which flies at high speeds with a powerful engine, has low-aspect-ratio wings and short wingspans.

**Wing TIPS**

Mathematically speaking, an aspect ratio is the square of the wingspan divided by the wing area.

Long, thin wings, such as a glider has, experience low induced drag. Wings designed in an elliptical shape have less drag than rectangular wings. The amount of an aircraft's induced drag is inversely proportional to its aspect ratio. That is, the greater the aspect ratio, the lower the induced drag. And the lower the aspect ratio, the higher the induced drag. Further, the lower the induced drag, the greater the lift, and vice versa.

Because gliders fly with no engine, their flight is relatively slow and their flight path is a constant decrease in altitude. This slow, steady descent gives them a small glide angle, *where the flight path intersects with the ground at an angle* (Figure 3.4).

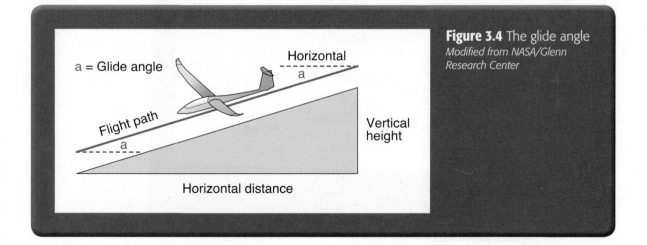

**Figure 3.4** The glide angle
*Modified from NASA/Glenn Research Center*

An aircraft's glide angle is a measure of its flying efficiency. A glider's high-aspect-ratio wings give it this efficiency because with the greater wing area, the wings generate plenty of lift. The wings' elliptical shape reduces drag and results in a long, slow flight.

The space shuttle with its low-aspect-ratio wings and thick chords from leading to trailing edge was a different case. This aircraft had to perform two very different duties. It used rocket engines to lift off, but it glided powerlessly back to Earth. On its return, the shuttle had a large glide angle because of its low-aspect-ratio wing shape and high speeds. This low aspect ratio and large glide angle meant it was a very poor glider.

### Wing Angles

Many planes have wings whose tips are higher than their roots—the part that attaches to the plane. *The angle that a wing makes with the aircraft's horizontal* is the dihedral angle (Figure 3.5). This angle gives an aircraft roll stability and helps it return to level flight.

Large commercial airliner wings have dihedral angles. Fighters, on the other hand, need to maneuver more, so they don't have dihedral. In fact, to increase fighters' ability to roll, engineers design many with *a negative dihedral* called anhedral (Figure 3.5). Aircraft with anhedral have wingtips closer to the ground than their wing roots. The Wright brothers' 1903 flyer had a slight anhedral.

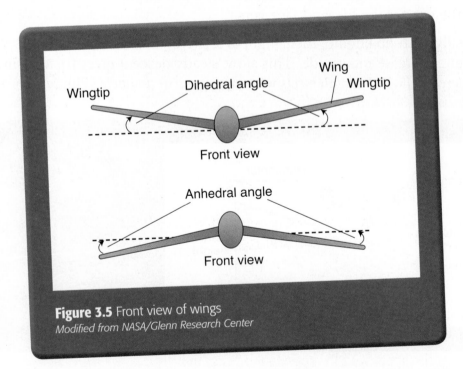

**Figure 3.5** Front view of wings
*Modified from NASA/Glenn Research Center*

## Winglets

A 1970s NASA invention called the winglet graces many modern wings. A winglet is *a plate attached to a wingtip or a wingtip bent upward*. The winglet increases efficiency by decreasing drag.

In the previous lesson you read about wingtip vortices that twirl the air into a whirlpool and create induced drag. Winglets reduce the strength of these tip vortices so the air flows more easily across the wing (Figure 3.6). This smoother airflow has other consequences: tests at NASA on a Boeing 707 with winglets showed a 6.5 percent decrease in fuel consumption.

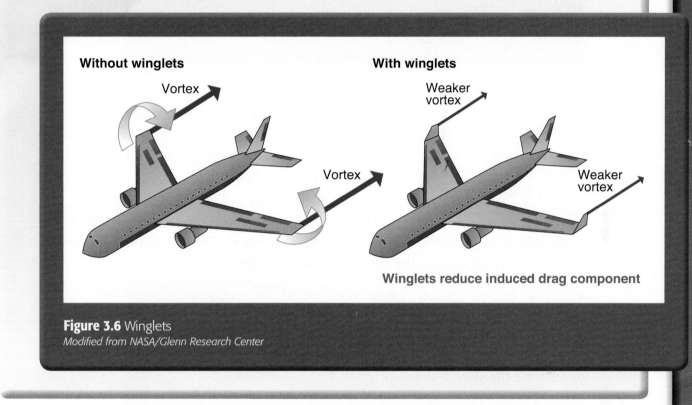

**Figure 3.6** Winglets
*Modified from NASA/Glenn Research Center*

## The Types, Purpose, and Function of Airfoil Design

Wing shape actually involves more than aspect ratios and angles. The relationship of an airfoil's upper and lower surfaces also plays a big role in generating lift.

### Airfoil Camber

In the opening lesson you read about the curve in an airfoil called the camber. The mean camber line is *a reference line from leading to trailing edge drawn an equal distance from the upper and lower wing surfaces*. In a symmetrical airfoil— *an airfoil in which the upper surface and the lower surface have the same shape*—

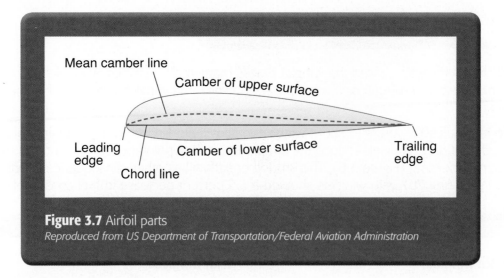

**Figure 3.7** Airfoil parts
*Reproduced from US Department of Transportation/Federal Aviation Administration*

the mean camber line and chord line will be the same. But in most airfoils the upper surface curves more than the lower surface, so the mean camber line will lie above the chord line (Figure 3.7).

You can also measure an airfoil's thickness. The thickness is the maximum distance between the upper and lower wing surfaces. To determine thickness you would measure the thickest vertical line of the wing, toward the rounded leading edge.

## Types, Purpose, and Function

Airfoils come in many combinations of camber and thickness because no one design meets every aircraft's needs. The aircraft's weight, speed, and purpose determine the wing's shape (Figure 3.8).

No one airfoil has been found that satisfies every flight requirement.

Engineers have found that concave lower surfaces—*hollow and rounded like an ice cream scoop*—produce the greatest lift at low speeds. However, this "scooped out" lower surface doesn't work well at high speeds.

Streamlined airfoils result in other tradeoffs. They may work at high speeds because they produce little wind resistance, but they don't create enough lift. So, modern aircraft have airfoils that find a happy medium between the two.

If you tried flying with a wing in the shape of a teardrop, you'd have a very streamlined airfoil, but it wouldn't create any difference in pressure between top and bottom and would have no lift at a zero angle of attack. The airflow over the top and bottom would be the same. If you chopped this teardrop in half lengthwise (along the chord line) so it was more in the shape of a typical airfoil (Figure 3.9), the top would be far more curved than the bottom. This would create the right conditions for lift.

**Figure 3.8** Airfoil designs
*Reproduced from US Department of Transportation/Federal Aviation Administration*

Early airfoil

Later airfoil

Clark "Y" airfoil (subsonic)

Laminar flow airfoil (subsonic)

Circular arc airfoil (supersonic)

Double wedge airfoil (supersonic)

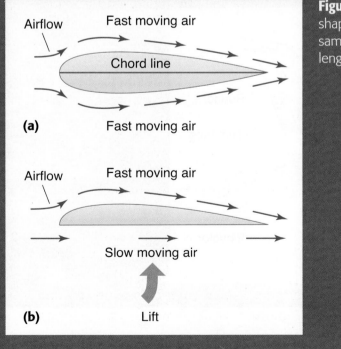

**Figure 3.9** (a) A teardrop-shaped airfoil; (b) the same airfoil cut in half lengthwise

Airflow

Fast moving air

Chord line

**(a)**  Fast moving air

Airflow

Fast moving air

Slow moving air

Lift

**(b)**

When airfoil curvature finds the right balance, the airstream striking the lower and flatter wing surface flows down to create an opposite upward lift. And the stream hitting the upper and more curved wing surface bends up.

Furthermore the pressure difference that a properly designed airfoil generates between its upper and lower surfaces is not enough to produce *all* of the lift force required (although it contributes greatly to it). With the right camber, the low pressure above the wing flows down to create downwash. Downwash is *a downward flow of air*. Meanwhile, the high-pressure air is flowing under the wing toward the trailing edge. When these two pressure areas meet, Newton's third law of action and reaction comes into play. The reaction of the downward backward flow of air results in an upward forward force on the airfoil.

## Wing TIPS

Stabilizers are wings. They just happen to sit at the plane's tail and are smaller than the main wings at the center of an aircraft.

## The Role of Stabilizers and Rudders

Another crucial aircraft part is the tail section. Pilots control their aircraft with the help of *stabilizers* and *rudders*. These two parts are attached to the rear of the fuselage at the tail (Figure 3.10). The vertical and horizontal stabilizers don't move; they are fixed in place. The rudder does move.

Stabilizers do just what it sounds like—they keep an aircraft stable so it can maintain a straight flight path. The vertical stabilizer prevents the nose from roving side to side. The horizontal stabilizers keep it from bobbing up and down.

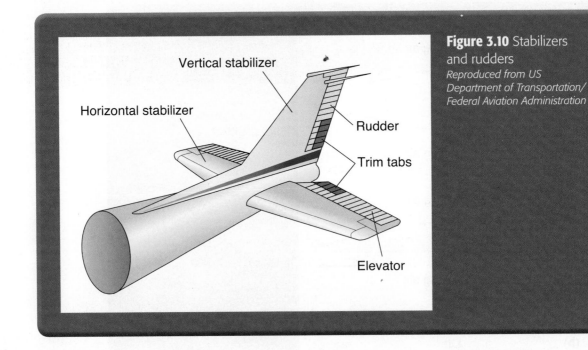

Vertical stabilizer

Horizontal stabilizer

Rudder

Trim tabs

Elevator

**Figure 3.10** Stabilizers and rudders
*Reproduced from US Department of Transportation/ Federal Aviation Administration*

The rudder, fastened with hinges to the back of the vertical stabilizer, lets the pilot steer the aircraft by moving the tail left and right. Another airplane part, called an *elevator*—attached with hinges to the back of the horizontal stabilizer—can direct the tail either up or down at the pilot's discretion. These movable parts at a stabilizer's rear change the force the stabilizers produce. Both the rudder and the elevator have *trim tabs*, also attached by hinges. Trim tabs are even smaller than the rudder and elevator and fine-tune the left-right, up-down movements by reducing pressure. They help "trim," or balance, the flight.

## The Positions of Flaps, Spoilers, and Slats on an Aircraft

An airplane's main fixed wings also have moving parts. These are flaps, spoilers, and slats (Figure 3.11). They play a major role during takeoff and landing when aircraft velocity is relatively slow.

A flap is *a hinged device at a wing's trailing edge that produces lift.* It moves in two directions: it bends downward and stretches aft (*toward the tail*). Flaps on large aircraft such as commercial airliners are more complex than those on smaller planes.

A spoiler is *a small, flat plate that attaches to the tops of the wings with hinges; it increases drag.*

At the front of the wings on some aircraft are slats. A slat is also *a movable, hinged part that pivots down to generate more force.* The next lesson will discuss the effects of flaps, spoilers, and slats on flight and motion.

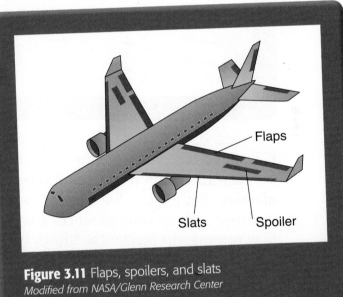

**Figure 3.11** Flaps, spoilers, and slats
*Modified from NASA/Glenn Research Center*

## How the Airflow and Airfoil Affect Flight Movement

The term *flow turning* describes much of what you've read so far in this lesson, most particularly airfoil design and wing controls such as flaps. The *amount* of lift an airfoil generates depends on how much it is able to change the path the airstream takes. How much an object can turn flow depends on its shape.

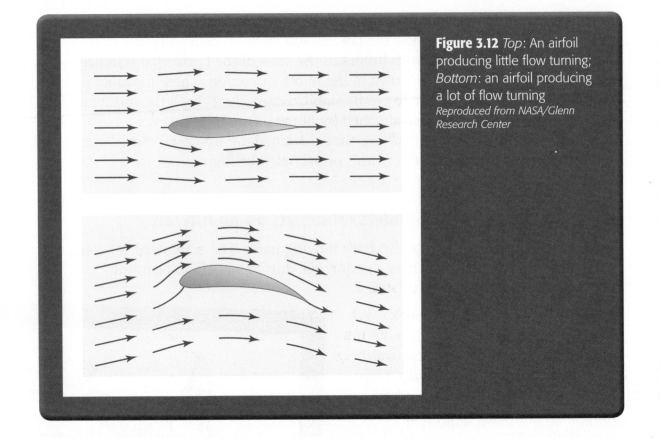

**Figure 3.12** *Top*: An airfoil producing little flow turning; *Bottom*: an airfoil producing a lot of flow turning
*Reproduced from NASA/Glenn Research Center*

If air flows evenly along both sides of a wing—as in that previous case of the symmetrical teardrop-shaped airfoil—the wing produces no lift at zero angle of attack. But if the airfoil curves dramatically toward its trailing edge, the flow is turned and the wing produces a large amount of lift. In short the greater the flow turning, the greater the lift. In Figure 3.12 the airfoils have identical leading edges. It is their trailing edges that determine whether an airfoil can turn flow and cause lift.

Control surfaces such as flaps, spoilers, and slats let pilots manipulate flow turning. With hinged devices down, the wing takes on a more sharply curved shape at its trailing edge to generate more lift. Extending flaps aft and slats forward also expands a wing's aspect ratio, which also helps lift. If the pilot desires less lift, he or she maneuvers the aft section up or activates the spoilers.

## The Purpose and Function of Propellers, Turbines, Ramjets, and Rocket Propulsion Systems

Propulsion systems provide the aircraft's thrust to maintain forward movement. The word propulsion has two Latin roots: *pro* for *forward* and *pellere* for *to drive*. Therefore, propulsion means *to drive forward or to push forward*.

A propulsion system is *an engine that generates thrust to move an aircraft forward*. These systems illustrate Newton's third law of action and reaction. Engines accelerate air—a gas—in one direction, and this action propels the aircraft in the opposite direction, which is the reaction. The engine pushes on the gas, and the gas pushes back. Propulsion systems are the machines that accelerate the gas. Propellers do this mechanically; jets do it both mechanically and through a chemical reaction.

> ### Wing TIPS
>
> Don't confuse the terms *gas* and *fuel*. Gas in this textbook refers to *working fluids*, such as air. Fuel, on the other hand, is the material that supplies power—it's what you put in your car's gas tank, the diesel in a truck, or the propellant in a rocket.

The amount of thrust an engine generates depends on how much gas it pushes through the propulsion system and the velocity at which the gas exits the system. The four main propulsion systems are propellers, turbines, ramjets/scramjets, and rockets. Each produces thrust in different ways.

Propulsion systems serve two roles. When an aircraft is in straight and level flight, the thrust an engine produces must balance the drag. When an aircraft is accelerating, the thrust produced must exceed the drag. The greater *the difference between thrust and drag*—called excess thrust—the faster the plane accelerates.

When choosing which type of propulsion system an aircraft will use, engineers consider what's more important for that particular plane: excess thrust, or high engine efficiency and low fuel usage. For an airliner, which generally travels at a steady, cruising speed over long distances, fuel efficiency takes top priority. But for a fighter, it's acceleration.

A C-130H3 taxies down the runway during a flight test at Edwards Air Force Base, California. It boasts eight-bladed NP-2000 propellers made for generating more thrust with greater efficiency.
*Courtesy of USAF/John Perry*

## Propellers

Many airliners and cargo planes use turboprops and high bypass turbofans because propellers and turbofans offer efficiency and low fuel usage.

Propellers have anywhere from two to eight blades attached to a hub. An engine provides the power to turn the blades. Propeller blades not only are long and thin like wings, but they also act like wings. They have camber, chord lines, and leading and trailing edges. They are subject to all the same forces that an airfoil is, including drag and lift. They generate thrust by accelerating the air that passes through them. Airfoil shape, the blades' angle of attack, and the engine's power all determine the amount of thrust produced.

Because a propeller is a rotating airfoil, it also generates lift as the blades move through the air. Propeller blades experience low- and high-pressure areas when producing lift. The low-pressure area forms behind the blades, and the high-pressure area sits in front of them, creating a lifting surface. This difference in pressure draws air through the propellers and pulls the plane forward.

As the blades spin, the tips have a faster linear speed than the hub because the tips have farther to travel in the same amount of time. To deal with this difference in speed along a blade and to produce uniform lift, blades are twisted from hub to tip. This twist introduces an angle of attack, and the angle of attack is lower at the tip than it is at the hub to accommodate the different speeds along the blade. That is, the blades are less twisted at the tip and more twisted at the hub. If the angle of attack were the same all along the blade, the hub would spin at a negative angle of attack and the tip would stall.

Airmen work on a C-130E propeller at their air base in Southwest Asia.
*Courtesy of USAF/SrA Laura Turner*

## Designing a Better Propeller

Frank Walker Caldwell (1889–1974) was with the US Army Air Service's Propeller Department of the Airplane Engineering Division at McCook Field in Dayton, Ohio, during World War I. Just a few years before, he had graduated from the Massachusetts Institute of Technology with a degree in mechanical engineering. It was up to him, as the Army Air Service's chief engineer, to design and test the best possible propellers for the Army *and* the Navy.

Just as World War II spurred advances in jet engines, World War I pushed advances in propellers. While the Wright brothers get credit for first discovering that propellers act like rotating wings, Caldwell made some very important observations and inventions as well.

The newer, more powerful aircraft rolling off the assembly lines beginning in World War I called for stronger, more-efficient propellers. Planes had fixed-pitch, wood propellers. The blade angle was always the same. But an efficient angle for a blade when an airplane is cruising isn't efficient at takeoff. What grew out of Caldwell's time as chief engineer were variable-pitch propellers that could be removed from the hub and their pitch changed with the plane on the ground and the engine off. These were ground-adjustable pitch propellers. They were also metal rather than wood.

In 1929, Caldwell left the military and joined the Hamilton Standard Propeller Corporation where he came up with a controllable-pitch propeller that could be adjusted during flight. This propeller had two positions, one good for takeoff, the other good for cruising. Later while still at Hamilton, Caldwell contributed to the Hydromatic constant-speed propeller, which automatically changed blade angle when the pilot changed engine speed. This propeller played a huge role in Allied aircraft during World War II.

## Turbines

Turbine (jet) engines can fly at higher speeds than propellers, although they also work at low cruising speeds. Aircraft manufacturers use gas turbines for everything from commercial airliners to military cargo planes to fighters engaged in air-to-air combat. One type of turbine engine is the turboprop, which boasts a turbine engine that turns a propeller.

A USAF C27J Spartan, a twin turboprop aircraft
*Courtesy of USAF*

These versatile engines—turboprop, turbojet, turbofan, and turboshaft—produce thrust by pushing gas out their exhaust at great velocity.

An F-16 Fighting Falcon has a typical jet engine. This F-16 is flying over Nevada in early 2011 on a training exercise.
*Courtesy of USAF/SSgt Benjamin Wilson*

Developed during World War II by Germany and Britain, gas turbines create hot exhaust gas and pass it through a nozzle to thrust an aircraft forward. Gas turbines depend on oxygen from the surrounding air for combustion, *the process of burning*. Many aircraft today use gas turbines. Airlines equip their passenger airliners with turbines because they often travel great distances at high speeds. Fighters need them because they must accelerate quickly in combat. And although cargo planes might not call for speed, they do carry heavy loads long distances.

## Ramjets

Ramjets are lighter and simpler than turbojets. Like a gas turbine, they combust fuel and derive thrust from the hot exhaust accelerated through the nozzle. The combustion pressure must be higher than the pressure at the nozzle for the engine to work. The working fluid is external air brought into the system—as with a gas turbine's operations. The system exerts a force on the external air to accelerate it, and the reaction produced is the thrust force on the system. Ramjets, like jet engines, rely more on gas exit velocity than on mass flow of gas through the engine.

Ramjets can't provide thrust unless an aircraft is already moving. Therefore, ramjets must operate with other propulsion systems, such as rocket engines, that provide an initial thrust. These engines are used for supersonic flight. But they can't combust fuel when traveling above the speed of sound unless the inlet slows the air to below the speed of sound. So aircraft with ramjets are capable of flying above the speed of sound, but the air must be slowed entering the engine for the ramjet to burn the fuel and create thrust. Aircraft designers use ramjets despite these complications because they are more efficient than other jet engines at these speeds. More-modern scramjets are an improvement over ramjets because the flow of air does not have to be reduced to less than the speed of sound. A scramjet allows supersonic combustion, improving the efficiency of the engine at high supersonic speeds.

## Rocket Engines

The working fluid for propellers, turbines, and ramjets is the surrounding air. A rocket engine's working fluid is hot rocket exhaust. This is because rockets carry their own oxygen to mix with fuel. This allows them to travel in space, where there is no oxygen. The other three engine types can't fly into space because they don't carry their own oxygen but instead draw on external air.

While rockets work for space travel, they can also fly in Earth's atmosphere. For instance, they work with ramjets for in-atmosphere flight. Engineers used them in the first aircraft to break the sound barrier, such as the Bell X-1 in 1947.

An X-15 flies over the desert in the 1960s. This aircraft flew with a rocket engine.
*Courtesy of NASA/USAF*

The two main types of rockets are liquid rockets and solid rockets. The first uses liquid propellant—*a mix of fuel and oxidizer* (an oxidizer is *a substance that includes oxygen to aid combustion*). A liquid rocket stores its liquids separately, then mixes them in a combustion chamber in the nozzle. The liquid is injected into the combustion chamber and ignited. The resulting hot gas ejects out the nozzle. A solid propellant is one where the ingredients are already mixed but they don't burn until ignited. The advantage of the liquid propellant is that the combustion can be turned on and off by no longer pumping the liquids into the combustion chamber. But a solid propellant, once lit, burns through unless you destroy its casing.

In general liquid rockets are heavier than solid rockets and more complex because of the pumps and storage tanks. Also you often can't load the combustible liquids into the tanks until right before launch. A solid propellant rocket, on the other hand, can sit in storage for a long time.

This lesson introduced you to airplane parts and their functions. In the next lesson, you'll read about aircraft motion and how to control it. After that, you'll study how aircraft engines work and the latest innovations in aircraft technology.

# ✓ CHECK POINTS

## Lesson 3 Review

Using complete sentences, answer the following questions on a sheet of paper.

1. A fuselage must be strong enough to withstand which force?

2. Why must a glider have high-aspect-ratio wings with long wingspans?

3. Where does the mean camber line usually lie in relation to the chord line?

4. At which speeds do wings with a concave-shaped lower surface produce the greatest lift?

5. What do a vertical stabilizer and a horizontal stabilizer do?

6. What does an elevator do?

7. When do flaps, spoilers, and slats play a major role?

8. Does a slat pivot up or down? Why?

9. How much an object can turn flow depends on what?

10. Which parts of a wing let a pilot manipulate flow turning?

11. How do turbine engines produce thrust?

12. What must already be happening for a ramjet to provide thrust?

# APPLYING YOUR LEARNING

13. Why is the air pressure both under and over a symmetrical airfoil the same?

### ✐ Quick Write

_____

_____

Colonel Lowe had the training, the experience, and the ability to stay calm in the middle of chaos. He used these qualities to rescue others. What are some ways you can develop those qualities in yourself?

### ✦ Learn About

- the axes of rotation and how the primary flight controls work
- the effects of flaps on flight
- the effects of slats on flight
- the effects of spoilers on flight
- the elements of controlled flight

**L**t Col Richard Lowe sat toward the back of a commercial flight taking off from Denver International Airport on 20 December 2008 when a powerful gust of wind slammed into the airliner. The plane lurched off the runway and, after several bounces into the air, came to a rough landing that damaged the fuselage and set the aircraft's right side on fire.

Lowe, a reservist, is a flight instructor with the 340th Flying Training Group. He's also a pilot with Continental Airlines. With 10,000-plus flying hours, the colonel had the experience needed to remain calm, assess the situation, and jump into action, even though he was simply hopping a ride aboard Continental Flight 1404 just like any other passenger.

With flames licking the aircraft, it was important to get everyone out as soon as possible because the plane would eventually explode. Lowe first helped a couple of passengers off the aircraft, then assisted two injured crew members. Lowe returned to the aircraft several times to aid others and check that no one was still onboard. The colonel later told a fellow Airman that on his last trip into the burning plane he could feel the hair stand up on the back of his neck. The plane exploded seconds after he exited.

For his courage, Colonel Lowe received the Airman's Medal at a ceremony in 2010 at the Air Force Reserve Command Joint Reserve Base in Fort Worth, Texas. The military awards the medal for "heroism not involving actual conflict with an armed enemy." It is the highest noncombat-related award granted by the Air Force. Lowe also received a presidential citation from the Air Line Pilots Association for his actions.

## Vocabulary

- attitude
- yaw axis
- yaw
- pitch axis
- roll axis
- roll
- ailerons
- control stick
- bank
- taxi
- headwinds
- tailwinds
- gradient

Maj Gen Frank Padilla (*left*), 10th Air Force commander, Air Force Reserve Command Joint Reserve Base, Fort Worth, Texas, presents the Airman's Medal to Lt Col Richard Lowe (*right*) during a ceremony in 2010.
*Courtesy of USAF/Don Lindsey*

## The Axes of Rotation and How the Primary Flight Controls Work

Flight takes place in three dimensions. A pilot's job is to control an aircraft's attitude in this three-dimensional space. Attitude is *an aircraft's orientation, or angle, in relation to the horizon.*

During flight, an aircraft rotates about its center of gravity. This rotation is described by a three-dimensional coordinate system made up of three principal axes. The center of gravity is the central point of this coordinate system, and the three axes starting at the center of gravity are perpendicular to one another (Figure 4.1).

### Wing TIPS

The terms roll, pitch, and yaw were originally nautical terms. Aircraft and ships have similar motions. Both *roll* along their longitudinal axis. Both *pitch* about their lateral axis (in aircraft, wingtip to wingtip). And both *yaw* about their vertical axis.

The yaw axis (vertical axis) is *a line that starts at the center of gravity, runs perpendicular to the wings, and is directed toward the aircraft's lower surface.* Yaw is *the side-to-side motion of an aircraft's nose.* The pitch axis (lateral axis) is *a line that starts at the center of gravity and runs from wingtip to wingtip.* (You read in Chapter 1, Lesson 1 that pitch is the up-and-down motion of an aircraft's nose.) The roll axis (longitudinal axis), too, is *a line that begins at the center of gravity, is perpendicular to the yaw and pitch axes, and runs from nose to tail.* Roll is *the up-and-down motion of an aircraft's wings.*

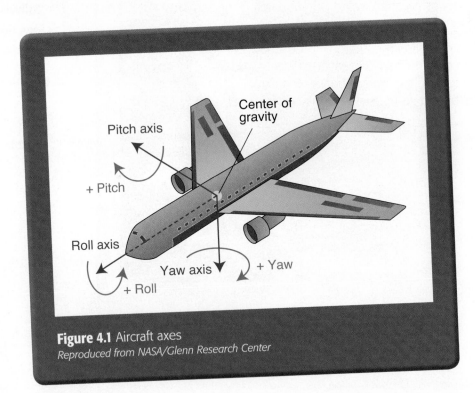

**Figure 4.1** Aircraft axes
*Reproduced from NASA/Glenn Research Center*

A pilot works with control surfaces to direct an aircraft's yaw, pitch, and roll. You read in the previous lesson about rudders, which control yaw, and elevators, which control pitch. The other control surface is the ailerons. An aileron is *a small hinged section on the outboard portion of a wing.*

These control systems respond differently, depending on airspeed. Their function is to change airflow and pressure distribution over and around the airfoil. Moving any one of the control surfaces changes the airflow and pressure. The changes in airflow and pressure distribution then affect the lift and drag produced by the combination of airfoil and control surface, and they let a pilot direct the aircraft about the three axes of rotation.

**Wing TIPS**

The Wright brothers' 1902 glider was the first aircraft to be controllable about all three axes.

## Ailerons

Pilots maneuver ailerons to control their roll about the longitudinal axis. Ailerons usually move in opposite directions. When one deflects up on one wing, the other deflects down on the other wing. For smaller aircraft, pilots move these hinged devices by means of a control stick—*a handle attached by cables, pulleys, or some other means to control surfaces for the purpose of controlling them.* Many times newer aircraft use advanced computer systems for control.

If a pilot flips the aileron up on his or her right wing, the lift decreases on that wing because the camber has decreased. At the same time, the aileron is deflecting down on the left wing, and the lift on that wing has increased because the camber has increased. The result is the left wing rises, the right wing drops, and the aircraft rolls to the right (Figure 4.2).

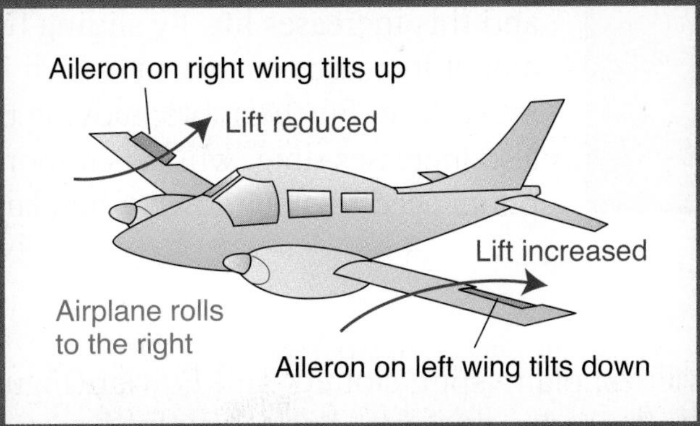

**Figure 4.2** Ailerons
*Reproduced from NASA/Virtual Skies*

Aileron on right wing tilts up

Lift reduced

Lift increased

Airplane rolls to the right

Aileron on left wing tilts down

## The Effects of Flaps on Flight

Maintenance technicians with the Maryland and Michigan National Guards install a wing flap on an A-10 Thunderbolt at Warfield Air National Guard Base in Essex, Maryland, in 2009.
*Courtesy of US Army/SSgt S. Patrick McCollum*

Rudders, elevators, and ailerons are an aircraft's *primary* control surfaces. They make an aircraft controllable and safe to fly. Aircraft also have *secondary* control systems composed of flaps, slats, and spoilers. They let a pilot maintain even more control over an aircraft's performance. As you read in the previous lesson, these secondary devices play a major role during takeoff and landing.

During takeoffs and landings, aircraft velocity is fairly low. Yet lift depends on sufficient velocity as well as airfoil shape and wing area. Takeoffs call for high lift and low drag. Landings require high lift and high drag. Engineers design wings to maintain high lift during such low-speed flight. By increasing wing area and altering wing shape with movable secondary control surfaces, an aircraft can get the lift it needs in challenging speed conditions. One airplane part manufacturers design for takeoffs and landings is the wing flaps, which you read about in the previous lesson.

Flaps, which sit at a wing's trailing edge, move down via hinges and move aft on metal tracks in the wing. By deploying the flaps down, a pilot increases the airfoil's camber and this increases lift. By sliding the flaps aft, a pilot increases wing area, which increases the lift surface. Furthermore, moving the flap aft also increases drag, which is important, because pilots need to slow down when landing.

### Types of Flaps

Wing flaps come in four varieties: plain, split, slotted, and Fowler (Figure 4.3). The *plain flap* is the simplest. It attaches at the trailing edge of the wing and, when deployed, increases camber and lift. It also increases drag because the surface bends into the main airflow, so the plane is subject to a nose-down pitching moment.

Orville Wright and J. M. Jacobs designed the *split flap* in 1920. The flap is hinged under the wing's trailing edge. It rotates down to generate lift, as the plain flap does, and increases drag. This helped a pilot descend toward the runway at a steeper rate than then-current wings would allow and thus made landing approaches easier.

The most commonly used flap is the *slotted flap*. The slotted flap sits in a groove carved into the underside of a wing's trailing edge. It generates more lift than plain and split flaps. When pivoted down, a duct forms between the lower surface of the wing's trailing edge and the flap's leading edge. High-energy air below the wing pours through this path to the flap's upper surface. It next accelerates the upper surface boundary layer and slows airflow separation, which gives the pilot more lift. This type of flap also generates needed drag but doesn't interfere with lift.

The fourth type of flap is the *Fowler flap*, which is a type of slotted flap. However, it doesn't have hinges, but instead uses metal tracks to slide backward and pivot down. When moderately extended, the Fowler flap increases lift by greater camber and wing area. When fully extended downward, however, the flap increases drag but provides little additional lift.

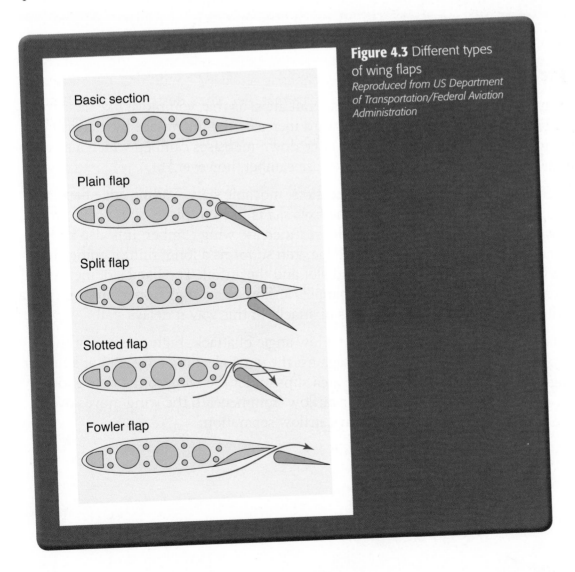

**Figure 4.3** Different types of wing flaps
*Reproduced from US Department of Transportation/Federal Aviation Administration*

Basic section

Plain flap

Split flap

Slotted flap

Fowler flap

## The Fowler Flap

Harlan D. Fowler, an American engineer with the US Army Air Corps and private aircraft firms, tried to improve the wing flap in the 1920s. Not many pilots used the trailing edge flap—even though it had been around since at least the 1910s—because they didn't think it made much difference in airplane performance. But Fowler thought differently.

In fact Fowler felt so strongly about the flap's usefulness that he spent his own time and money to develop it. His flap was different from previous ones because it slid back on a track under the wing to increase the wing area and, therefore, the lift surface. This extra lift grew increasingly important as planes could fly faster and faster and carried more weight.

Despite his flap's effectiveness and a test wing that he installed on airplanes from 1927 to 1929, Fowler couldn't get anyone to buy into his discovery. During the Great Depression, he worked as a salesman to continue funding his private research. Finally he caught a break in 1933 when aircraft manufacturer Glenn L. Martin gave him a job and installed his flaps on a bomber. Another big boost came in 1937 when the Lockheed 14 commercial airliner used his invention. In addition, a major player during World War II, the B-29 bomber, also adopted the Fowler flap.

## The Effects of Slats on Flight

At the front of the wings on some aircraft are slats that you move like flaps to generate more lift. Sliding slats forward increases the lift surface by increasing wing area. Rotating a slat's leading edge down increases camber, which also helps with lift. (Note that not all slats increase camber, however.)

Aircraft use four types of slats: fixed slots, movable slats, leading edge flaps, and leading edge cuffs (Figure 4.4). The *fixed slot* is fixed in place, so it doesn't move, swivel on hinges, or slide, and it doesn't increase wing camber. It is also a *fixed* distance from the airfoil's leading edge, and so forms a long, thin opening along the wing's length between the fixed slot and the airfoil. The fixed slot increases lift because, like the slotted flap, it channels airflow to a wing's upper surface to delay airflow separation at higher angles of attack. In this way it delays stall.

*Movable slats* slide along tracks. At a low angle of attack, high pressure at the wing's leading edge pins the slats against the wing's leading edge. But at a higher angle of attack, the high-pressure area slips under the wing so the slats glide forward. When the slats open, the airflow from beneath the wing moves over the wing's upper surface and delays airflow separation.

*Leading edge flaps* are yet another invention. They increase lift and wing camber and decrease the size of the nose-down pitch produced by trailing edge flaps. They extend down and forward from a hinge under the wing's leading edge. Extended just a bit, they increase lift more than drag. Fully extended, drag increases more than lift.

The fourth type of slat is the *leading edge cuff*. Manufacturers and pilots slip these fixed devices onto a wing's leading edge either during or after assembly to increase lift and wing camber. They curve the wing's leading edge down and forward. The airflow attaches to the wing's upper surface at higher angles of attack and lowers the stall speed. So with a leading edge cuff installed, an aircraft can assume a higher angle of attack without reaching a critical angle of attack when landing at relatively low speeds. Leading edge cuffs can decrease efficiency at cruising speeds, although improved technology has erased some of this drawback.

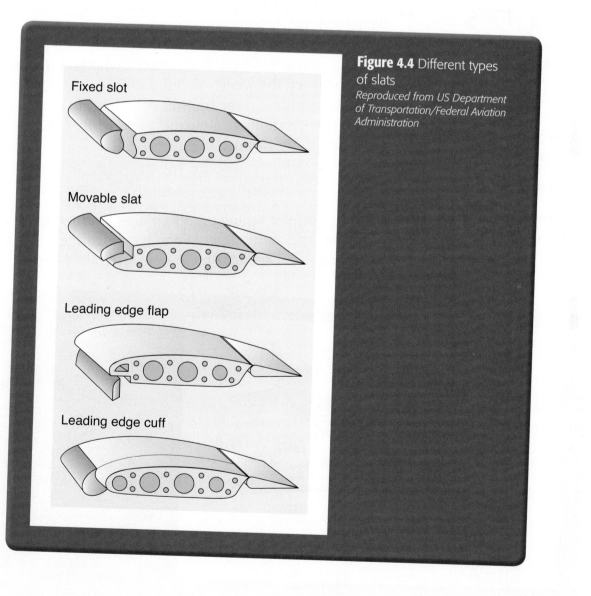

**Figure 4.4** Different types of slats
*Reproduced from US Department of Transportation/Federal Aviation Administration*

# The Effects of Spoilers on Flight

**Wing TIPS**

A spoiler's hinges are at the front of the plate. When a pilot raises a spoiler, the aft flips up into the airstream.

Spoilers are small, flat plates that attach to the tops of the wings with hinges. When a pilot deploys a spoiler, it pivots up into the airstream. A spoiler's purpose is to "spoil" the airflow, increasing drag and decreasing lift. Spoilers have a different role than flaps and slats, which pilots use to increase lift. Designers refer to spoilers as high-drag devices.

## Spoilers Deployed on Both Wings

Raising spoilers on both wings slows an aircraft in any phase of flight. Pilots also use the spoilers to "dump" lift, which forces an aircraft to descend more rapidly.

**Wing TIPS**

Pilots use spoilers along with the ailerons to increase the roll rate (the rate at which the aircraft turns). When used alone and in tandem, spoilers act as speed brakes in any phase of flight.

Once on the runway, pilots raise their spoilers to keep the aircraft on the runway by erasing lift. This also improves the efficiency of the brakes by shifting the aircraft's weight from the wings to the wheels. Friction forms between the wheels and the runway because of the loss of lift and the transfer of weight. In addition, the spoilers continue to slow the plane while rolling to a stop on the ground (Figure 4.5).

## Spoiler Deployed on One Wing

Raising spoilers on only one wing causes a rolling motion. Pilots use this method to bank—*to roll or tilt sideways*—an aircraft in flight. When banking, one wingtip falls and the other climbs. If you deploy a spoiler on the right wing, this decreases

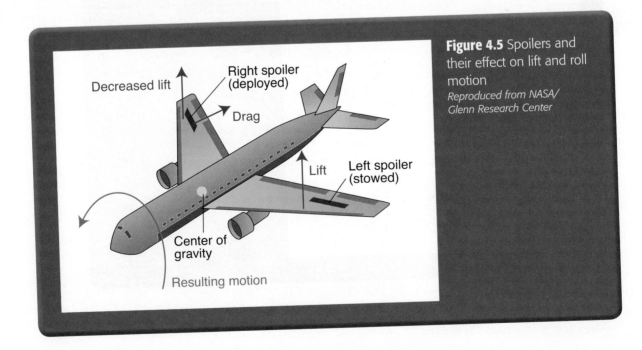

**Figure 4.5** Spoilers and their effect on lift and roll motion
*Reproduced from NASA/ Glenn Research Center*

the lift and raises the drag on the right wing because the spoiler has disturbed the airflow over that wing. As a result, the right wing dips down, the left wing rises, and the aircraft banks and yaws right.

Spoilers cause torque, just as rudders, elevators, and ailerons do. Torque results in rotation. The *net torque*—the difference in forces—is what causes the aircraft to rotate about the center of gravity. For instance, after a pilot deploys a spoiler on the right wing, an aircraft will roll clockwise to the right. But if a pilot tilts a spoiler on the left wing, the aircraft will roll counterclockwise to the left.

## The Elements of Controlled Flight

Flight consists of a number of phases: takeoff, climb, cruise, descent, and landing. Each of these stages of flight applies the forces and control surfaces you've read about in this and previous lessons.

*Wing* TIPS

Manufacturers provide *performance charts* with their aircraft, which let a pilot know how an aircraft will perform during takeoff, climb, cruise, descent, and landing.

### Takeoff, Climb, Cruise, Descent, and Landing

Pilots taxi their aircraft along a runway or taxiway before and after a flight. To taxi is *to move slowly on the ground before takeoff and after landing*. Takeoff begins from a standstill and accelerates to takeoff speed to get into the air. Landing involves touching down on the runway at a landing speed and slowing down to zero speed.

#### Takeoff

During the takeoff phase of flight, engines provide the thrust that gets the aircraft from zero speed to a speed sufficient for takeoff. The thrust is usually set to a maximum at this stage. The runway must be long enough for the aircraft to reach takeoff speed. A safe takeoff speed is comfortably above stall speed and gets an aircraft into the climb phase with satisfactory aircraft control.

Some other factors affect takeoff. These include aircraft weight, wind, and runway slope and condition, which all help determine how long a runway an aircraft needs to take off safely. As weight increases, requirements change: an aircraft must accelerate to a higher takeoff speed, and it must combat an increase in drag and friction with the ground.

Wind conditions are also important. Headwinds reduce takeoff distance because they increase the rate of airflow over the wings from leading to trailing edge and therefore contribute to lift. Tailwinds increase takeoff distance for just the opposite reason. (A headwind is *wind blowing against the direction of travel*. A tailwind is *wind blowing from behind*.)

How much a runway slopes and the condition of the surface (wet, icy, etc.) also affect takeoff and landing distance. The slope influences how long it will take an aircraft to get off the ground or come to a stop. The runway condition affects such factors as ground roll and brake efficiency.

## Runway Conditions

A pilot generally wants to see a paved, level, relatively smooth, and dry runway surface. But runway surfaces differ from location to location. Some surfaces are concrete, whereas others are asphalt, gravel, dirt, or grass. A pilot needs to know what conditions he or she is flying into or taking off in, as these factors affect how long a runway will be needed, among other things. The Federal Aviation Administration offers digital online Airport/Facility Directories that include all kinds of information about airports, including runway data.

Runways that aren't hard and smooth increase ground roll during takeoff. Tires can't roll smoothly if the runway has potholes, the ground is muddy, or the runway is simply a soft, grassy field. Mud and wetness decrease the friction between tires and ground, which makes for less-efficient braking. Mud and wetness can also have the opposite effect by acting as obstructions and reduce landing distance.

Runway length also plays a role in takeoff and landing. If surface conditions aren't ideal (wet and muddy, for instance), the pilot must consider those factors when calculating how long a runway he or she will need to take off or come to a stop when landing. A runway's gradient—or *slope*—may change as well. For example, every 100 feet of a runway may rise by 3 percent. When taking off on an upsloping runway, an aircraft takes longer to accelerate and needs a longer runway. But landing on an upsloping runway reduces the length of runway needed. Conversely, a downsloping runway increases acceleration and shortens the takeoff distance. Landing on a downsloping runway interferes with deceleration and lengthens landing distance.

## Climb

Once an aircraft has accelerated to a sufficient takeoff speed along the runway, the pilot raises its nose and the plane becomes airborne. The aircraft now enters the climb phase (Figure 4.6).

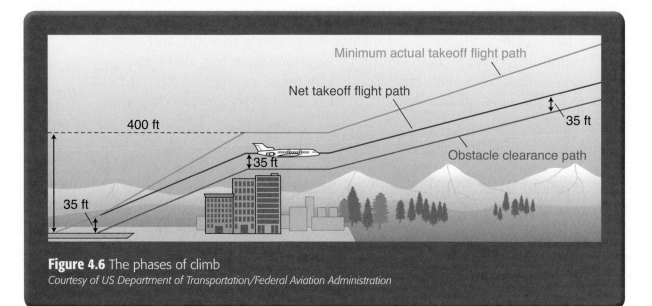

**Figure 4.6** The phases of climb
*Courtesy of US Department of Transportation/Federal Aviation Administration*

As soon as the aircraft reaches a positive rate of climb, the pilot raises the landing gear and accelerates out to the speed at which he or she retracts the flaps. The pilot raises the flaps and accelerates to climb speed. He or she then sets the engines on climb power, which is usually below full power. Thereafter, the pilot climbs until reaching the assigned level-off altitude.

## Cruise

Once the pilot gets to cruising altitude, the aircraft remains there until arriving near the destination airport. Sometimes the aircraft must change altitude because of weather or turbulence. The flight crew members now concern themselves with weather conditions, avoiding other aircraft, fuel consumption, and perhaps passenger issues, among other things. When the aircraft gets near the destination, the pilot and crew prepare to descend and land.

## Descent and Landing

An aircraft descends from its cruising altitude by decreasing thrust and/or engaging the secondary control systems. During the final phases of descent, a pilot will lower the aircraft's landing gear to prepare for eventual contact with the runway.

When landing, the pilot continues to engage the flaps, slats, and spoilers to generate the high lift and high drag that landings require. The minimum safe speed is that which is above stall. The speed must also be enough to provide the pilot sufficient control and ability to abort a landing, climb, and circle around for another try.

The YAL-1A, a modified Boeing 747-400F known as the Airborne Laser, lands at Edwards Air Force Base, California.
*Courtesy of USAF*

**Wing TIPS**

All these elements are crucial, but pilot technique is a critical factor in aircraft performance.

Adequate runway length for a landing depends on wind, weight, and runway characteristics and slope, among other considerations. According to federal regulations, the landing distance is that length of runway needed to land and come to a complete stop from a point 50 feet above the threshold end of the runway (where the runway begins from the aircraft's approach direction). The wheels generally won't hit the runway until about 1,000 feet into the runway distance. The rules adjust depending on the aircraft, however.

Landings can be rough on brakes and tires, so an aircraft takes advantage of aerodynamic drag to slow down. Once the aircraft decelerates sufficiently, drag is no longer great enough to be of much use, so the pilot has to rely on the brakes for any continued deceleration. (Some aircraft also use their engines to slow down by using a device called a "thrust reverser," which reverses an engine's thrust.)

This lesson covered yaw, pitch, roll, primary and secondary control surfaces, and the different phases of flight. Engines usually provide the thrust to help make all these motions and controls possible. The next lesson will look at a number of different engine types and the scientific laws that explain how they operate.

# ✔ CHECK POINTS

## Lesson 4 Review

Using complete sentences, answer the following questions on a sheet of paper.

1. During flight, an aircraft rotates about which point?

2. What must a pilot work with to direct an aircraft's yaw, pitch, and roll?

3. How can an aircraft get the lift it needs in challenging speed conditions?

4. Who designed the split flap in 1920?

5. Sliding slats forward does what?

6. Rotating a slat's leading edge down does what?

7. Once on the runway, why do pilots raise their spoilers?

8. What kind of motion will result by raising spoilers on only one wing?

9. What are some other factors that affect takeoff?

10. Adequate runway length for a landing depends on which considerations?

# APPLYING YOUR LEARNING

11. Describe how a pilot would deploy the secondary control surfaces during a descent (which ones would he or she deflect up, down, aft, forward, or some combination) and what effect each deployment would achieve.

# LESSON **5**
## ✈ Flight Power

### Quick Write

_____

_____

Which is the more significant achievement— being the first to invent something, or the first to make it practical? Or are both equally important?

### Learn About

- the principles of Boyle's law, Charles's law, and Gay-Lussac's law
- the characteristics of internal combustion engines
- the mechanical, cooling, and ignition systems of reciprocating engines
- how the different types of jet engines work
- the role of reversers and suppressors used in jet aircraft
- reaction engines
- the development of new engine technology

**D**r. Hans Joachim Pabst von Ohain of Germany designed the first operational jet engine. But he didn't get credit for being the first to invent the jet engine. Great Britain's Frank Whittle, who registered a patent for the turbojet engine in 1930, received that recognition—although he did not perform a flight test until 1941.

Ohain was born on 14 December 1911 in Dessau, Germany. He came up with his theory of jet propulsion in 1933 while pursuing his doctorate in physics.

Ohain received a patent for his turbojet engine in 1936. He also joined the Heinkel Company in Rostock, Germany. By September 1937 he had built and tested a demonstration engine. By 1939 he developed an operational jet aircraft, the He 178. Soon after, Ohain directed the construction of the He S.3B engine. This engine was installed in the He 178 airplane. It made the world's first jet-powered aircraft flight on 27 August 1939.

Ohain developed an improved engine, the He S.8A, which first flew on 2 April 1941. This engine design, however, was less efficient than one designed by Anselm Franz. That engine powered the Me 262, the first operational jet fighter aircraft.

Ohain came to the United States in 1947 and became a research scientist at Wright-Patterson Air Force Base. In September 1963 he was appointed chief scientist of the Aerospace Research Laboratories. In 1975 he became chief scientist of Wright's Aero Propulsion Laboratory. There he was responsible for maintaining the technical quality of research in air-breathing propulsion, power, and petrochemicals. After retiring in 1979, he became a consultant to the University of Dayton Research Institute.

During his 32 years of US government service, Ohain published more than 30 technical papers and registered 19 US patents. His many honors and awards included the Goddard Award of the American Institute of Aeronautics and Astronautics, the Air Force Systems Command Award for Meritorious Civilian Service, and the Department of Defense Distinguished Civilian Service Award. He was enshrined in the International Aerospace Hall of Fame and the Engineering and Sciences Hall of Fame. In 1990 he was inducted into the National Aviation Hall of Fame. In 1991 Ohain was honored by the US National Academy of Engineering with the Charles Stark Draper Prize as a pioneer of the jet age.

Ohain died on 13 March 1998 at his home in Melbourne, Florida.

## Vocabulary

- piston
- crankshaft
- spark plugs
- baffle
- magnetos
- compressor
- reaction engine
- emissions
- nacelle
- composite

# The Principles of Boyle's Law, Charles's Law, and Gay-Lussac's Law

Centuries ago chemists and physicists were discovering many of the scientific principles that operate in aircraft engines today. Some of those laws are Boyle's law, Charles's law, and Gay-Lussac's law. These help explain how certain engines work and create thrust.

## Boyle's Law

The types of engines you will read about in this lesson—including turbines and ramjets—derive thrust by doing work on gases. The scientific laws listed above explain the relationships among properties of gas. Those properties include pressure, temperature, mass, and volume. If the value of any two of the properties is constant, you can determine the nature of the relationship between the other two.

In 1662 a British chemist named Robert Boyle examined the relationship between the pressure and volume of a confined gas held at a constant temperature. Boyle noted that the product of the pressure and volume was nearly constant. When pressure increased, the volume decreased, and when the pressure decreased, the volume of the confined gas increased. This inverse relationship between pressure and volume is Boyle's law (Figure 5.1).

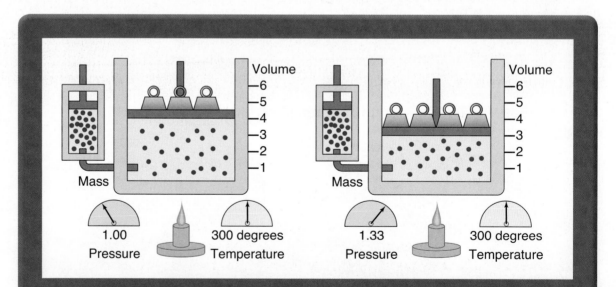

**Figure 5.1** Boyle's law in operation: for a given mass at a constant temperature, the pressure times the volume is a constant.
*Reproduced from NASA/Glenn Research Center*

# The Three States of Matter

Of all the three common states of matter, gases have a unique set of properties. In solids, *intermolecular forces* (the forces between molecules) are strong enough to keep molecules vibrating about fixed positions so that solids have a definite shape and volume. As the temperature is raised, the molecules can gain enough kinetic energy to overcome some of these attractive forces and the substance melts. At this point the liquid molecules have enough attraction to have a definite volume, but not enough to stay in a fixed position. Liquids take the shape of their container. Raise the temperature high enough and the molecules move fast enough to overcome nearly all the remaining intermolecular forces so that the substance becomes a gas. Gases both fill their container and take its shape. In the case of many small molecules like oxygen, methane, or carbon dioxide, the forces between molecules are so small that these substances are already gases at room temperature.

Because the motion of these widely spaced gaseous molecules is random, they collide with each other and with the walls of their containers, creating pressure. Because gas molecules are so far apart, gases are compressible and readily mix. Unlike solids and liquids, a gas can undergo significant changes in pressure and volume. The mathematical expressions that describe these changes are called the gas laws (Figure 5.2).

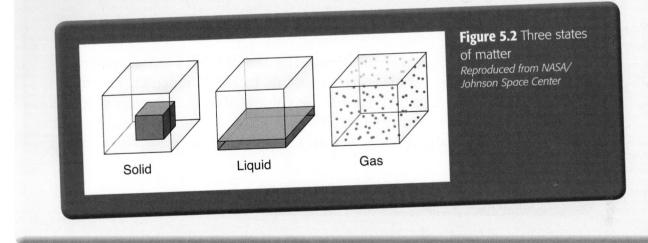

**Figure 5.2** Three states of matter
*Reproduced from NASA/ Johnson Space Center*

Solid    Liquid    Gas

Imagine gas confined in a jar with a piston on top. A piston is *a metal device that moves back and forth inside a cylinder*. The volume of the gas is 4.0 cubic meters, and the pressure is 1.0 kilopascal (*kilo* is equal to a thousand, and *pascal* is a unit of pressure). The temperature and mass are constant. When you add weights to the top of the piston, thus compressing the gas, the pressure increases to 1.33 kilopascals and the volume decreases to 3.0 cubic meters. If you multiply the original values of pressure and volume (4.0 × 1.0) and compare that product with the new product of values (3.0 × 1.33), you will see that the product remains constant.

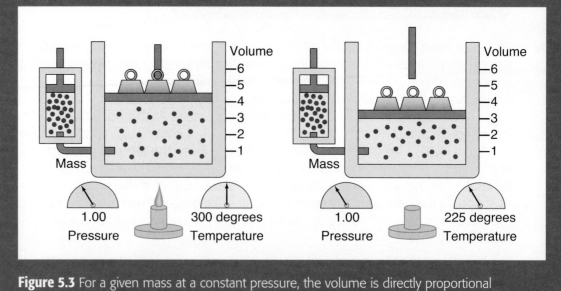

**Figure 5.3** For a given mass at a constant pressure, the volume is directly proportional to temperature.
*Reproduced from NASA/Glenn Research Center*

## Charles's Law

Another law of gas properties considers the relationship between temperature and volume, with a constant mass and pressure. In 1787 French chemist and physicist Jacques Charles observed that the volume of a gas is directly proportional to its temperature. If the temperature of a gas rises, its volume increases, and if the temperature falls, the volume decreases (Figure 5.3).

### Absolute Temperature

It is important to note that the temperature scale used must be *absolute temperature*. An absolute temperature scale such as the Kelvin scale starts with zero at the coldest temperature possible, which is absolute zero. The Celsius (C) scale has zero at a point well above *absolute zero* (–273 degrees C), so a change from 10 degrees C to 20 degrees C would not be a true doubling of temperature. The Kelvin temperature scale uses the same size degrees as the Celsius scale, and because it starts with absolute zero as zero, it has no negative temperatures. The units are called Kelvins (not degrees Kelvin), and the abbreviation is K (not °K). Add 273 to the Celsius temperature to convert to Kelvins; 0 degrees C is equal to 273 Kelvins.

## Gay-Lussac's Law

A second French chemist and physicist named Joseph Louis Gay-Lussac also studied gases and reactions. In 1802 he confirmed Charles's law with his own publication on the relationship of temperature and volume. He later proposed a relationship between the pressure of a gas and its absolute temperature, when the volume is constant. He found that they are directly proportional: pressure rises when temperature rises, and pressure falls when temperature falls. This law about the relationship of temperature and pressure is known as Gay-Lussac's law.

## The Characteristics of Internal Combustion Engines

One engine that works on gases is the internal combustion engine. For the 40 years following the Wright brothers' first flight, airplanes used internal combustion engines to turn propellers, which generate thrust. Today most general aviation or private airplanes are still powered by propellers and internal combustion engines, much like the automobile engine.

The combustion process of an internal combustion engine takes place in an enclosed cylinder where chemical energy (fuel) converts to mechanical energy (the movement of engine parts to turn the propeller). Inside the cylinder a moving piston compresses a mixture of fuel and air before combustion. The piston is then forced back down the cylinder following combustion.

The Wright brothers built an internal combustion engine in 1903. It was a very simple engine by today's standards, which makes it a good model to study. This type of internal combustion engine was a four-stroke engine because it included four movements (strokes) of the piston before the entire engine firing sequence repeated. The four strokes, in order, are *intake*, *compression*, *power (or ignition)*, and *exhaust*. The next section explains these movements.

**Wing TIPS**

The series of four strokes is referred to as the Otto cycle. In 1876 a German named Nikolaus Otto patented the Otto cycle four-stroke engine consisting of intake, compression, power, and exhaust.

## The Mechanical, Cooling, and Ignition Systems of Reciprocating Engines

An internal combustion engine is a reciprocating engine because of the burning process that takes place in the cylinders. The name *reciprocating engine* derives from the back-and-forth, or reciprocating, movement of the pistons, which produces the mechanical energy necessary to do work. Engineers design most small aircraft with reciprocating engines.

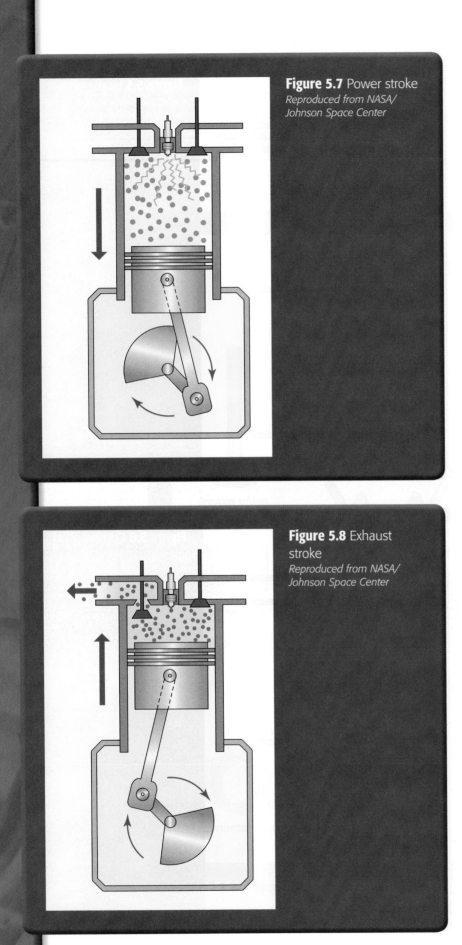

**Figure 5.7** Power stroke
*Reproduced from NASA/ Johnson Space Center*

**Figure 5.8** Exhaust stroke
*Reproduced from NASA/ Johnson Space Center*

3. *Power or ignition stroke*— The third stroke. As the piston nears the top, the system sends a surge of high-voltage current to the spark plug. This produces a high-energy spark, which ignites the compressed fuel-air mixture. The fuel rapidly combines with the oxygen (that is, it burns) and produces carbon dioxide gas and water vapor. These hot gases expand and exert tremendous force on the piston, driving the piston down the cylinder and turning the crankshaft. The crankshaft turns the aircraft propeller (Figure 5.7).

4. *Exhaust stroke*—The fourth stroke. Once the piston has reached the bottom and starts back up the cylinder, the exhaust stroke begins. The exhaust valve opens, residual heat is released, and the pressure returns to atmospheric conditions. The piston pushes the waste gases out of the cylinder, and the process is ready to begin again (Figure 5.8).

The intake and exhaust cycles don't actually produce any power. Only the compression and power strokes determine the work available from the reciprocating engine.

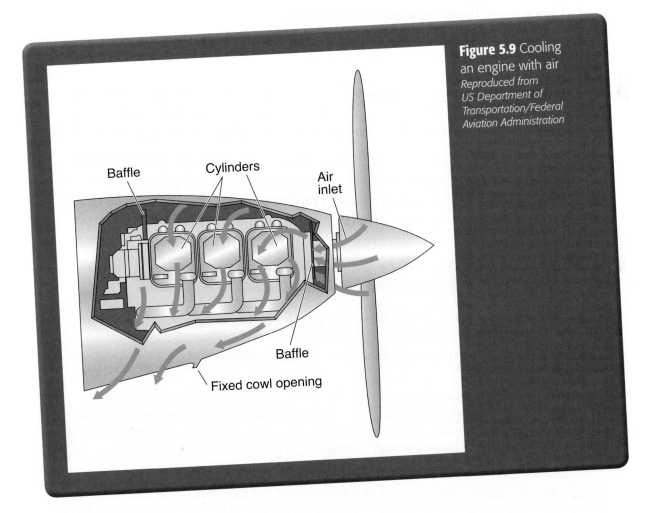

**Wing TIPS**

Even when an engine is operating at a fairly low speed, the four-stroke cycle takes place several hundred times each minute.

## Cooling

The burning fuel inside the cylinders produces intense heat. Exhaust systems expel much of the heat. Still, enough remains that engines can overheat. Designers use either air or liquid to address this danger.

In an air-cooled system, air flows through openings at the front of the engine cowling, or cover, into the engine compartment. A number of devices called a baffle (*a partition that changes the airflow direction*) route air over fins attached to the cylinders and other parts of the engine where the air absorbs the heat. The hot air leaves the engine compartment through one or more openings in the lower, aft portion of the engine cowling (Figure 5.9).

**Figure 5.9** Cooling an engine with air
*Reproduced from US Department of Transportation/Federal Aviation Administration*

Baffle    Cylinders    Air inlet

Baffle

Fixed cowl opening

These turbine engines offered a much better power-to-weight ratio than piston engines could. They consist of an air inlet, a compressor (*a device that compresses— or increases pressure on—air or gas*), combustion chambers, a turbine section, and exhaust. The turbine engine draws in air and compresses it, adds fuel and burns it, and the hot gases expand out the rear of the engine to push the aircraft forward. By increasing the velocity of the air flowing through the engine, you can increase the thrust. Some of these exhaust gases turn the turbine, which drives the compressor.

Engineers have developed a number of different turbine engines for use depending on the needs of a particular type of aircraft.

> ## Wing TIPS
>
> The turbine engine is still an internal combustion engine. But rather than turning a propeller for forward momentum, it expels a high-velocity exhaust out a nozzle for thrust.

## Turbojets

Turbojets were the first type of turbine engines developed. All the thrust comes through the turbine and nozzle, which are the *core* of the engine. These are what people commonly refer to as jet engines.

The turbojet engine is a departure from the standard piston engine. Instead of burning fuel in a confined space that depends on precise timing of ignition, the turbojet engine is essentially an open tube that burns fuel continuously. According to Newton's third law, as hot gases expand out the rear of the engine, the engine accelerates in the opposite direction. The engine consists of three main parts, the compressor, the combustion chamber, and the turbine, along with the inlet, shaft, and nozzle (Figure 5.10).

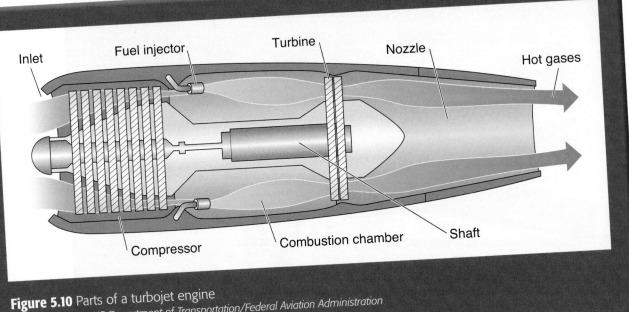

**Figure 5.10** Parts of a turbojet engine
*Reproduced from US Department of Transportation/Federal Aviation Administration*

## Afterburners

Only supersonic high-performance aircraft use afterburners, and they use them for only short periods. Fuel is injected into the hot exhaust stream after the hot gases have passed through the turbine to produce additional thrust, allowing for high speed at a cost of high fuel consumption.

The afterburner uses compressor air not burned in the combustor. It doesn't burn fuel as efficiently as the combustion chamber, which is why an afterburner dramatically increases fuel consumption. It is generally only used on fighter aircraft to gain short bursts of speed such as for short takeoffs and in dogfighting. Otherwise, the aircraft could quickly run out of fuel.

A large mass of air enters the engine through the inlet and is drawn into a rotating compressor. The compressor raises the pressure of the air entering the engine by passing it through a series of rotating and stationary blades on its way to the combustion chamber. As the compressor forces the gas into smaller and smaller volumes, the gas pressure increases. The gas also heats up as the compressor decreases its volume.

Next, a fuel injector injects fuel into the combustion chamber, where it ignites. The energy of the gas rises as its temperature increases. The gas accelerates toward the turbine due to the high pressure created by the compressor. Next, the heated gas passes over the turbine blades causing them to rotate and, in turn, to rotate a shaft connected to the compressor. The turbine removes some energy from the flow to drive the compressor, but there remains sufficient energy in the gas to do work as it exits the nozzle.

The nozzle's purpose is to convert chemical energy into mechanical energy, thus producing thrust. The nozzle allows the flow of hot gases to exit the rear of the engine. Most nozzles restrict the flow somewhat before allowing it to expand. This creates additional pressure and, thus, additional thrust. It also controls the mass flow through the engine, which, along with the velocity, determines the amount of energy the engine produces.

### Wing TIPS

Turbine engines can operate at much higher temperatures and can produce much more thrust than propeller engines. However, they are less efficient at low speeds and low altitudes.

## Turbofans

A *turbofan* is a modified version of a turbojet engine. Both share the same basic core of an inlet, compressor, combustion chamber, turbine, and nozzle. However, the turbofan has an additional turbine to turn a large, many-bladed fan located at the front of the engine. This is a *two-spool* engine. One spool powers the compressor and the other turns the large fan (Figure 5.11).

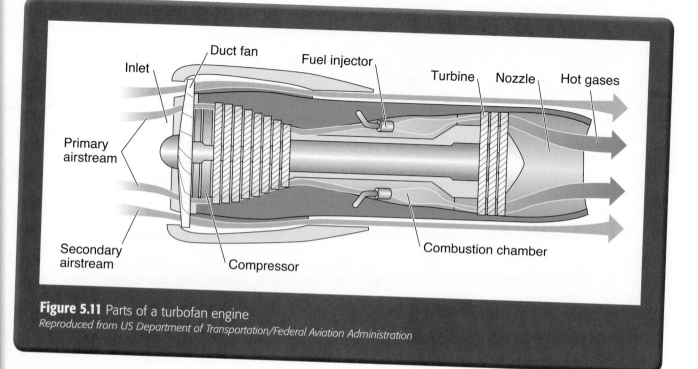

**Figure 5.11** Parts of a turbofan engine
*Reproduced from US Department of Transportation/Federal Aviation Administration*

Some of the air from this large fan enters the engine core, where fuel burns to provide some thrust. But up to 90 percent of the air goes around—or bypasses—the engine core. It is this bypass stream of air that is responsible for the term *bypass engine*. As much as 75 percent of the engine's total thrust comes from the bypass air. This air passes through a fan, which acts as a low-pressure compressor.

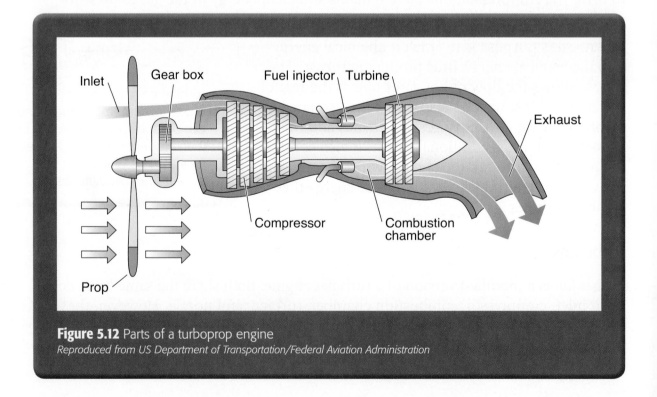

**Figure 5.12** Parts of a turboprop engine
*Reproduced from US Department of Transportation/Federal Aviation Administration*

**CHAPTER 1** How Airplanes Fly

It is then ejected directly as a "cold" jet or mixed with the exhaust to produce a "hot" jet. In addition, the rear end of the bypass area is narrower than the front end, thus creating more thrust.

## Turboprops

This engine is a hybrid of a turbojet and a propeller engine. It has at its heart a turbojet core to produce power but with two turbines. The first turbine powers the compressor while the second powers the propeller through a separate shaft and gear reduction (a gear reduction *reduces* the speed of something, in this case, the propellers). The gears are necessary to keep the propeller from going supersonic and losing efficiency (Figure 5.12).

## Ramjets and Scramjets

Chapter 1, Lesson 3 introduced you to ramjets, which work in conjunction with another power source for initial thrust. A rocket is one such initial power plant. Once up to sufficient speed, the ramjet operates by combusting fuel in a stream of air compressed by the aircraft's forward motion, as opposed to a normal jet engine in which the compressor section (the fan blades) compresses the air. The airflow through a ramjet engine is subsonic, or less than the speed of sound. Ramjet-propelled vehicles operate from about Mach 3 to Mach 6 (Figure 5.13).

*Wing* **TIPS**

The speed of sound is referred to as Mach. Mach 1 is the speed of sound, Mach 2 is twice the speed of sound, Mach 3 is three times the speed of sound, and so forth. The speed of sound is not a constant; it depends on temperature and the air's composition. On a standard day, the speed of sound is about 750 miles per hour.

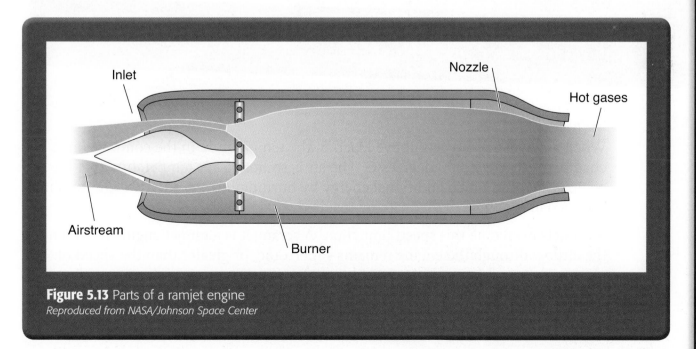

**Figure 5.13** Parts of a ramjet engine
*Reproduced from NASA/Johnson Space Center*

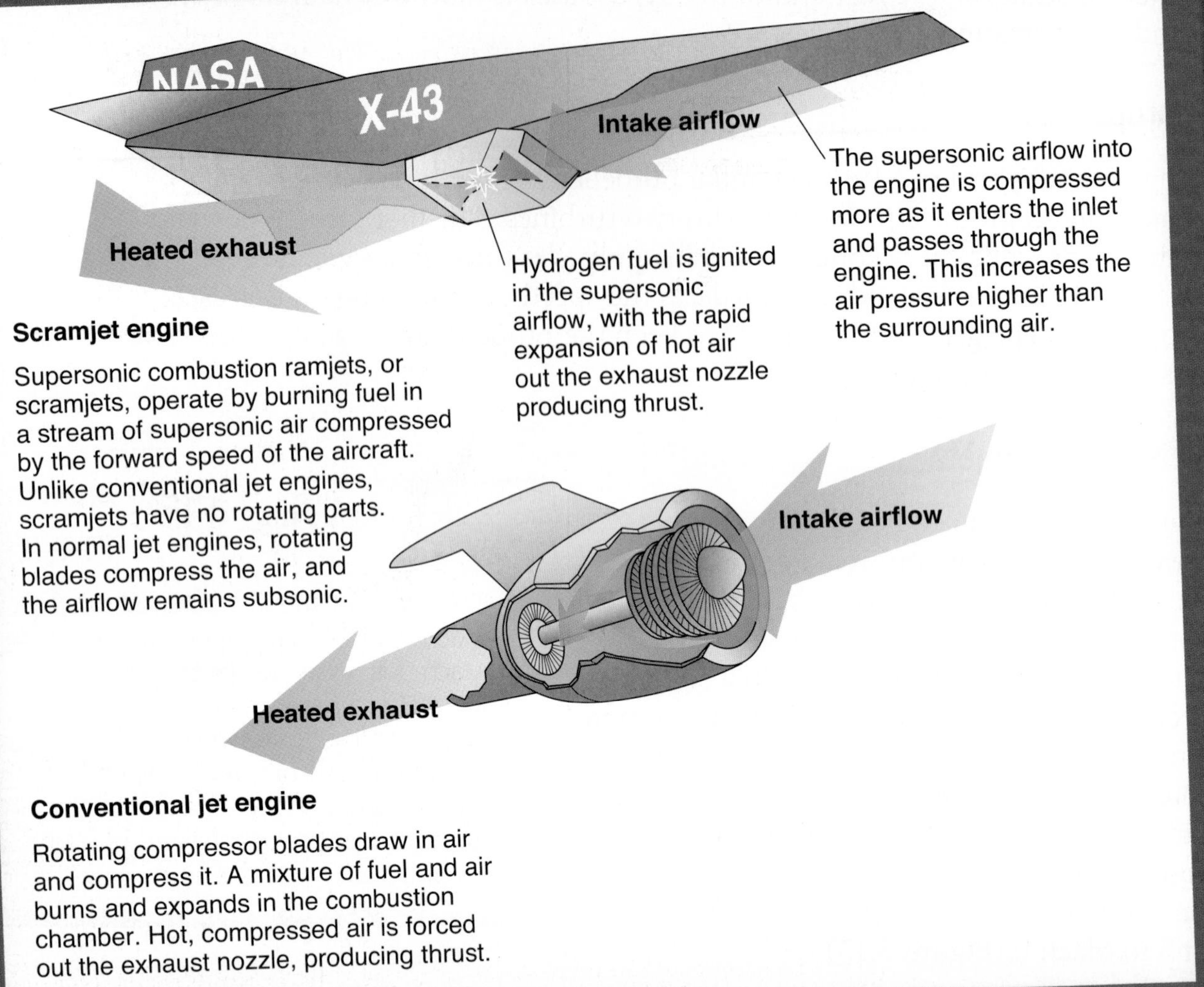

**NASA X-43**

**Intake airflow**

The supersonic airflow into the engine is compressed more as it enters the inlet and passes through the engine. This increases the air pressure higher than the surrounding air.

**Heated exhaust**

Hydrogen fuel is ignited in the supersonic airflow, with the rapid expansion of hot air out the exhaust nozzle producing thrust.

**Scramjet engine**

Supersonic combustion ramjets, or scramjets, operate by burning fuel in a stream of supersonic air compressed by the forward speed of the aircraft. Unlike conventional jet engines, scramjets have no rotating parts. In normal jet engines, rotating blades compress the air, and the airflow remains subsonic.

**Intake airflow**

**Heated exhaust**

**Conventional jet engine**

Rotating compressor blades draw in air and compress it. A mixture of fuel and air burns and expands in the combustion chamber. Hot, compressed air is forced out the exhaust nozzle, producing thrust.

**Figure 5.14** Comparison of a scramjet and conventional jet engine
*Reproduced from NASA's Dryden Flight Research Center*

## Scramjets

When a ramjet's speed gets above Mach 5, the temperature in the combustion chambers exceeds 2,000 degrees C. The air is so hot at this temperature that the engine can't gain much additional energy by burning fuel. In addition, the extreme speeds damage some of the material inside the engine.

Scramjets overcome this speed limitation. A scramjet is a ramjet engine in which the airflow through the engine remains supersonic, or greater than the speed of sound. The word "scramjet" is an abbreviation for *supersonic-combustion ramjet*. A scramjet can't operate at subsonic or even low supersonic speeds, so it needs another engine or vehicle to accelerate it to operating speed (Figure 5.14).

**CHAPTER 1** How Airplanes Fly

In 2004 NASA flew its X-43A scramjet at Mach 9.6. A B-52 carried to 40,000 feet the first stage of a Pegasus rocket with an X-43A attached. It released the Pegasus, which ignited and transported the scramjet to 110,000 feet, where it released the X-43A to fire its scramjet engine.

In 2010 the US space agency test-flew another scramjet, the X-51A. Its air-breathing engine burned for more than 200 seconds and accelerated to Mach 5, or five times the speed of sound. This broke the X-43A's previous record for longest scramjet burn.

## The Role of Reversers and Suppressors Used in Jet Aircraft

As jet aircraft increased speed and altitude, a couple of new problems arose. First, aircraft coming in at a greater speed when landing are harder to stop; and second, they generate lots of noise that bothers communities near airports.

### Thrust Reversers

As you read in earlier lessons, jet aircraft have several ways to come to a stop when landing: aerodynamic braking (spoilers, flaps, and slats), wheel brakes, and the *thrust reverser* (a device that diverts thrust to the opposite direction of the aircraft's motion). Pilots use a combination of these methods when landing.

The thrust reverser can be especially important in difficult landing conditions, such as a wet, slippery runway where wheel braking won't be as effective as on a dry surface.

One type of thrust reverser design that pilots use to change the direction of the exhaust stream is the *clamshell reverser*. It fits over the engine nozzle. When engaged, the reverser opens up like a clamshell. Each half swings back until the two halves meet at the nozzle exit. By forming a shield at the back of the nozzle, the reverser deflects the exhaust so it no longer acts to produce forward thrust.

A clamshell reverser
*Courtesy of Dan Brownlee*

On a large, commercial airliner, according to a design study in the 1990s by General Electric Aircraft Engines, each thrust reverser weighs about 1,500 pounds. They require extra maintenance and increase fuel consumption by as much as 1 percent. Boeing Commercial Airplane Company estimated in another study during the same period that purchasing, installing, and using thrust reversers on a Boeing 767 cost about $125,000 (not adjusted for inflation) per year for each airplane. Boeing's study further showed that while thrust reversers decrease the wear and tear on the wheel brakes, thrust reversers still end up being the more expensive way to slow down an aircraft. Academic and industry researchers continue to work on developing more effective wheel brakes.

A second type of reverser is the *cascade reverser*. A series of airfoils with a high degree of camber opens up to change the airflow's direction. Some engines use just one of these reverse thrust designs; others employ both cascade and clamshell reversers.

## Noise Suppressors

Jet engine noise is also a modern problem, as you might have noticed when a passenger airliner has flown overhead. This is particularly a problem during periods when the aircraft is relatively low to the ground and has high-powered settings. These generally occur during takeoff and climb or during approaches to landing. To protect communities surrounding airports, laws regulate how much noise aircraft can make.

**Wing TIPS**

Airplane noise raised so many complaints from the public that Congress first put noise regulations into law in 1969.

Chevrons are teeth cut into a nozzle's edge to reduce jet exhaust noise.
*Copyright © Boeing. All Rights Reserved.*

The flow of exhaust creates much of the racket. Aircraft engines generate power from expanding gases. Turbine blades move large volumes of air backward to push aircraft forward. All this activity pushes on the surrounding air, causing compression and rarefaction (a thinning out) of the air molecules. This produces pressure waves, which you perceive as sound if they are strong enough and at the right frequencies.

Engineers have designed a number of different noise suppressors over the years to deal with this issue. One invention is the chevron. These are teeth cut into a nozzle's edge to reduce jet exhaust noise. Chevrons change the way engine exhaust mixes with the surrounding air. Another idea is the corrugated noise suppressor. Something that's corrugated has ridges. These ridged nozzles break up the low-frequency noise pouring out in a large exhaust flow.

Yet another invention attaches many smaller tubes (*multi-tube-type*) at the nozzle exit to fracture the exhaust flow. The fourth concept, referred to as an *ejector-type* noise suppressor, directs surrounding air so it mixes with the high-velocity exhaust. Mixing external air with the exhaust slows down the exhaust's velocity and reduces noise. Noise is still a problem, however, and engineers are working to find even better solutions (Figure 5.15).

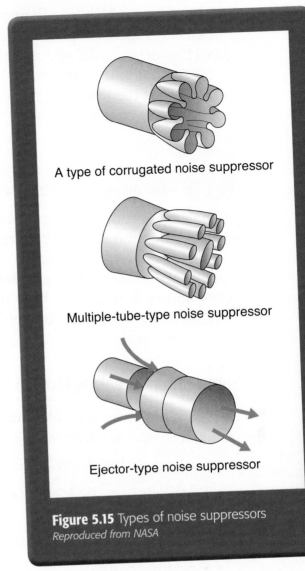

A type of corrugated noise suppressor

Multiple-tube-type noise suppressor

Ejector-type noise suppressor

**Figure 5.15** Types of noise suppressors
*Reproduced from NASA*

## Reaction Engines

The same high-velocity exhaust that makes so much noise is also what makes a jet engine a reaction engine. A reaction engine is *an engine that develops thrust by its reaction to a substance ejected from it; specifically, an engine that ejects a jet or stream of gases created by the burning of fuel within the engine.*

A reaction engine operates according to Newton's third law of motion that to every action, there is an equal and opposite reaction. As you've read, a jet engine produces a high-velocity exhaust that shoots out a nozzle and propels the aircraft in the opposite direction.

Rocket engines, which launch missiles and spacecraft, are also reaction engines. Most rocket engines are internal combustion engines and likewise propel the rocket in the opposite direction of the exhaust flow. Ion engines are yet another kind of reaction engine. They create ions (charged particles) and then eject the ions at high speeds to push vehicles forward. Whereas most rocket engines use chemical reactions for power, ion engines use electric fields.

## The Development of New Engine Technology

Jet engines consume many millions of tons of fuel each year and release harmful gases into the atmosphere. One of these harmful emissions (*a discharge of gases*) is carbon dioxide ($CO_2$), which comes from burning fossil fuels. Another is nitrogen dioxide ($NO_2$), which contributes to the yellow-brown haze you see hanging over cities. Aerospace engineers are working on new engine technologies that aim to cut fuel use and reduce such emissions.

### Some Groundbreaking Moves

Pratt & Whitney, an aerospace manufacturer, recently developed a geared turbofan engine that reduces:

- Fuel consumption (by 12 percent to 15 percent)
- Emissions (e.g., by 1,500 tons of carbon dioxide per plane each year)
- Engine noise (by 50 percent)
- Operating costs.

The company spent 20 years researching, developing, and ground- and flight-testing its new engine, the PurePower PW1000G. This engine allows the fan and the turbine to operate separately, a real breakthrough in technology. This is important because fans, which draw air into an engine for the combustion process, run more efficiently at speeds lower than a turbine's optimum operating speed. Turbines, which exhaust the high-velocity gases exiting the combustion chambers, perform far better at higher speeds. Until Pratt & Whitney's engine, the two parts turned at the same speed. Now that they can work independently of one another, this engine burns less fuel, makes less noise, and breathes out fewer unhealthy emissions.

**CHAPTER 1** How Airplanes Fly

Meanwhile, General Electric (GE) and NASA's Glenn Research Center have teamed up to develop an *open rotor* engine. This open rotor engine is a jet engine with two high-speed propellers outside the nacelle (*a streamlined casing around an engine*). The concept dates from the 1970s when GE and NASA first joined forces on open rotor research to find ways to save fuel. Oil prices were high then, which inspired 20 years of research. Now that oil costs are rising once more, study has resumed.

An open rotor engine
*Courtesy of NASA/Glenn Research Center*

GE and NASA's renewed work on open rotor engine technology is focusing on fuel efficiency, reduced emissions, and noise reduction. Today's more sophisticated technology means advances are even more likely. NASA's test rig, for instance, allows one propeller to spin one way and the other to turn in the opposite direction. Furthermore NASA says that the airfoil shapes of fan blades can now be custom designed to get the best performance.

GE is also working to reduce fuel consumption by using lighter-weight engine material. One avenue of research is the ceramic fan blade for turbines. GE Aviation refers to the material as ceramic matrix composites. Anything that is composite is *made up of a combination of materials.*

GE Aviation says these composites have two main advantages for jet engines. First they are lightweight—in general, one-third the density of pure metals—and so increase fuel efficiency because the aircraft is carrying less weight. Second the composites are tough and more heat resistant than metals, requiring less cooling and so improving engine efficiency and/or thrust.

## Another Breakthrough: Thrust Vectoring

Although not a twenty-first century invention, thrust vectoring is still a groundbreaking engine technology. First tested in the early 1990s, the thrust vector engine has nozzles that turn to redirect thrust. This lets aircraft maneuver with greater precision in the very slow speed, very high angle of attack regime. The aim of this technology is maneuverability, not fuel efficiency.

The F-15 ACTIVE flies over the Mojave Desert. The twin-engine fighter is equipped with Pratt & Whitney nozzles that can turn up to 20 degrees in any direction, giving the aircraft thrust control in the pitch and yaw directions.
*Courtesy of NASA/Jim Ross*

One of the earliest aircraft to fly with thrust vectoring technology was the F-15 ACTIVE (Advanced Control Technology for Integrated Vehicle). NASA conducted flight tests with this aircraft using thrust vector engines built by Pratt & Whitney. Today Pratt & Whitney continues to produce engines with vectored thrust. Two of the manufacturer's F119 engines drive the F-22 Raptor, a fifth-generation fighter.

Many other private and government engineers are working on various advances in engine technologies. Their hope is that future engines will further increase efficiency, reduce noise and pollution, and work with other exotic, lightweight materials.

This lesson has taken you through a comparison of different types of airplane engines, from the internal combustion engine to jet engines to a look at new engine technologies. The next lesson will explore what drives development of aerospace technology and explore some of the new technologies.

# ✔ CHECK POINTS

## Lesson 5 Review

Using complete sentences, answer the following questions on a sheet of paper.

1. What relationship did Robert Boyle examine, and what did he note about it?

2. What did French chemist and physicist Jacques Charles observe in 1787?

3. Where does the combustion process of an internal combustion engine take place?

4. Name the four strokes of an internal combustion engine in order.

5. Which two strokes determine the work available from a reciprocating engine?

6. How many magnetos do most engines have?

7. What happens to propeller tips when they approach the speed of sound?

8. What was the first type of turbine engine developed?

9. When can a thrust reverser be especially important?

10. When is jet engine noise particularly a problem?

11. What is a reaction engine?

12. A reaction engine operates according to which law?

13. What did GE and NASA hope to find when they first joined forces on open rotor research in the 1970s?

14. What does a thrust vector engine have?

# APPLYING YOUR LEARNING

15. What other ways can you think of—not discussed in this lesson— that could further reduce how much fuel an aircraft uses?

# LESSON 6

## Aviation Innovation

### Quick Write

_____

_____

Why do you think Richard Whitcomb's inventions were such important advances in aircraft design?

### Learn About

• the latest topics of aviation research
• the use of remotely piloted aircraft
• the most recent innovations in aircraft design

Without Richard T. Whitcomb, supersonic flight wouldn't be such a smooth ride. Whitcomb was an aviation pioneer whom NASA has dubbed "the most significant aerodynamic contributor of the second half of the twentieth century." The celebrated engineer made three major discoveries that reduced drag on aircraft during his career with the aeronautics and space agency.

Whitcomb started at what is now NASA's Langley Research Center in 1943—during World War II. He had just graduated with highest honors from Worcester Polytechnic Institute with a degree in mechanical engineering. One of his first challenges was how to achieve transonic (_aircraft speeds nearing the speed of sound_) and supersonic flight. When aircraft neared the speed of sound, they were hit with a sharp increase in drag. Whitcomb and other scientists spent years trying to understand what caused this drag.

One day in the 1950s, while sitting with his feet propped up on his desk, an idea struck Whitcomb. He had learned from a visiting German scientist that although air flows easily around a plane at lower speeds, the area of the airflow remains constant at higher speeds and can no longer flow smoothly around the airplane structure. Whitcomb had been working on streamlining the wings and fuselage as independent areas of a plane, but now the notion came to him to consider the whole aircraft body at once—wings, fuselage, and tail. In particular, he focused on an aircraft's cross-sectional area where wings and fuselage meet. He used the curved shape of a Coke bottle as his model. His newly shaped fuselages tested beautifully in the wind tunnels. The increased drag disappeared. This is often referred to as _Whitcomb's area rule_.

In the 1960s Whitcomb struck aerodynamic gold again. He invented the *supercritical wing* (airfoil) for jet liners. These airfoils were flatter on top and rounder on the bottom, with a downward curve on the trailing edge. This innovative shape delayed the formation of the supersonic shock wave that forms just below and just above the speed of sound. This increased fuel efficiency as the airliners neared Mach 1.

In the next decade Whitcomb came up with the *winglet*, which you first read about in Lesson 3 of this chapter. Other engineers had thought of adding plates to wingtips to reduce drag. But an article Whitcomb read about birds sparked a more complete idea. "It is a little wing. That's why I called them winglets," he said. "It's designed with all the care [with which] a wing was designed." The winglet reduces drag as well as fuel consumption.

Whitcomb retired in 1980 and passed on in 2009 at the age of 88. Yet his three aerodynamic contributions continue to impress. "Dick Whitcomb's three biggest innovations have been judged to be some 30 percent of the most significant innovations produced by NASA Langley through its entire history," said Langley chief scientist Dennis Bushnell, who worked with the legendary engineer.

## Vocabulary

- transonic
- biofuel
- acoustic
- decibel
- air traffic control
- autonomous
- prototype
- micro-UAV
- nanotechnology
- nano-UAV
- cure
- tensile strength
- compressive strength

Richard Whitcomb looks at a model that sports his supercritical wing concept.
*Courtesy of NASA*

## The Latest Topics of Aviation Research

Aircraft research is taking dramatic turns in the twenty-first century. As you read in the previous lesson, some of the main goals driving innovation include cutting fuel consumption, emissions, and noise. Engineers are studying lighter materials and new fuels, among other possibilities. Shorter travel time that incorporates green goals is another big aim.

### Hypersonic Aircraft

If you've ever traveled long distances—East Coast to West Coast, North America to Europe, or Los Angeles to Tokyo, for instance—then you know how very long and tiring a long-distance flight can be. Imagine cutting the travel time from Los Angeles to New York from six hours to no more than 35 minutes!

**Figure 6.1** The HyperSoar would bounce along Earth's atmosphere, giving the aircraft improved fuel efficiency and shorter flight times.
*Reproduced from the Lawrence Livermore National Laboratory/DOE*

Building on decades of research in hypersonic flight, engineers at the US government's Lawrence Livermore National Laboratory in California have come up with a futuristic concept that would do just that. It's called the HyperSoar. While still just in the conceptual phase, this aircraft would be capable of taking off from any airport, traveling at Mach 10 (10 times the speed of sound), and climbing to 40 kilometers (about 24.8 miles) to skip along Earth's atmosphere to its destination (Figure 6.1).

Once the aircraft reaches the 40-kilometer mark, its engines would be turned off and it would coast up to about 60 kilometers. At that point it would fall back down to about 35 kilometers, well inside the atmosphere's upper level. Here it would be pushed up by aerodynamic lift, and the engines would fire briefly, propelling it back into space. Outside the atmosphere, the engines would shut off and the process would repeat. Each skip would take about two minutes.

Besides saving time, HyperSoar would burn liquid hydrogen, which emits water vapors. Liquid hydrogen is a clean fuel. Also, by spending so much of its travel distance out of Earth's atmosphere, HyperSoar should be able to radiate any heat it generates into space. To deal with heat issues, engineers generally add heavy, strong materials to engines and airframes (another name for an aircraft body). This added weight cuts into fuel efficiency and decreases the amount of cargo an aircraft can carry. HyperSoar's trajectory along the top of Earth's atmosphere addresses these problems. HyperSoar could also have many other uses: move passengers and cargo, deliver satellites to space, or bomb enemy targets. These other uses would bring down the aircraft's cost.

In 2004 NASA and the Air Force successfully tested the scramjet engines such an aircraft would need with the X-43A hypersonic technology demonstrator. Another successful flight took place in 2010, when a scramjet engine burned for more than 200 seconds to power the Air Force's X-51 vehicle to five times the speed of sound (Mach 5).

## New Fuels

Another area of aviation research concerns new fuels for powering aircraft. In 2008 Boeing and several international partners flew a small motor-glider powered by hydrogen fuel cells. They made history with that manned flight.

### Hydrogen Fuel Cells

This two-seat motor-glider used a combination of hydrogen fuel cells and lithium-ion batteries to power an electric motor that turned a propeller. Once the plane reached 3,300 feet—cruising altitude—the pilot disconnected the batteries. For the next 20 minutes, the plane flew 62 miles per hour using only power generated by the fuel cells.

In 2008 Boeing Research & Technology Europe test-flies a manned airplane powered by hydrogen fuel cells, which exhaust only heat and water.
Copyright © Boeing. All Rights Reserved.

Hydrogen fuel cells are electrochemical devices that convert hydrogen (a gas) into electricity and heat. They do not produce any of the typical products of combustion such as carbon dioxide. Instead, they exhaust only heat and water.

## Biofuels

Researchers in both the Air Force and private industry are exploring the use of biofuel, *a fuel made from plants.* In March 2011, an Air Force F-22 Raptor successfully flew at Mach 1.5 on a 50/50 blend of conventional petroleum and biofuel derived from *camelina*, a weedlike plant not used for food.

The flight was a milestone in the Air Force's goal to obtain 50 percent of its domestic fuel requirement using alternative fuel blends. These blends should come from domestic sources produced in a way that is "greener"—more environmentally friendly—than fuels produced from conventional petroleum.

In February 2011, the Air Force certified the entire C-17 Globemaster fleet for flight operations using a biofuel blend. The C-17 was the first Air Force aircraft to receive such certification.

Several aircraft manufacturers have also combined efforts to study biofuel. In early 2008 Boeing, Virgin Atlantic, and GE Aviation all took part in the first commercial airline flight with a biofuel mix. Virgin Atlantic airlines flew a Boeing 747 from London to Amsterdam with a 20 percent biofuel/80 percent kerosene mix in one of four engines.

In its *2009 Environment Report*, Boeing listed among its reasons to study biofuels a desire to be less dependent on fossil fuels, especially when oil prices can be unpredictable. The company also said that the types of plants it's investigating would not compete with food crops (unlike ethanol, which uses corn), and it argued that plants absorb carbon dioxide, a greenhouse gas. The report further stated that a plant such as algae could produce up to 2,000 gallons of oil per year per acre.

## It's Not Just Science Fiction Anymore

Engineers at NASA have built an amazing machine that's reminiscent of the replicator from *Star Trek: Next Generation*. That fictional replicator let Captain Picard order a cup of his favorite tea, and out it would pop from a hole in the wall. NASA's real-life contraption requires only a drawing and the right materials—and the Electron Beam Freeform Fabrication, or EBF3, does the rest.

"You start with a drawing of the part you want to build, you push a button, and out comes the part," says Karen Taminger, technology lead for the research project that's a branch of NASA's Fundamental Aeronautics Program in Virginia.

The technique is called *rapid prototyping*, a technology that first became available in the late 1980s. It's been evolving ever since.

The EBF3 actually manufactures objects one layer at a time. It works in a vacuum chamber in which an electron beam focuses on a source of metal constantly being fed in. The beam melts the metal, which is then applied one layer at a time on top of a rotating surface. The drawing supplied to the EBF3 must be very detailed and three-dimensional. The object created must also be made out of materials compatible for use with an electron beam.

The Electron Beam Freeform Fabrication made this structural metal part.
*Courtesy of NASA*

Once perfected, aircraft manufacturers could use the EBF3 to build major structural parts of a plane for much less per pound than on a typical manufacturing floor. The electron beam concept is also more environmentally friendly. For instance, rather than having to trim a 6,000-pound block of titanium to a 300-pound part, with all the rest left as recyclable waste, the EBF3 could work with a 350-pound initial block and have only 50 lbs. of recyclable material left over.

## Noise Reduction

As you read in the last lesson, researchers are trying to reduce aircraft noise. Noise suppressors such as chevrons are one way to muffle the racket. But engineers are working on a number of other ideas as well.

NASA acoustic researchers have been tinkering with metallic foam made from stainless steel that they install around engines. (Acoustic is *having to do with sound*.) The foam is firm, tightly packed like a honeycomb, gently abrasive, and lighter than you might expect. While just about any foam will absorb noise, most would catch fire from engine heat. Metallic foam solves that problem. NASA has joined forces with a Michigan-based company called Williams International to refine the concept in a way that doesn't add weight or cost, or hurt performance.

Taking a wider view, NASA is also crafting a plan for noise-reduction strategies 10, 20, and 30 years out. The space agency refers to this as "next" generation technology, or N+1 for 10 years out, N+2 for 20 years into the future, and N+3 for anything 30 years and beyond. For instance, N+1 aircraft would look just like today's aircraft with a tube-shaped fuselage. But engineers could modify the noisy parts, such as flaps and slats, by doing away with their sharp edges. They could also slow fan speeds or streamline landing gear.

N+2 aircraft might meld wings and body into one smooth unit and mount engines on top of the "blended or hybrid wing body." By placing the engines on top, the airplane body itself could block noise from reaching communities below.

This test rig evaluates the noise reduction of a newly developed metallic foam liner at NASA's Glenn Research Center in Cleveland.
*Courtesy of NASA*

A model of an N+2 aircraft with top-mounted engines and a blended wing body
*Courtesy of NASA*

No one really knows what an N+3 aircraft will look like. One of the trickiest aspects to inventing new technologies is making sure they don't improve flight in one way while damaging it in another. As NASA aerospace engineer Edmane Envia says, "What makes our job very hard is that we are asked to reduce the noise but limit as much as possible any negative impacts on the overall performance characteristics of the airplane, including fuel burn, aircraft weight, and range."

## Wing TIPS

A decibel is *a measure of loudness*. A busy city street produces about 70 decibels of noise. A Boeing 737 churns at about 85 decibels to 100 decibels. NASA's goal for N+1 aircraft is to reduce noise by about 32 decibels from a Boeing 737's decibel level.

## Air Traffic Control

How efficiently planes can take off, fly, and land also impacts noise pollution, emissions levels, and fuel use. Since World War II, the United States has relied on radars to manage its military and commercial air traffic. This system currently services a huge number of planes. For instance, at any given moment, 5,000 airplanes—civilian and military—are in the air over the United States. In 2009 they carried 689 million passengers around the country. In addition, planes move an average of 36 billion pounds of cargo each year. The Federal Aviation Administration (FAA), which monitors aviation, expects air traffic to grow 50 percent by 2025.

Certain authorities such as the FAA are working on new technologies to improve air traffic efficiency and safety, particularly as these numbers continue to grow. They have formed a plan called NextGen that uses satellite technology, much like the GPS people use in their cars. Introducing NextGen is an ongoing process. Its satellite-based system should provide information to pilots and air traffic control (*a management system for coordinating air traffic at airports and in the air*) on the ground in real time. This would allow planes to safely fly closer together, take more direct routes, and be more aware of their position relative to other aircraft. These more precise flight paths should result in fewer delays for passengers and thus lower fuel consumption, emissions, and noise.

The effects are noticeable on the ground as well. Because of this still-evolving system's improved precision, airports can install runways closer together. For instance, in 2008 O'Hare (Chicago), Dulles (Washington, DC), and Seattle airports installed additional runways within their existing airfield footprints. With these new strips, the three airports, together, can handle 300,000 additional flights per year. This speeds the rate of takeoff and landing patterns, which decreases delays, fuel use, noise, and emissions.

### Continuous Descent Approach

NASA has also been developing a fuel-saving idea called *continuous descent approach*. It tested this idea in 2007 and 2009 at different US airports. Today when planes land, they follow an arrival path that looks like a set of stairs. It is not very efficient because it requires planes to frequently change altitude, direction, and speed to keep a safe distance from other aircraft. NASA's continuous descent approach allows airplanes to coast during their final flight stages, which uses less power. Planes coming in at a continuous, gliding descent at low engine power cut fuel burn and all the emissions and noise that go with it.

**CHAPTER 1**   How Airplanes Fly

Of course, the new landing approach requires careful coordination. Using continuous descent approach, planes must follow specific landing routes at specific speeds for safety's sake. NASA is researching a system called Efficient Descent Advisor, which will be a tool for air traffic controllers. The agency estimates that this system could cut fuel use for a large aircraft by up to 3,000 pounds per flight and carbon dioxide emissions by about 10,000 pounds. That's 27 percent less than normal. Research is continuing on the system.

## The Use of Remotely Piloted Aircraft

Another heavily researched area is the unmanned aircraft system (UAS), also referred to as unmanned aerial vehicle, or UAV. A UAS consists of one or more aircraft, along with equipment and operations and maintenance personnel. The US military and intelligence services use UASs for reconnaissance and combat. (You may sometimes see them referred to in the press as *drones*.) They've flown many missions in recent years, including into Afghanistan, Iraq, and Pakistan. Because pilots control the aircraft remotely—from the ground rather than from the cockpit—UASs are also called *remotely piloted aircraft*, or RPAs.

One of the more familiar RPAs is the Predator, which has seen a lot of action overseas. It is propeller driven; conducts intelligence, surveillance, and reconnaissance (ISR); and is armed. The earliest armed Predator generation—designated the MQ-1—first flew in 1994 and has around 920,000 hours in the air. The Air Force received its very last MQ-1 in March 2011 from manufacturer General Atomics Aeronautical Systems. Each costs about $5 million, and the Air Force has bought 268 of them in all.

### Wing TIPS

The letters and numbers in MQ-1 mean something. The *M* stands for multirole (intelligence, surveillance, reconnaissance, and kill capability), the *Q* stands for unmanned aircraft system, and the *1* indicates the aircraft is the first of the series of remotely piloted aircraft systems. Some aircraft have the initials *RQ*. The *R* stands for reconnaissance; these aircraft do not carry weapons.

General Atomics has also built a larger and more powerful generation of Predator called the MQ-9 Reaper for the Air Force. It has a turboprop engine. Like the MQ-1, it can be taken apart and shipped in a single container for duty anywhere in the world. Each MQ-9 costs about $13 million. The company has recently manufactured yet another Predator generation called the Predator C with a jet engine.

The RQ-4 Global Hawk is another well-known UAS. It is a large-size UAS, compared with the MQ-9 and MQ-1, which are both midsize. Its function is solely intelligence gathering, so it flies at high altitudes. It went to work in November 2001 to support the global war on terror. The Global Hawk's power plant is a turbofan. Northrop Grumman is the prime contractor, and each Global Hawk costs anywhere from $55 million to $81 million.

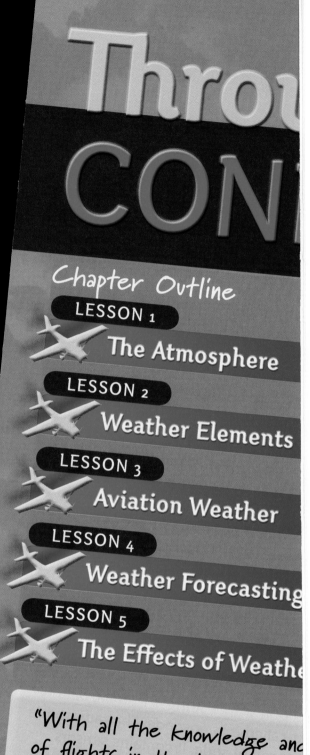

Chapter Outline

"With all the knowledge and of flights in the last ten yea today of making my first fligh in a twenty-seven mile wind, the machine had already been

# LESSON 1

## ✈ The Atmosphere

### Quick Write

_____

_____

What are some of the advantages to using low-tech equipment? What are some of the disadvantages?

### Learn About

• the atmosphere's regions
• the roles of water and particulate matter in the atmosphere
• the primary causes of atmospheric motion
• the types of clouds
• how the atmospheric layers impact flight

E ven in the twenty-first century some science tools remain low-tech. Weather offices have been using balloons since 1918 to measure different atmospheric qualities. These measurements ultimately help pilots fly more safely through the atmosphere because forecasters use the data to identify severe weather.

The balloons are filled with lighter-than-air gases, such as hydrogen, helium, or natural gas. Latex balloons form nice round spheres and rise fast and uniformly into the sky. Neoprene rubber balloons flatten out the more they rise. They also climb more slowly and less uniformly than latex balloons.

The National Weather Service, a part of the US government's National Oceanic and Atmospheric Administration (NOAA), releases balloons twice each day from about 100 locations across the United States. This adds up to nearly 75,000 balloons per year. The balloons ascend more than 100,000 feet (20 miles) into the atmosphere. Over the course of their two-hour flight, they can drift up to 200 miles from their launch site. These balloons provide a meteorologist (_a person who forecasts the weather_) data to predict everything from thunderstorms and hurricanes to aircraft icing, jet stream positions, and temperatures.

Attached to each balloon with a string is a small, battery-powered instrument called a _radiosonde_. The radiosonde is barely larger than a can of soda. It contains sensors that measure humidity, temperature, air pressure, wind speed, and wind direction. Ground-based radar tracks the radiosonde and collects and processes the weather data.

Rain/ T'Storm

Snow

Rain

Rain/ Sno

Rain/ T'Storms

Rain/ T'Storms

Weather Forecast for Tue, Sep 27, 201
DOC/NOAA/NWS/NCEP/Hydrometeor
Prepared by Ryan based on HPC, SPC

This NOAA map shows the weathe
Note the large low-pressure system
thunderstorms are possible.
_Courtesy of NOAA_

Not only do local weather offices use this information to forecast the weather, but *all* of the data also goes to a supercomputer in Washington, DC, which generates a computer model of the atmosphere.

When a balloon has risen as far as it can tolerate, it pops. A small parachute opens, which gently floats the radiosonde back to Earth. If you ever find one of these instruments tethered to a deflated balloon, it should have the message "Harmless Weather Instrument" printed on it. It may also smell like a rotten egg and make some strange noises, but don't be alarmed! That's just the nature of this equipment. Each instrument has an addressed, postage-paid return mailbag attached to it, so you can mail it back to the National Weather Service. Currently, the weather service only gets back 20 percent of the radiosondes it sends into the upper-air atmosphere each year. It has to replace the other 80 percent.

NOAA weather balloon launch
*Courtesy of NOAA*

## Vocabulary

- meteorologist
- atmosphere
- jet stream
- evaporation
- sublimation
- condensation
- deposition
- humidity
- relative humidity
- dew point
- saturated
- precipitation
- particulate matter
- nucleus
- atmospheric pressure
- latitude
- ceiling
- cumulonimbus clouds
- wind shear
- cabin altitude

## The Atmosphere's Regions

The atmosphere is *a blanket of air that surrounds Earth and consists of a mixture of gases.* It extends more than 350 miles from Earth's surface, but the air grows thinner the farther away you get from the planet. This mixture is in constant motion, like an ocean with its waves, swirls, and eddies.

Life on Earth is supported by the atmosphere as well as by solar energy and the planet's magnetic fields. The atmosphere absorbs energy from the sun, recycles water and other chemicals, and works with electrical and magnetic forces to provide a moderate climate that can support organic life. It also protects Earth from high-energy radiation and the frigid vacuum of space.

Nitrogen accounts for 78 percent of the atmosphere's gases; oxygen represents about 21 percent; and argon, carbon dioxide, and traces of other gases account for the remaining 1 percent (Figure 1.1). The atmosphere also contains some water vapor, which varies from 0 percent to about 5 percent by volume. This small amount of water vapor is responsible for major changes in the weather.

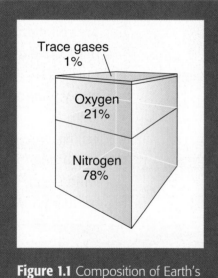

Trace gases
1%

Oxygen
21%

Nitrogen
78%

**Figure 1.1** Composition of Earth's atmosphere
*Reproduced from US Department of Transportation/Federal Aviation Administration*

## Five Distinct Layers

Five distinct layers compose Earth's atmosphere. Scientists identify each one by differences in temperature, chemical composition, movement, and density.

The lowest layer of the atmosphere, the *troposphere*, reaches from sea level up to anywhere from about four miles (eight km or 20,000 feet) at the poles to about nine miles (14.5 km or 48,000 feet) at the equator. This is a dense stretch of atmosphere where most weather takes place, along with clouds, storms, and temperature changes. The temperature drops at a rate of about 3.5 degrees F (2 degrees C) for every 1,000 feet gained in altitude. The pressure decreases as well at a rate of about one inch per 1,000 feet altitude gain. A boundary called the *tropopause* caps the troposphere and traps moisture and weather in the troposphere.

# The Jet Stream

The jet stream is *a strong current of air that generally sits atop the troposphere and flows from west to east.* Scientists often refer to it as a "river" of air because of its shape—up to thousands of miles long, a few hundred miles wide, and only a few miles deep. This stream can move along at more than 275 miles per hour.

The jet stream signals the boundary between hot and cold air in the atmosphere. The Northern and Southern Hemispheres sometimes have two jet streams each, the *subtropical jet* around 30 degrees North and South and the *polar jet* at 50 degrees to 60 degrees N/S (Figure 1.2). These are the latitudes where the temperature changes are often greatest. In the United States, the subtropical jet flows along the country's southern border, while the polar jet flows overhead somewhat following its northern border. Depending on the season and conditions, though, the jet streams can change latitudes and altitudes, split into narrower streams, or even disappear for a time.

Pilots of big commercial and military aircraft that fly at high altitudes keep track of the jet streams. If a plane is traveling from west to east, the jet stream provides a tailwind that can help the airplane make good time. But if the plane is traveling from east to west, the jet stream becomes a headwind that increases flying time and cuts fuel efficiency, so pilots try to avoid it when they can.

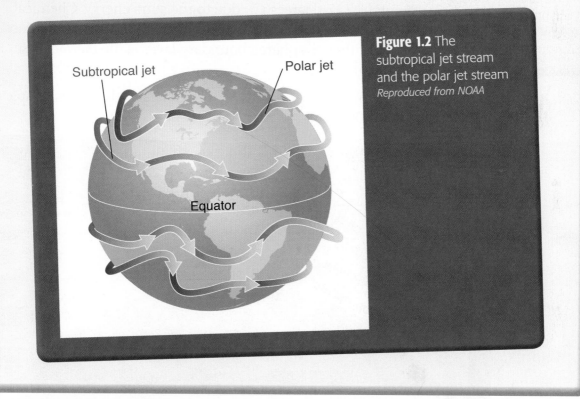

**Figure 1.2** The subtropical jet stream and the polar jet stream
*Reproduced from NOAA*

The *stratosphere* is the next layer of the atmosphere. It starts above the troposphere and extends to about 30 miles (50 km or 160,000 feet) high. This second layer is drier and less dense than the troposphere. The temperature actually begins to rise again in this region to about 26.6 degrees F (−3 degrees C) as you gain altitude because the stratosphere absorbs ultraviolet radiation from the sun. The ozone layer, which absorbs and scatters solar ultraviolet radiation, resides here. The stratosphere has a boundary layer called the *stratopause* that separates it from the next layer.

The *mesosphere* runs from above the stratosphere to about 53 miles (85 km or 280,000 feet) high. Temperatures dip once more, as low as −135.4 degrees F (−93 degrees C), with increases in altitude. The chemicals in this layer exist in an excited state because they absorb energy from the sun. The boundary layer at the top of the mesosphere is the *mesopause*.

The *thermosphere* starts above the mesosphere and extends to more than 350 miles (560 km). The temperature takes yet another swing, going as high as 3,140.6 degrees F (1,727 degrees C), due to the sun's energy. Chemical reactions occur much faster here than on Earth's surface. The thermosphere's boundary layer is the *thermopause* (Figure 1.3).

Beyond the thermosphere is the *exosphere*, which stretches to the edges of space at around 6,200 miles (10,000 km).

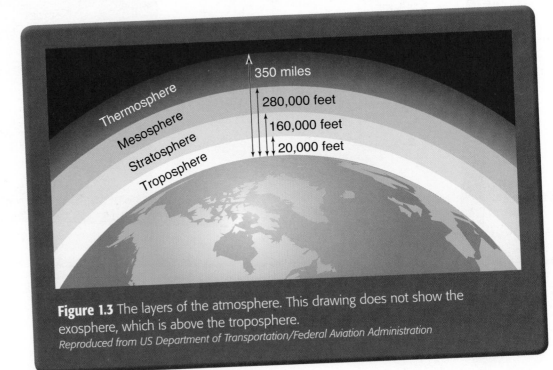

**Figure 1.3** The layers of the atmosphere. This drawing does not show the exosphere, which is above the troposphere.
*Reproduced from US Department of Transportation/Federal Aviation Administration*

# The Roles of Water and Particulate Matter in the Atmosphere

As you read earlier, the atmosphere contains water in the form of water vapor. Temperature determines how much moisture the atmosphere holds. A rise in temperature increases the amount of moisture the air can hold, just as a drop in temperature decreases that amount.

Water in the atmosphere takes three forms: liquid, solid, and gas. Each form can change into another form. When they do change, a heat exchange occurs. These changes come about through the processes of:

- Evaporation—*The transformation of a liquid to a gaseous state, such as the change of water to water vapor*

- Sublimation—*The process by which a solid changes to a gas without going through the liquid state*

- Condensation—*A change of state of water from a gas—water vapor— to a liquid*

- Deposition—*The process by which a gas changes to a solid without going through the liquid state*

- Melting

- Freezing.

However, water vapor enters the atmosphere only through evaporation and sublimation.

## Evaporation and Sublimation

Oceans hold 97 percent of the world's water, and they are the source of 86 percent of the evaporation taking place around the planet. Besides moisture, however, evaporation requires heat. This heat destroys the bonds between water molecules and allows them to evaporate. When water evaporates from the oceans, the air loses heat when the liquid water changes into vapor. This is known as the *latent heat* of evaporation. It cools ocean surfaces (just as sweat cools your skin).

The warm water vapor enters the atmosphere and generally condenses onto tiny particles of dirt, dust, or pollution in the atmosphere because of the colder temperatures aloft. The condensing water then forms clouds. The water in these clouds can fall as rain, snow, or in some other form over the oceans or land. Moisture falling over the land seeps into the soil or groundwater through *infiltration*. Evaporation takes place at other sources, too: lakes, rivers, trees, and even soil (in the case of plants and soil, the process is called *transpiration*— Figure 1.4).

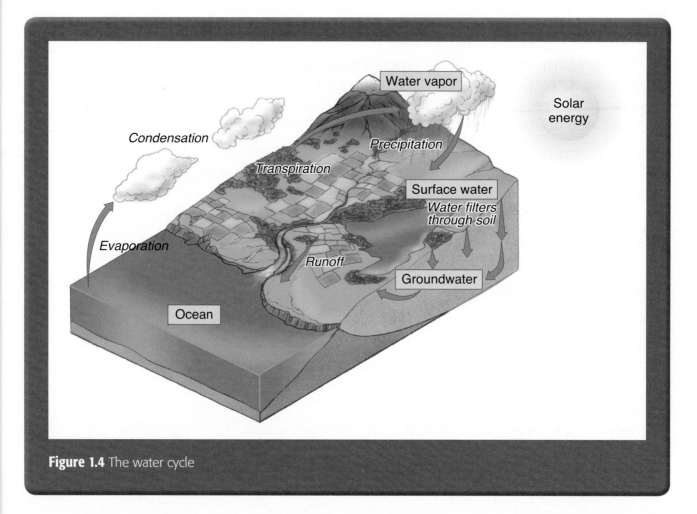

**Figure 1.4** The water cycle

Sublimation contributes far less water vapor overall to the atmosphere than the evaporation process does. It generally takes place in colder climates than evaporation, as well. For instance, the temperatures will often be too cold on a mountaintop to melt snow and ice to liquid water, but intense sunlight sometimes coupled with strong winds can transform that snow and ice into water vapor. In the American West, scientists have a name for a wind so strong and so dry in the region that it turns water solids straight into gas with no melting in between: Chinook, after an Indian tribe that lives along the Pacific Northwest coast.

## Humidity

Humidity is *the amount of water in the atmosphere at a given time*. Relative humidity is *the actual amount of moisture in the air compared with the total amount of moisture the air could hold at that temperature*. For example, if the current relative humidity is 65 percent, the air is holding 65 percent of the total amount of moisture that it is capable of holding at that temperature and pressure. While much of the western United States rarely sees days of high humidity, relative humidity readings of 75 percent to 90 percent are common in the southeastern United States during the warmer months.

**CHAPTER 2**   Working Through Flight Conditions

## Dew Point

The relationship of dew point and temperature further explains relative humidity. Dew point is *the temperature at which air can hold no more moisture*. When temperatures fall to the dew point, the air becomes saturated—or *as full of moisture as something can get*—and the water in the atmosphere condenses as fog, dew, frost, clouds, rain, hail, or snow. However, rain is the most common form of precipitation—*any or all forms of water particles that fall from the atmosphere and reach Earth's surface*.

Air can reach its complete saturation point by four methods. First, when warm air moves over a cold surface, the air temperature drops and reaches the saturation point. Second, when cold air mixes with warm air. Third, when air cools at night through contact with the cooler ground. And fourth, when air rises upward in the atmosphere, it uses heat energy to expand. As a result, the rising air loses heat rapidly, and so arrives at its saturation point.

## Particulate Matter

Water vapor goes through a complex process before it can fall as rain. Water vapor and cloud droplets make up a cloud. Both are very, very small. However, when you look at a cloud, what you see is not the invisible water vapor but the cloud droplets. Even so, these droplets aren't big enough or heavy enough to fall as rain. In fact, it takes millions of cloud droplets to form one raindrop.

A shelf cloud rolls over Oklahoma in 2008.
*Courtesy of Sean Waugh NOAA/NSSL*

To create a single raindrop, water vapor must first cling to particulate matter (*material suspended in the air in the form of minute solid particles, such as dust, salt, or smoke particles*) to condense into cloud droplets. The bits of particulate matter are smaller than the water vapors, yet they are necessary to the formation of raindrops because they act as the nucleus—*or core*—of the raindrop. The cloud droplets collide with other droplets and continue to grow until they are heavy enough to fall to Earth.

## The Primary Causes of Atmospheric Motion

In addition to affecting moisture levels, heat also causes air to circulate around Earth's surface. Scientists refer to this as *atmospheric circulation*. While other factors play a role in moving the air about, a main factor is energy radiation from the sun.

When the sun heats the atmosphere, air molecules spread apart and the warm air rises. As the air expands, it becomes less dense and lighter than the surrounding air. Cooler air, with tightly packed air molecules, sinks and replaces the warmed air. This process of rising warm air and heavy, sinking cool air results in the atmosphere's circular motion.

Furthermore, because Earth rotates on a tilted axis while orbiting the sun, some regions of the planet receive more heat at any given time than others. This also affects atmospheric circulation. For instance, because more heat from the sun reaches the equator, the air there is less dense and rises. This warm air flows toward the poles where it cools and sinks.

### Atmospheric Pressure

The unequal heating of Earth's surface also affects air pressure. The invisible gas particles that make up the atmosphere create atmospheric pressure. These gas particles—or air molecules—have weight and take up space. Therefore, atmospheric pressure is *the weight of air molecules*. Altitude—along with temperature and air density—determines how much pressure these particles apply.

Pressure is greatest at sea level where the gas particles are close together. The weight of the air above compresses the molecules below and so increases the pressure. As altitudes increase, the molecules begin to spread apart so that there's more space between them, causing the pressure to drop.

The actual pressure at a given place and time changes depending on altitude, temperature, and air density. These conditions are important for flight, particularly during takeoff, climb, and landing. Scientists measure atmospheric pressure with instruments called *barometers*. When the pressure is high, the weather should be good. But when the barometer shows a drop in pressure, weather conditions will deteriorate. You'll read more about air pressure's effect on weather in the next lesson.

## Coriolis Force

In 1835 a French scientist named Gustave-Gaspard Coriolis came up with a theory to further explain atmospheric circulation. His discovery—called the *Coriolis force*—describes how Earth's rotation affects the motion of air.

Low-pressure areas lie at Earth's warm equator regions. High-pressure areas sit over the cold polar regions. If Earth were stationary, these high-pressure areas would flow along Earth's surface straight toward the equator. But Earth isn't still. Its rotation about its axis creates the Coriolis force, which affects large bodies such as air masses.

This force deflects air to the right in the Northern Hemisphere. The air takes a curved path rather than flowing in a straight line. In the Southern Hemisphere, the Coriolis force pushes the air left. The air once again curves rather than flowing in a straight line.

The size of the curve depends on latitude (*a line north or south from Earth's equator and parallel to it*) and the moving body's speed. The deflection is greatest at the poles and decreases to zero by the time you reach the equator. And the greater the air's speed, the more the curve increases.

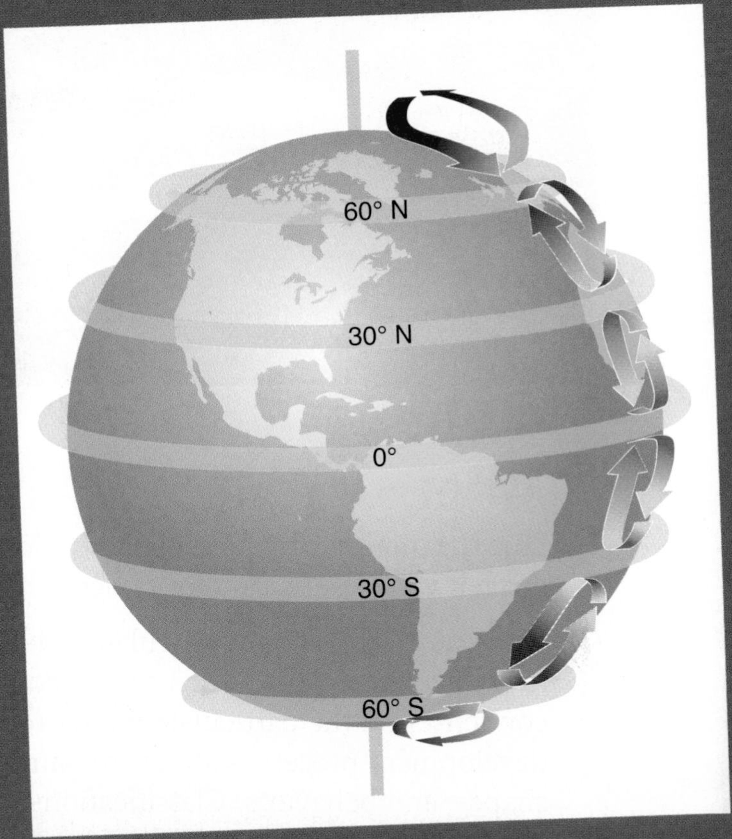

The speed of Earth's rotation breaks up the flow of air into three cells at different latitudes in each hemisphere (Figure 1.5). In the Northern Hemisphere, these cells are from the equator to 30 degrees N, from 30 degrees N to 60 degrees N, and from 60 degrees N to the North Pole. From the equator to 30 degrees N, warm air rises from the equator, moves north, and curves east due to Earth's rotation. The air cools as it moves, sinks at around 30 degrees N to form a high-pressure area, flows south along Earth's surface, and bends right due to the Coriolis force. This creates northeasterly trade winds from the equator to 30 degrees N. The other latitudes also experience circulation cells. The continental United States, for instance, sees westerly winds.

### Wing TIPS

Actually, the Coriolis force affects everything, even humans walking on Earth's surface, but the effects of the force are only noticeable over long distances and on objects of great size, such as air masses and oceans.

**Figure 1.5** Circulation cells in the Northern and Southern Hemispheres due to Earth's rotation
*Reproduced from US Department of Transportation/Federal Aviation Administration*

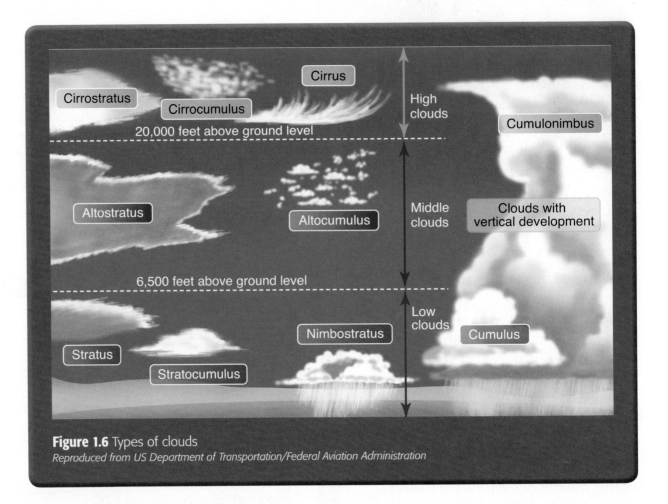

**Figure 1.6** Types of clouds
*Reproduced from US Department of Transportation/Federal Aviation Administration*

## The Types of Clouds

Like the motion of the atmosphere, clouds are an important weather factor. They can tell you about current weather conditions as well as what to expect in the future.

You read earlier about particulate matter. Clouds form from water vapor clinging to particulate matter. When the air cools to its saturation point, the water vapor condenses on the particulate matter to form visible cloud droplets. While the development process is about the same for each, clouds come in many heights, shapes, and behaviors. Classifications depend on the height of a cloud's base—low, middle, or high—as well as its vertical development (Figure 1.6).

*Low clouds* form near Earth's surface up to about 6,500 feet. Water droplets are their main ingredient. Sometimes they include *supercooled water droplets* (water droplets that have cooled to below freezing but are still in a liquid state), which can produce hazardous aircraft icing. Typical low clouds are *stratus*, *stratocumulus*, and *nimbostratus*. Fog is also a type of low cloud. All have a low ceiling (*the height above Earth's surface of the lowest layer of clouds*) and can change quickly. They can also make visibility difficult, so pilots must rely on instruments to fly through them.

*Middle clouds* start around 6,500 feet and reach up to about 20,000 feet. They are made up of water, ice crystals, and supercooled water droplets. Typical middle clouds are *altostratus* and *altocumulus*. You might spot these if you sit in a window seat when flying at a higher altitude while traveling cross-country. Altostratus clouds pose a moderate icing hazard and can create turbulence. Altocumulus clouds, which can form when altostratus clouds break up, may offer light turbulence and icing.

*High clouds* start above 20,000 feet. They usually form only in stable air, that is, air where the ride is smooth and little weather is present. High clouds contain ice crystals, yet pose no real risk of turbulence or icing. Typical high clouds are *cirrus*, *cirrostratus*, and *cirrocumulus*.

Clouds with lots of vertical development are *cumulus* clouds. This type of cloud has a flat base that forms in the low or middle cloud regions. Its plump, billowing vertical development stretches up into the high cloud zone. Cumulus clouds, particularly cumulonimbus clouds—*thunderstorms*—can mean turbulent weather ahead. Cumulonimbus clouds contain large amounts of moisture and unstable air, which can produce hazardous weather such as lightning, hail, tornadoes, gusty winds, and wind shear (*a sudden, drastic shift in wind speed, direction, or both that may take place in the horizontal or vertical plane*). These can be the most dangerous type of cloud to encounter in flight.

Cumulus clouds as seen from a NOAA Research Aircraft DC-6 40C
*Courtesy of NOAA/AOML/Hurricane Research Division*

## Fog

Fog is a cloud that begins within 50 feet of Earth's surface. It typically forms in a stable air mass when the air's temperature near the ground has cooled to the air's dew point. At this point, water vapor in the air condenses and appears as fog.

Fog comes in several varieties, depending on how it forms. These include:

- *Radiation fog*—This type of fog develops on clear nights when little or no wind is present. It forms in low-lying areas such as mountain valleys. This type of fog occurs when the ground cools rapidly and the surrounding air temperature reaches its dew point. As the sun rises and temperature increases, radiation fog lifts and eventually burns off. Wind quickens the dissipation process.

- *Advection fog*—This fog occurs when a layer of warm, moist air moves over a cold surface. It requires wind up to 15 knots to form. Above 15 knots, the fog lifts and forms low stratus clouds. Advection fog is common along the coast where sea breezes blow air over cooler landmasses.

- *Upslope fog*—This variety of fog forms when moist, stable air moves up sloping land features such as mountain ranges. Like advection fog, it needs wind to form and survive. Upslope and advection fog can last for days (unlike radiation fog, which burns off with the morning sun). It also reaches greater heights than radiation fog.

- *Steam fog*—Also called *sea smoke*, this fog develops when cold, dry air moves over warm water. As water evaporates from the water surface, it rises and looks like smoke. This type of fog is common over water during the coldest times of year. Steam fog produces low-level turbulence and icing.

- *Ice fog*—This is also a cold weather fog. It occurs when the temperature is a good deal below freezing (usually –25 degrees F or more). The water vapor forms directly into ice crystals. Otherwise, it requires the same conditions as radiation fog. It generally forms in the arctic regions, although sometimes the middle regions get ice fog.

Radiation fog
© iStockphoto/Thinkstock

Advection fog rolls over the Golden Gate Bridge in San Francisco. This type of fog occurs when a layer of warm, moist air moves over a cold surface.
*Courtesy of NOAA/National Weather Service*

Some very faint sea smoke, or steam fog, over the Bering Sea west of Alaska, as taken from onboard the NOAA ship *Surveyor*.
*Courtesy of NOAA*

**LESSON 1** ■ The Atmosphere

## Cloud Classifications

Clouds come in many shapes, sizes, and functions. If a pilot knows how to read them, they can help him or her fly more safely. The cloud types are:

- *Cumulus*—Heaped or piled clouds
- *Stratus*—Formed in layers
- *Cirrus*—Ringlets, fibrous clouds, also high-level clouds above 20,000 feet
- *Castellanus*—Common base with separate vertical development, castlelike
- *Lenticularus*—Lens-shaped, formed over mountains in strong winds
- *Nimbus*—Rain-bearing clouds
- *Fracto*—Ragged or broken
- *Alto*—Meaning high, also middle-level clouds existing at 5,000 feet to 20,000 feet.

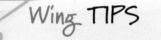

## Wing TIPS

Some aircraft fly in the stratosphere for the advantages it offers, such as safety from the effects of stormy weather and increased fuel efficiency.

## How the Atmospheric Layers Impact Flight

The different layers of the atmosphere present their own set of challenges in flight. The troposphere, however, is where most flight takes place. Three atmospheric factors to take into account are air density, pressure, and temperature. These characteristics change in part due to changes in altitude.

### Density

Altitude, in fact, affects every aspect of flight from aircraft performance to human health. At lower altitudes, the density of the air increases, and at higher altitudes, air density decreases. As air becomes less dense:

- Engines and propellers are less efficient because they take in less air (although jet engines operate better with cold intake air)
- Engines and propellers generate less thrust
- Lift decreases because the thin air exerts less force on the airfoils
- Drag also decreases
- Takeoff and landing distances increase because it takes longer to create enough airflow over the airfoils for lift.

## Pressure

As you climb through the troposphere, the increasing altitude also affects atmospheric pressure. As you read earlier in this lesson, air pressure is greatest at sea level and decreases as you move higher. Because atmospheric pressure changes with time and location, scientists developed another means to measure pressure. They came up with *standard conditions*, a reference for the standard atmosphere at sea level, where the surface pressure is measured as 29.92 inches of mercury ("Hg), or 1,013.2 millibars (mb). A millibar is one-thousandth of a bar, a measure of pressure developed by a British meteorologist in 1909 (Figure 1.7).

Under what are called *standard conditions* at sea level, the weight of the atmosphere exerts an average pressure of 14.7 pounds per square inch (psi) of surface, or 1,013.2 millibars (mb). The higher the altitude, the less air you have above; therefore, the atmosphere's weight at 18,000 feet is half what it is at sea level (Figure 1.8).

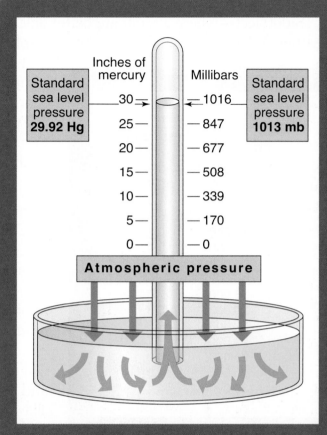

**Figure 1.7** Standard sea level pressure: inches of mercury versus millibars
*Reproduced from US Department of Transportation/Federal Aviation Administration*

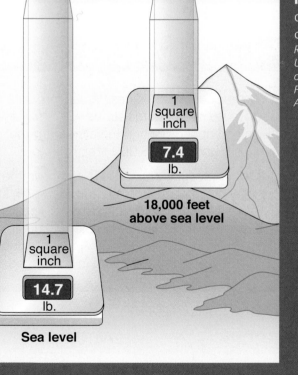

**Figure 1.8** Weight of atmosphere at different altitudes
*Reproduced from US Department of Transportation/ Federal Aviation Administration*

At sea level, the atmospheric pressure is just right for people to lead healthy lives. But by 18,000 feet—where aircraft are better able to conserve fuel and avoid bad weather and a lot of turbulence—the atmospheric pressure drops to a level that can be fatal. This is because air molecules spread out more and more with greater altitudes. People can't get enough oxygen to survive. (You'll read more about flight and health in Chapter 3.)

To counter these ill effects, aircraft have pressurized cabins. The sealed fuselage holds air under a pressure that's higher than the atmospheric pressure outside the airplane. Government regulations require a cabin altitude—*cabin pressure equal to what it would be at the same altitude above sea level*—of 8,000 feet (Figure 1.9). It keeps the passengers and crew comfortable and safe. The system also releases air from the fuselage through an outflow valve and admits a fresh supply through an inflow valve.

The altitude at which the standard air pressure is equal to 10.9 psi can be found at 8,000 feet.

At an altitude of 28,000 feet, standard atmospheric pressure is 4.8 psi. By adding this pressure to the cabin pressure differential of 6.1 psi difference (psid), a total air pressure of 10.9 psi is obtained.

| Atmospheric Pressure | |
| --- | --- |
| Altitude (ft) | Pressure (psi) |
| Sea level | 14.7 |
| 2,000 | 13.7 |
| 4,000 | 12.7 |
| 6,000 | 11.8 |
| 8,000 | 10.9 |
| 10,000 | 10.1 |
| 12,000 | 9.4 |
| 14,000 | 8.6 |
| 16,000 | 8.0 |
| 18,000 | 7.3 |
| 20,000 | 6.8 |
| 22,000 | 6.2 |
| 24,000 | 5.7 |
| 26,000 | 5.2 |
| 28,000 | 4.8 |
| 30,000 | 4.4 |

**Figure 1.9** Standard atmospheric pressure chart
*Reproduced from US Department of Transportation/Federal Aviation Administration*

## No April Fool's Joke: Lost Cabin Pressure

On 1 April 2011 Southwest Airlines passengers on a flight from Phoenix to Sacramento heard a loud noise and then felt strong winds sweep through the cabin. A hole about five-feet long had ripped open in the aluminum roof of the Boeing 737-300, leaving a view of the sky above. At 36,000 feet, cabin pressure fell to dangerous levels.

Oxygen masks immediately dropped from above the passenger seats. Some people passed out before they could get their masks on, however, and two people had minor injuries. To get the plane back to a safer altitude where passengers could breathe, the pilot quickly descended to 11,000 feet. The plane then made an emergency landing at Yuma Marine Corps Air Station. The Federal Aviation Administration ordered emergency inspections of about 80 737-300 airplanes operated in the United States.

**CHAPTER 2** Working Through Flight Conditions

## Temperature

Temperatures also drop dramatically in the troposphere with altitude (Figure 1.10). Earlier you read that they fall at a rate of about 3.5 degrees F (2 degrees C) for every 1,000 feet gained. This drop continues until somewhere around 36,000 feet, where the temperature sits at about −65 degrees F (−55 degrees C) up to 80,000 feet (by which point, you're in the stratosphere and temperatures begin to slowly rise).

Today's airplanes have systems that regulate temperature and heat cabins. Passengers and crew can travel in comfort. But back in World War II, bomber crews endured wretched conditions during missions. Air systems and pressurized cabins were not yet the norm.

"Breathing was possible only by wearing an oxygen mask—cold and clammy, smelling of rubber and sweat—above 10,000 feet in altitude," writes historian and author Stephen E. Ambrose in his book about World War II B-24 bomber crews, *The Wild Blue: The Men and Boys Who Flew the B-24s Over Germany*. "There was no heat, despite temperatures that at 20,000 feet and higher got as low as 40 or even 50 degrees [F] below zero.... The oxygen mask often froze to the wearer's face. If the men at the waist touched their machine guns with bare hands, the skin froze to the metal."

Ambrose also details how wind whipped through the cabin whenever the bombardiers opened the bomb bay doors to deliver their loads on targets below. Frigid air also got through windows that the waist gunners used to shoot at enemy planes. Further, the lack of pressurization meant that any food the crew ate could create excruciating "pockets of gas in a man's intestinal tract." This list of lack of comforts goes on.

By contrast, today's Airmen sit in cockpits and wear suits that protect them from the elements. For instance, the F-22, a mainstay of the US Air Force, can fly well into the stratosphere with the aircraft's ceiling at 50,000 feet. Its maximum speed is Mach 2. These two characteristics expose pilots to many dangers, such as below-freezing temperatures and pressure changes. But the F-22's life-support systems overcome these challenges.

| Standard Atmosphere | | | |
|---|---|---|---|
| Altitude (ft) | Pressure (Hg) | Temperature °C | °F |
| 0 | 29.92 | 15.0 | 59.0 |
| 1,000 | 28.86 | 13.0 | 55.4 |
| 2,000 | 27.82 | 11.0 | 51.9 |
| 3,000 | 26.82 | 9.1 | 48.3 |
| 4,000 | 25.84 | 7.1 | 44.7 |
| 5,000 | 24.89 | 5.1 | 41.2 |
| 6,000 | 23.98 | 3.1 | 37.6 |
| 7,000 | 23.09 | 1.1 | 34.0 |
| 8,000 | 22.22 | −0.9 | 30.5 |
| 9,000 | 21.38 | −2.8 | 26.9 |
| 10,000 | 20.57 | −4.8 | 23.3 |
| 11,000 | 19.79 | −6.8 | 19.8 |
| 12,000 | 19.02 | −8.8 | 16.2 |
| 13,000 | 18.29 | −10.8 | 12.6 |
| 14,000 | 17.57 | −12.7 | 9.1 |
| 15,000 | 16.88 | −14.7 | 5.5 |
| 16,000 | 16.21 | −16.7 | 1.9 |
| 17,000 | 15.56 | −18.7 | −1.6 |
| 18,000 | 14.94 | −20.7 | −5.2 |
| 19,000 | 14.33 | −22.6 | −8.8 |
| 20,000 | 13.74 | −24.6 | −12.3 |

**Figure 1.10** Standard atmosphere properties
*Reproduced from US Department of Transportation/Federal Aviation Administration*

An F-22A Raptor flies over Fort Monroe, Virginia.
*Courtesy of USAF/TSgt Ben Bloker*

The fighter's cockpit comes equipped with an onboard oxygen generation system (OBOGS) so pilots can breathe normally. If a pilot has to eject into water during a mission, he or she will be wearing a suit capable of maintaining body heat even when submerged in water that's only 32 degrees F (0 degrees C) for up to two hours. Ejecting from an aircraft like the F-22 can mean exposing a pilot to winds of 600 knots. So the fighter's escape system shields the pilots from the heat and ripping effects these winds could inflict. The F-22's life-support systems also address hazards such as high altitudes, high speeds and acceleration, and chemical and biological situations, among others.

This lesson has discussed the atmosphere's makeup. In the process, you've also gotten an idea of the hazards a pilot faces from the atmosphere at great altitudes. The more you understand the atmosphere and how to read its signs, the more safely you will be able to fly. In fact, the next lesson will delve further into the atmosphere's components and how they affect weather.

# ✔CHECK POINTS

## Lesson 1 Review

Using complete sentences, answer the following questions on a sheet of paper.

1. Which is the lowest layer of the atmosphere?

2. How far does the stratosphere extend from Earth's surface?

3. Water vapors enter the atmosphere through only two processes. What are they?

4. How many cloud droplets does it take to form one raindrop?

5. What is the process that results in the atmosphere's circular motion?

6. In which direction does the Coriolis force deflect air in the Northern Hemisphere?

7. What can clouds tell you?

8. What do cumulonimbus clouds contain?

9. What are two things that happen when air becomes less dense?

10. What can't people get enough of to survive at higher altitudes?

## APPLYING YOUR LEARNING

11. What types of conditions and challenges would you expect when flying in the stratosphere?

## ✈ Weather Elements

### ✎ Quick Write

_____

_____

Aircraft and flight instruments have made great gains since the days when Charles Lindbergh and others took daring trips over oceans and continents. Do you think today's pilots still need to worry about weather conditions, even with their more advanced gear? What kinds of dangers might weather continue to pose to modern aircraft?

### Learn About

- types of air masses and fronts
- the factors that impact air mass
- how high- and low-pressure systems are key factors in wind and atmospheric motion
- how fronts are boundaries between air masses
- the terrain factors that affect weather

**W**eather challenged the famous aviator Charles A. Lindbergh many times during his career. It delayed flights. It posed dangers such as sleet and ice once up in the air. Conditions such as fog made finding where he was going difficult.

A couple of weeks before his historic transatlantic (*across the Atlantic Ocean*) flight from New York to Paris in 1927, the 25-year-old pilot gave his plane a test run. A lengthy trial journey in the single-engine monoplane, *Spirit of St. Louis*, would additionally give Lindbergh a chance to prove that he was really up to the cross-ocean trip. Some questioned whether his age and experience qualified him to make the dangerous, nonstop flight to Paris. For his trial run, he would fly from San Diego to New York, a transcontinental (*coast-to-coast*) flight. But day after day the weather was so bad that he couldn't take off.

Pressures were everywhere to get this test flight off the ground as soon as possible so he could move on to the even bigger journey across the Atlantic. For one thing, the first person to make the New York to Paris flight would win $25,000. For another, some pilots were preparing to attempt the flight while Lindbergh was still twiddling his thumbs back in San Diego.

The young pilot consulted frequently with the US government's Weather Bureau in San Diego to check conditions along the first leg of his route, San Diego to St. Louis. He wrote about his daily visits to the bureau in his autobiography *The Spirit of St. Louis*, "Each time it has been the same story—a low-pressure area covering the route I want to follow—mountaintops in clouds—low visibility passes—heavy rain—local reports of ice and hail.

Three days wasted on account of weather, waiting for the clouds to clear. Why does flying have to be so dependent on the eyesight of a pilot?"

(Low-pressure systems pose hazards to aviators such as clouds and rain, or worse, rapid changes in wind speed and direction. These were the years before the introduction of sophisticated flight instruments that guide today's pilots through bad conditions when they can't see much out their windows. What a pilot could see was critical to a safe flight. Otherwise, if the pilot got completely disoriented, he or she might crash into the side of a mountain or plow straight into the ground.)

The weather cleared, however, enough for Lindbergh to head to St. Louis and then New York—in record time no less. Once again he started consulting with the Weather Bureau for the flight from New York to Paris. Clouds, fog, rain, and storms threatened all along his route. So as not to risk every danger, Lindbergh's course was going to keep him over land as much as possible. He would fly over Nova Scotia, Newfoundland, and only then face the 15-hour journey over the ocean to Ireland and then a far safer last leg over land to Paris. He adds in his autobiography, "I wouldn't be so concerned about weather if the moon weren't already past full. Soon it won't be any use to me." He needed light from the moon to help him navigate, as he didn't have much in the way of equipment other than a compass.

Finally, Lindbergh got word that the weather was clearing. A high-pressure system, which brings good weather with it, was pushing away a low-pressure system that had been hovering over Newfoundland. Lindbergh wrote, "The time has come, at last, for action." He headed for Roosevelt Field on Long Island to prep his plane with the help of friends and ready it for takeoff.

## Vocabulary

- transatlantic
- transcontinental
- barometric pressure
- air masses
- fronts
- maritime
- continental
- wind shifts
- warm front
- visual flight rules
- stratiform
- cold front
- cirriform
- squall line
- stationary front
- occluded front
- terrain
- orographic lift
- windward
- leeward
- rain shadow

According to the Weather Bureau, Lindbergh should have waited at least another 12 hours before taking off on 20 May 1927. As a result he faced fog and rain, even an ice cloud. "Ice clouds toss you in their turbulence, lash you with hailstones, poison you with freezing mist," he said. A July 1927 edition of *Boys' Life* magazine reported, "All that night, his faithful engine beating faultlessly, he kept the plane headed for Europe, sometimes trying to fly below a fog, on one occasion rising to ten thousand feet to get over the sleet...."

Despite the weather obstacles, not to mention the difficulty of flying alone with no sleep, Lindbergh did what he set out to do. He flew 3,610 miles in 33.5 hours to land at Le Bourget field in France on 21 May. He became one of the most famous aviators of all time.

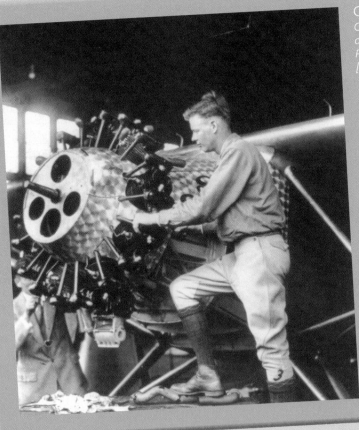

Charles A. Lindbergh
*Courtesy of Library of Congress, Prints & Photographs Division [LC-USZ62-51834]*

## Types of Air Masses and Fronts

Pilots watch the weather because it affects how their aircraft will perform, from the runway to the air and back down again for landing. *Weather* refers to temperature, moisture, wind velocity (calm or stormy), visibility (clear or cloudy), and barometric pressure (*air pressure as measured by a barometer*) at a specific location. These are all conditions of the atmosphere, some of which you read about in the previous lesson.

Two formations in which these atmospheric conditions take place are air masses and fronts. An air mass is *a large body of air having fairly uniform properties of temperature and moisture.* A front is *the boundary between two different air masses.*

### Air Masses

Scientists classify air masses according to where they form. They refer to these areas as *source regions.* The bodies of air adopt the temperature and moisture traits of their source region if they remain over the area long enough. Source regions generally maintain their temperature and moisture levels for days at a time, which allows air masses to assume those same values.

Flat, uniform areas where the temperature and moisture don't change rapidly make ideal source regions. These include Polar Regions, such as Canada and Siberia, and tropical oceans such as the South Atlantic. Therefore, air masses are either polar or tropical depending on temperature traits, and either maritime (*of or relating to the sea*) or continental (*of or relating to land*) depending on moisture content.

Air masses generally fall into four categories based on a mix of these source regions. Each is assigned an *acronym*, which is a name made up of initials. If an air mass starts over a cold, polar region, it gets a capital *P* for *polar*. If the air mass source region is warm and tropical, it uses a capital *T* for *tropical*. An air mass over land, which will be dry, gets a lowercase *c* for *continental*. And an air mass forming over water uses a lowercase *m* for *maritime*. A fifth category sometimes appears: in winter an air mass that forms over extremely cold, dry land gets a capital *A* for *arctic*. These initials combine to form the following types of air masses (Figure 2.1):

- *cP*—Cold, dry air mass
- *mP*—Cold, moist air mass
- *cT*—Warm, dry air mass
- *mT*—Warm, moist air mass
- *cA*—Extremely cold and dry air mass.

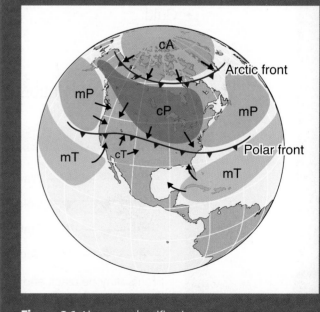

**Figure 2.1** Air mass classifications
*Reproduced from NOAA/National Weather Service*

Air mass classifications are further refined by how cold or warm an air mass is when compared with the temperature of the land or water surface over which it's moving. The lowercase initial *k* stands for an air mass that's *colder than surface temperature*. The lowercase initial *w* is for an air mass that's *warmer than surface temperature*.

A cold, dry air mass that's colder than surface temperature, for example, would read as cPk. A warm, moist air mass that's warmer than the surface temperature carries the initials mTw. An air mass that's warmer than the surface (mTw) produces more stable conditions, while an air mass that's colder than the surface (cPk) leads to less stable conditions. That's because the air mass interacts with the surface below it. A cPk air mass becomes destabilized as it interacts with heat rising from below. An mTw air mass becomes more stable as it interacts with the cold below it.

## Fronts

Air masses eventually meet other air masses of different temperature and moisture characteristics. This contact creates fronts. As the boundary layer between these two types of air masses, fronts always signal a change in weather.

Fronts are divided into four types based on the temperature of the advancing air compared with the temperature of the air it is replacing. In other words, fronts are named according to which type of air mass (warm or cold) is taking the place of another air mass (Figure 2.2). The four types are:

- Warm
- Cold
- Stationary
- Occluded.

You'll read more about fronts later in this lesson.

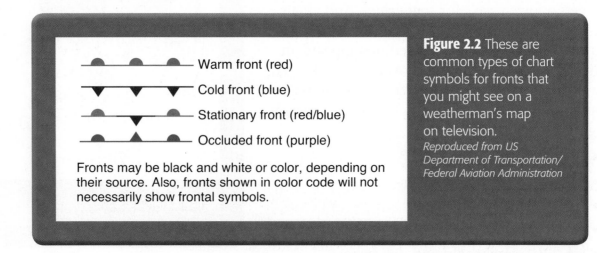

Warm front (red)

Cold front (blue)

Stationary front (red/blue)

Occluded front (purple)

Fronts may be black and white or color, depending on their source. Also, fronts shown in color code will not necessarily show frontal symbols.

**Figure 2.2** These are common types of chart symbols for fronts that you might see on a weatherman's map on television.
*Reproduced from US Department of Transportation/ Federal Aviation Administration*

## The Factors That Impact Air Mass

The different air masses that meet to form fronts are themselves shaped by outside factors. These include source region, the qualities of the surface over which air masses travel, and season.

You've already read that an air mass gets its initial temperature and moisture characteristics from the source region over which it forms. Eventually most air masses will drift away from their source region. As they do they end up adopting new temperature and moisture traits from the land or water surfaces they pass over. These changes may be drastic or slight. This process is sometimes referred to as *modification.*

An air mass passing over a warmer surface is warmed from below. This produces updrafts in which warm air rises from the surface and cooler air rushes in to take its place. An unstable air mass results that offers good surface visibility because the updraft gets rid of any smoke, dust, or other particles in the air mass. The moist, unstable air produces cumulus clouds, showers, and turbulence.

Now reverse the above example with an air mass passing over a colder surface. It won't exhibit updrafts. Instead, it produces a stable air mass with poor surface visibility. The smoke, dust, or other particles can't get out of the air mass. They are trapped at Earth's surface. A stable air mass sometimes produces low stratus clouds and fog.

Seasons (created by the tilt of Earth's axis as it orbits the sun) also affect air masses. For instance, summer temperatures can warm and inject moisture into an otherwise cold, dry (cP) air mass traveling south from a northern region such as Canada. Winters can chill and dry out a warm, moist (mT) air mass drifting north over land from the subtropics.

## How High- and Low-Pressure Systems Are Key Factors in Wind and Atmospheric Motion

Air flows from areas of high pressure into areas of low pressure because air always seeks out lower pressure. As you read in the previous lesson, air pressure, temperature changes, and the Coriolis force work together to create motion in the atmosphere. In fact they set the atmosphere in motion in two ways—a vertical movement of rising and falling *currents* and a horizontal movement in the form of *wind.* Wind is the result of having a low-pressure area next to a higher-pressure area because air molecules in the higher-pressure zone will migrate to the low-pressure area's "roomier" surroundings.

Pilots must be familiar with currents (which will be discussed in detail in the next lesson) and wind because they affect takeoff, landing, and cruising. Even more importantly, currents and wind—or atmospheric circulation—can change the weather. For instance, when an air mass moves away from its source region, it can bring cold, warm, wet, or dry weather conditions with it.

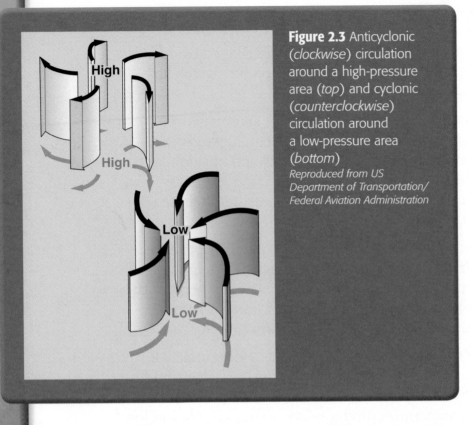

**Figure 2.3** Anticyclonic (*clockwise*) circulation around a high-pressure area (*top*) and cyclonic (*counterclockwise*) circulation around a low-pressure area (*bottom*)
*Reproduced from US Department of Transportation/ Federal Aviation Administration*

## Wind Patterns

In the Northern Hemisphere the flow of air from areas of high to low pressure swerves to the right to produce a clockwise circulation around a high-pressure area. Scientists call this *anticyclonic circulation*. The flow of air around a low-pressure area deflects as well and produces a counterclockwise circulation referred to as *cyclonic circulation* (Figure 2.3).

High- and low-pressure systems create weather. High-pressure systems generally produce dry, stable, descending air. You'll often enjoy good weather because of a high-pressure system. On the other hand, low-pressure systems often bring in bad weather with increasing cloudiness and precipitation. This is because the air flowing into a low-pressure system to replace rising air tends to be unstable.

An Expedition 27 crew member aboard the International Space Station took this photo of a low-pressure system in the eastern North Pacific Ocean on 20 March 2011.
*Courtesy of NASA*

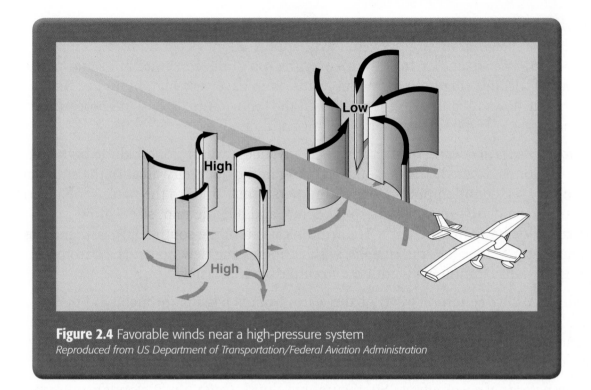

**Figure 2.4** Favorable winds near a high-pressure system
*Reproduced from US Department of Transportation/Federal Aviation Administration*

Pilots who have a good understanding of high- and low-pressure wind patterns can take advantage of tailwinds. When flying from west to east, a pilot would find favorable winds on the northern side of a high-pressure system or the southern side of a low-pressure system (Figure 2.4). On the return flight from east to west, the favorable winds would be along the southern side of the same high-pressure system or the northern side of the low-pressure system. In addition, the pilot familiar with these systems could also more accurately predict the kind of weather he or she is likely to run into.

These theories of circulation and wind patterns work for large-scale atmospheric motion. However, they don't take into account changes to circulation on a local scale. Local conditions, land features, and other factors can change wind direction and speed close to Earth's surface.

## How Fronts Are Boundaries Between Air Masses

As noted earlier in this lesson, high- and low-pressure systems meet in fronts. Their winds move in opposite directions where they make contact—the winds rotate clockwise in a high-pressure system and counterclockwise in a low-pressure system. Fronts are the boundaries between these two areas of pressure.

Because of the opposite natures of these pressure systems, the wind continually changes direction. Cold fronts produce the most dramatic wind shifts, which are *abrupt changes in the direction of wind by 45 degrees or more in less than 15 minutes with winds of at least 10 knots.* The following paragraphs break down the types of fronts, their phases, and their characteristics.

A warm front is *a boundary area that forms when a warm air mass advances and replaces a colder air mass.* It moves slowly, at only 10 miles per hour to 25 miles per hour. The advancing front slides over the top of the cooler air mass and pushes it out of the area (Figure 2.5). Warm fronts often have very high humidity. As the warm air flows up and over the cooler air, the temperature drops and moisture condenses. Warm fronts cause low ceilings and rain.

*Before passage of a warm front*: Signs of an advancing warm front include high-level cirrus clouds, layered stratus clouds, and fog in the boundary area. In the summer, you may see cumulonimbus clouds (thunderstorms) prior to the passage of a warm front. Precipitation may be light to moderate and include rain, sleet, snow, or drizzle. Visibility may be poor. The wind blows from the south-southeast, and the temperatures are cool or even cold, with an increasing dew point. The barometric pressure falls continuously until the front passes completely.

*During passage of a warm front*: As the warm front is passing through, you may see layered stratus clouds and some drizzle. Visibility is still poor, although it can clear up depending on the wind. The temperature rises steadily because warmer air is flowing into the area. Pressure levels off and the dew point remains steady.

*After passage of a warm front*: Stratocumulus clouds (high, puffy clouds) command the sky and rain showers are possible. Visibility improves at this stage of the warm front, but conditions may remain hazy for a bit. The wind blows from the south-southwest. As the temperature climbs, the dew point also rises and then levels off. Finally, the barometric pressure rises just slightly only to fall again.

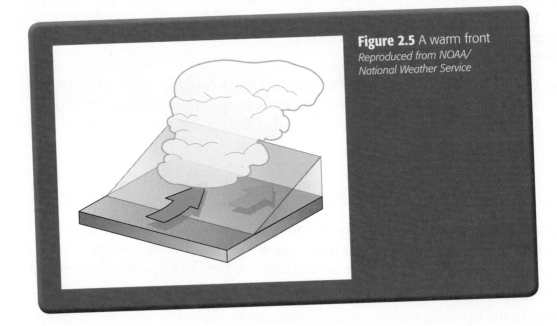

**Figure 2.5** A warm front
*Reproduced from NOAA/ National Weather Service*

# Flying Into a Warm Front

You're on a westward flight departing from Pittsburgh for St. Louis. The weather is good enough for visual flight rules (*also referred to simply as VFR, these rules are in play when the weather is clear enough that a pilot can operate an aircraft by sight as opposed to relying on aircraft instruments*). As you near Columbus, Ohio, and closer to the warm front, the clouds gets thicker and appear more stratiform (*of or pertaining to all types of stratus clouds*) with a ceiling of 6,000 feet. Visibility decreases to six miles, and the barometric pressure falls (Figure 2.6).

As you approach Indianapolis farther west, the weather grows more unstable with broken clouds at 2,000 feet and only three miles visibility with rain. Fog is also likely because the temperature and dew point are the same. As you leave Indianapolis, you can no longer continue VFR. You must switch to your instruments. Finally, you reach St. Louis. The sky is overcast, the clouds are low and it is drizzling, and visibility is down to just one mile.

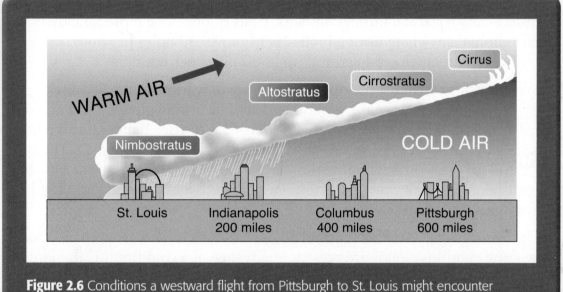

**Figure 2.6** Conditions a westward flight from Pittsburgh to St. Louis might encounter in face of a warm front advancing eastward
*Reproduced from US Department of Transportation/Federal Aviation Administration*

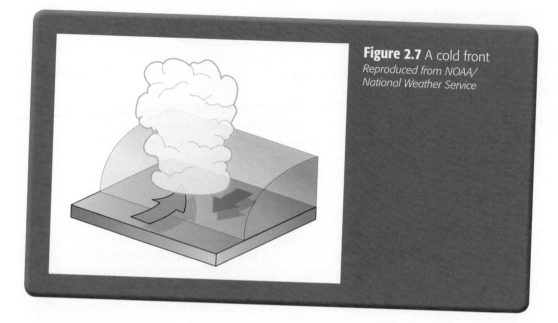

**Figure 2.7** A cold front
*Reproduced from NOAA/
National Weather Service*

A cold front is *a boundary area that forms when a cold, dense, and stable air mass advances and replaces a warmer air mass.* Cold fronts move more quickly than warm fronts, anywhere from 25 miles per hour to 30 miles per hour. Extreme cold fronts can move even faster at up to 60 miles per hour.

Cold fronts move in just the opposite manner of warms fronts. Because they're dense, they stick close to the ground and act like a snowplow by sliding *under* the warmer air and thereby forcing the less dense air to rise. Temperatures decrease quickly in the rapidly rising air, and this creates clouds (Figure 2.7). The type of cloud depends on how stable the warmer air is. Cold fronts in the Northern Hemisphere normally take up a northeast to southwest direction and can be several hundred miles long.

*Before passage of a cold front:* Cirriform (*of or relating to all types of cirrus clouds*) or towering cumulus clouds are present in advance of a cold front. Cumulonimbus clouds are possible. You might also see rain showers and haze as these clouds quickly develop. The wind from the south-southwest replaces the warm temperatures with colder air. A high dew point and falling barometric pressure are also signs that a cold front is coming.

*During passage of a cold front:* You'll continue to see towering cumulus or cumulonimbus clouds dominating the sky at this stage. Depending on how intense the cold front is, you might get heavy rain showers accompanied by lightning, thunder, and/or hail. Even more severe cold fronts can produce tornadoes. Visibility is poor, with winds variable and gusty, and the temperature and dew point drop rapidly. Once the quickly falling barometric pressure has bottomed out during a cold front's passage, it then begins to gradually rise.

*After passage of a cold front*: The towering cumulus and cumulonimbus clouds begin to dissipate to cumulus clouds and a decrease in precipitation. Visibility improves with winds from the west-northwest. The temperatures remain cooler and the barometric pressure continues to rise.

## Flying Into a Cold Front

Once again you're flying west from Pittsburgh to St. Louis into a weather front, only this time it's a cold front. The weather is VFR as you take off with three miles visibility. You encounter smoke and a scattered layer of clouds at 3,500 feet. As you near Columbus and edge closer to the cold front, the clouds show some vertical development with a broken layer at 2,500 feet. Visibility is now up to six miles but in haze with a falling barometric pressure (Figure 2.8).

By the time you approach Indianapolis, the weather has deteriorated to overcast clouds at 1,000 feet and visibility has decreased again to three miles with thunderstorms and heavy rain showers. The weather improves in St. Louis with 10 miles visibility and scattered clouds at 1,000 feet.

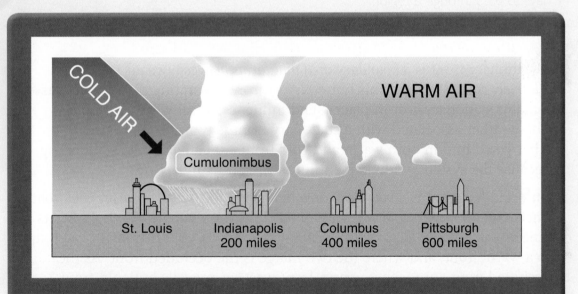

**Figure 2.8** Conditions a westward flight from Pittsburgh to St. Louis might encounter in face of a cold front advancing eastward
*Reproduced from US Department of Transportation/Federal Aviation Administration*

## Two Very Different Characters: Cold Fronts Versus Warm Fronts

Cold fronts and warm fronts have very different personalities. They also pose different hazards to pilots and the public. They contrast in speed, shape, the weather they bring, and how easy it is to predict a front's approach.

Cold fronts, at 20 to 35 miles per hour, move much more quickly than warm fronts, which travel only 10 to 25 miles per hour. Further, a cold front's face slopes far more steeply than a warm front's. (Note the gentle angle of the warm front in Figure 2.6 and the cold front's sharply angled slope in Figure 2.8.)

Cold fronts bring violent weather, such as sudden storms, gusty winds, turbulence, and sometimes hail or tornadoes. Generally, this rough weather takes place along the frontal boundary, not in advance. However, a squall line (*a narrow band of active thunderstorms*) can form in summer as far as 200 miles ahead of a severe cold front. In comparison, warm fronts serve up low ceilings, poor visibility, and rain.

Cold fronts approach quickly with little or no warning. They radically change the weather in just a few hours. The weather clears rapidly after passage and drier air with unlimited visibility prevails. Warm fronts, on the other hand, provide advance warning and can take days to pass through a region.

A stationary front is *a boundary area that forms between two air masses of relatively equal strength.* This boundary or front separating the two air masses remains stationary—or in place—and influences local weather for days. The weather associated with a stationary front is typically a mixture of what you'd find in both warm and cold fronts. It can also lead to flooding, because the system, with its plentiful rain, doesn't move for days.

An occluded front is *a boundary area that forms when a fast-moving cold front catches up with a slow-moving warm front.* This is only logical, as cold fronts tend to move faster than warm fronts. Eventually a cold front will "catch up" to a warm front, and where the two merge is the occluded front. As the occluded front approaches, warm-front weather prevails but cold-front weather immediately follows it.

Occluded fronts come in two varieties, depending on the temperatures of the colliding fronts. A *cold front occlusion* takes place when a fast-moving cold front is colder than the air ahead of the slow-moving warm front. When this occurs, the cold air replaces the cool air and pushes the warm front up into the atmosphere. Generally, if the air is stable, this creates a mix of weather found in both warm and cold fronts (Figure 2.9).

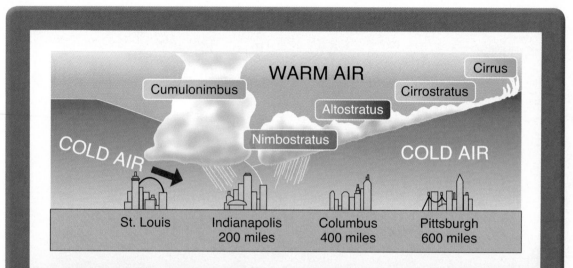

**Figure 2.9** A cold front occlusion forms. The warm front slopes over the prevailing cooler air and produces warm front-type weather. Before passage of the occluded front, cirriform and stratiform clouds prevail with light to heavy precipitation, poor visibility, steady dew point, and falling barometric pressure. During passage, nimbostratus and cumulonimbus clouds rule the skies with towering cumulus clouds also possible. Once again, precipitation is light to heavy and visibility is poor. But winds are variable and the barometric pressure levels off. After the front passes, nimbostratus and altostratus clouds form, precipitation decreases then clears, and visibility improves.
*Reproduced from US Department of Transportation/Federal Aviation Administration*

A *warm front occlusion* comes about when the air ahead of the warm front is colder than the cold front. In this case, the cold front rides up and over the warm front. If the air forced up by the warm front occlusion is unstable, the weather is more severe than the weather in a cold front occlusion. Thunderstorms, rain, and fog result.

## The Terrain Factors That Affect Weather

Like fronts and air masses, terrain—*the landscape, anything from mountains to coastlines and lakeshores*—also influences the weather. This is because the shape of the land can force air masses in new directions.

Mountains are one type of landscape feature that affects weather. When air approaches a mountain, the land's upward slope pushes the air up where the water vapor cools and condenses into clouds. Orographic lift is the name of this *event that forces air to rise and cool.* If the air cools enough, it either rains or snows on the mountain's windward side—*the side facing the wind.*

Orographic lift of moist air coming off the Pacific Ocean produces clouds along the Santa Lucia Mountains south of Monterey, California.
*Courtesy of Robert Schwemmer/NOAA/NOS/CINMS*

As the air reaches a mountain's peak and starts down its leeward side (*the side facing away from the wind*), it warms up, and rain and snow slow or stop. This zone is also referred to as a mountain's rain shadow, that *area on a mountain's leeward side with reduced precipitation because of warming air and decreasing clouds.* As a result, a mountain's leeward side tends to have less vegetation than its moist windward side. Furthermore, valleys at the base of mountains receive less precipitation because of these effects. For instance, San Jose at the foot of the Santa Cruz Mountains in California receives only one-quarter as much rain as the mountains do.

Big bodies of water that encounter cold, dry air create weather as well. For instance, the Great Lakes produce what's called *lake effect snow*. Lake effect snow is the result of a cold air mass passing over a warmer body of water (Figure 2.10). As relatively warm water vapors rise from the lake, the arctic or polar air mass cools the moisture until it condenses into clouds of snow that fall wherever the winds happen to be blowing. Parts of New York along Lake Erie and Lake Ontario get more than 200 inches of snow each year because of this lake effect.

The uneven heating of large bodies of water and land kick up two more types of weather known as *sea breeze* and *land breeze*. During the day, land heats up much faster than water. This difference in temperature produces a sea breeze that blows from off the cool ocean surface onto the warmer land. Recall from earlier in this lesson that air always seeks a lower pressure, which during the day is over the warm land.

Land breezes operate in just the opposite fashion. Come night, land cools off faster than water. Therefore, a land breeze flows from the cool land surface to the warmer ocean surface. Again, air is seeking a lower-pressure region.

This lesson has given you an overview of weather and some of its components such as air masses and fronts. The next lesson will consider how weather affects flight and how pilots deal with some types of severe weather. You'll read more, for instance, about how a pilot familiar with land and sea breezes can better handle an aircraft when approaching a landmass from over water or leaving a landmass to fly over a large body of water.

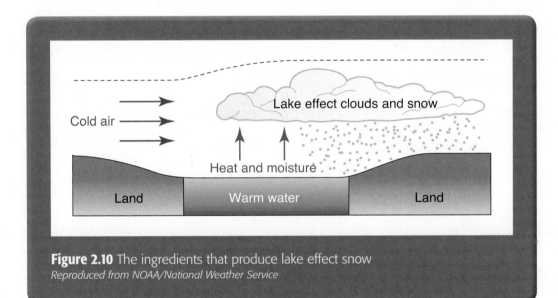

**Figure 2.10** The ingredients that produce lake effect snow
*Reproduced from NOAA/National Weather Service*

# ✔ CHECK POINTS

## Lesson 2 Review

Using complete sentences, answer the following questions on a sheet of paper.

1. How do scientists classify air masses?

2. What do fronts always signal?

3. What are the outside factors that shape the different air masses?

4. What can summer temperatures do to an otherwise cold, dry (cP) air mass traveling south from a northern region such as Canada?

5. Why does air flow from areas of high pressure into areas of low pressure?

6. When flying from west to east, a pilot would find favorable winds on which side of a high-pressure system? And on which side of a low-pressure system?

7. Where do high- and low-pressure systems meet?

8. Before passage of a warm front, what are some of the signs you might see?

9. What happens when air approaches a mountain?

10. Which two types of weather does the uneven heating of large bodies of water and land kick up?

## APPLYING YOUR LEARNING

11. It's a hot summer day. A pilot is flying a small aircraft over Lake Erie when he or she reaches Ohio. Which is warmer, the lake surface or the Ohio land surface, and how will the aircraft react to the change in air pressure when crossing from water to land?

## 🖉 Quick Write

_____

_____

Colonel Duckworth was an expert in instrument flying. Why was this skill important when flying through a hurricane?

## ✦ Learn About

- causes of atmospheric instability
- types and causes of turbulence
- how types of severe weather affect aviation

**C**ol Joe Duckworth slipped into a single-engine, propeller-driven North American AT-6 "Texan" trainer and flew into the heart of a hurricane. It was 1943—the middle of World War II—and he agreed to the flight because of a dare.

Duckworth and another Army Air Corps pilot challenged each other to do what no one had ever done before— fly through a hurricane _on purpose_. At the time of the dare, Duckworth was stationed at Bryan Field in Bryan, Texas. He trained veteran British pilots how to fly using instruments rather than relying solely on sight. Most of these men were aces—pilots who had shot down at least five enemy airplanes. Therefore, they'd already seen plenty of action in the skies over Europe. What prompted Duckworth to accept the challenge were complaints from the aces about the quality of their AT-6 trainer aircraft. They wanted to learn instrument flying in fighters that were the best of the best, not in some flimsy trainers.

Early on 27 July 1943 a hurricane was heading across Galveston, Texas, on the Gulf of Mexico, and north across Galveston Bay toward Houston. With his dare made that morning, Duckworth set out to prove how sturdy the snubbed AT-6 trainer actually was. The aircraft had two seats, so he asked the only navigator on base at that hour— 2d Lt Ralph O'Hair—to join him.

As they neared the hurricane, Duckworth and O'Hair flew between 4,000 feet and 9,000 feet. They encountered a lot of turbulence. According to O'Hair, they were "tossed about like a stick in a dog's mouth." Updrafts and downdrafts knocked their aircraft about, the sky around them was dark, and the rain gushed down.

But then they entered the hurricane's eye. The turbulence fell away. A hurricane is a low-pressure system that develops over tropical or subtropical waters, produces winds of at least 74 miles per hour, and has a well-defined circulation. Its core is referred to as its *eye*. If you've ever seen a satellite image of a hurricane, you might have noticed that in the middle of the storm is a small, round center. The winds are calm in the center. The winds die, and the sun shines. But a very turbulent eyewall surrounds this eye. This is what Duckworth and O'Hair had just broken through to reach the eye.

O'Hair told historians with the National Oceanic and Atmospheric Administration (NOAA) that he remembered the eye was cone-shaped and leaned with its bottom lagging behind its top. The pair from Bryan Field took a quick tour of the eye, which was about 10 miles across. Then they headed back into the tumultuous, black eyewall and made their return to Bryan Field.

Duckworth and O'Hair had just completed a historic flight. But Duckworth's day wasn't over. A weather officer named 1st Lt William Jones-Burdick asked for his chance to see a hurricane firsthand. So Duckworth headed back up. The air base hadn't sanctioned the flights. In fact, no one asked for permission for fear they'd be denied. But the flights were a first, and they convinced the British aces that perhaps the AT-6 wasn't a wimpy plane after all.

## Wing TIPS

Today the 53rd Weather Reconnaissance Squadron with the Air Force Reserve studies hurricanes and other storms that threaten American shores. Members of the squadron are more popularly called *Hurricane Hunters*. The pilots fly WC-130J aircraft.

## Causes of Atmospheric Instability

When people talk about weather, they often discuss its more spectacular forms, such as thunderstorms, lightning, and snow. What changes weather from the calm, good conditions that most hope for to the more violent conditions that occupy meteorologists has to do with how stable the atmosphere is.

The atmosphere's stability depends on its ability to resist vertical motion. A stable atmosphere impedes and destroys any small vertical disturbances. An unstable atmosphere, on the other hand, allows small vertical movements to grow until they produce turbulent airflow and air circulation. Instability can lead to lots of turbulence, massive vertical clouds, and severe weather.

Rising air expands and cools due to the decrease in air pressure as altitude increases. Descending air acts just the opposite way: as air falls it compresses and heats up because of the increase in atmospheric pressure. Two terms that describe this temperature change are adiabatic heating—*the process of heating dry air through compression*—and adiabatic cooling—*the process of cooling air through expansion*.

The adiabatic process takes place in all upward and downward moving air. When air rises into an area of lower pressure, it expands to a larger volume. As the air molecules spread out, the air temperature lowers. As a result, when a parcel of air rises, pressure decreases, volume increases, and temperature decreases. When air descends, the opposite is true: pressure increases, volume decreases, and temperature increases. *The rate at which temperature decreases with an increase in altitude* is its lapse rate (Figure 3.1). Recall from Chapter 2, Lesson 1 that as air rises, the average rate of temperature loss is 3.5 degrees F (2 degrees C) per 1,000 feet.

![Wing TIPS]

Warm air holds more water vapor than cool air.

Water vapor is less dense than air. Therefore, moisture decreases air density and causes air to rise. Conversely, a decrease in moisture makes air denser and causes it to sink. Moist air cools at a slower rate than dry air. Further, moist air is less stable than dry air because the moist air must rise higher than dry air before its

### Why Air Cools When It Rises and Warms Up When It Falls

The adiabatic process requires energy, which either cools or heats air. For instance, when air rises into an area of lower pressure and expands, this expansion of air molecules requires energy. This process actually takes energy from the air. This is why air cools as it rises.

The reverse is also true. As air falls into an area of higher pressure, the pressure compresses the air molecules. This compression heats the air as it sinks. This is why air gets warmer as it falls.

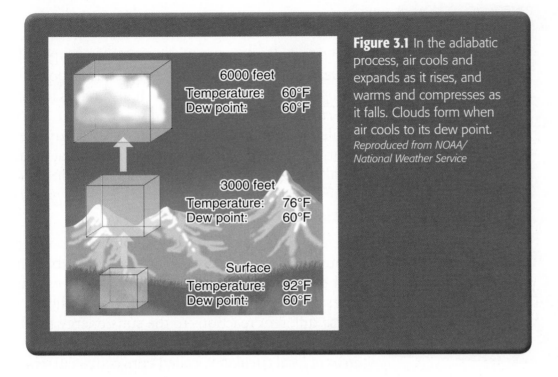

**Figure 3.1** In the adiabatic process, air cools and expands as it rises, and warms and compresses as it falls. Clouds form when air cools to its dew point.
*Reproduced from NOAA/ National Weather Service*

Contents of image:

6000 feet
Temperature:    60°F
Dew point:    60°F

3000 feet
Temperature:    76°F
Dew point:    60°F

Surface
Temperature:    92°F
Dew point:    60°F

temperature cools to that of the surrounding air. The dry adiabatic lapse rate (that is, the rate at which the temperature of dry air decreases with an increase in altitude) is 5.4 degrees F (3 degrees C) per 1,000 feet. The moist adiabatic lapse rate (that is, the rate at which the temperature of moist air decreases with an increase in altitude) varies from 2 degrees F to 5 degrees F (1.1 degrees C to 2.8 degrees C) per 1,000 feet.

The combination of moisture and temperature determines air stability and the resulting weather. Cool, dry air is very stable and resists vertical movement, which leads to good and generally clear weather. Warm moist air produces the greatest instability, the type you might see in tropical zones in summer. Typically, thunderstorms appear on a daily basis in these tropical regions due to the surrounding air's instability.

## Types and Causes of Turbulence

Unstable atmospheres produce turbulence, which is a sign of rapidly rising and falling currents of air. Pilots often have to deal with turbulence, which can take place close to the ground or high up in the sky. It can be dangerous and make for a very uncomfortable, rough flight as a plane bounces this way and that. This section discusses three types of turbulence and what causes them.

**Wing TIPS**

In 1996 President Clinton encountered turbulence during a flight at 33,000 feet over Texas in *Air Force One*. A thunderstorm had produced updrafts and downdrafts so wild that the ensuing rough ride ended up injuring a passenger on the president's official plane.

## Thermal Turbulence

*Thermal turbulence* is one of the three main types of turbulence. Something that's thermal has to do with *the property of being warm or hot*. It also refers to *the rising current of warm air that causes turbulence*. Thermal turbulence is the circulation of air, particularly the rise of warm air, taking place on a local scale, that is, in a small area. Convection is *the atmospheric circulation that results when warm air rises and cooler air moves in to replace it* (Figure 3.2).

Different surfaces radiate, or give off, heat in varying amounts. Plowed ground, rocks, sand, and barren land emit a large amount of heat. Water, trees, and other green growing things tend to absorb heat. These different reactions to the sun's energy result in the uneven heating of the air, which creates *small areas of local circulation* called convective currents.

**Figure 3.2** When the sun warms the ground, warm air rises and cooler air moves in to replace it in a process called convection.
*Reproduced from NOAA/Earth System Research Laboratory*

Convective currents create the bumpy, turbulent air that pilots sometimes experience when flying at lower altitudes during warmer weather. This is particularly an issue for pilots of small aircraft. On a low-altitude flight over varying surfaces, pilots may run into updrafts over pavement or barren places, and downdrafts over water or forests. To avoid these turbulent conditions, they can fly at a higher altitude, even above cumulus clouds (Figure 3.3). Cumulus clouds are actually one of several signs that convective currents are present. But sometimes pilots just can't avoid them—like during takeoffs or landings.

**Figure 3.3** The aircraft flying above the cumulus clouds is avoiding the turbulence that the plane below it is facing.
*Reproduced from US Department of Transportation/Federal Aviation Administration*

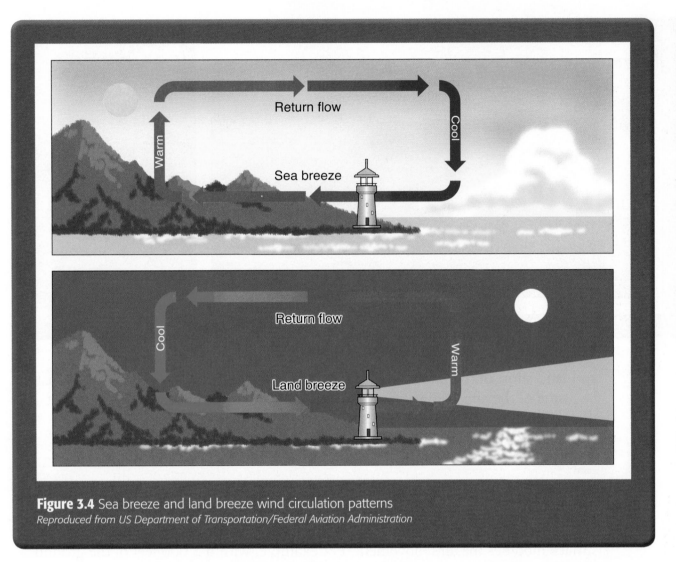

**Figure 3.4** Sea breeze and land breeze wind circulation patterns
*Reproduced from US Department of Transportation/Federal Aviation Administration*

Pilots may particularly notice convective currents when flying over an area with a landmass right next to a large body of water, such as an ocean or large lake. During the day the land heats faster than the water so the air over the land becomes warmer and less dense. This air rises, and cooler, denser air flowing in from over the water takes its place. This causes the onshore wind you read about in the previous lesson called a *sea breeze*. Conversely, at night when land cools faster than water, the air over the land also cools faster than the air over the water. The warmer air over the water rises and is replaced by cooler, denser air from the land. This creates the offshore wind called a *land breeze* (Figure 3.4). This reverses the local wind circulation pattern. Convective currents can occur anywhere the Earth's surface experiences uneven heating.

Convective currents close to the ground can affect a pilot's ability to control an aircraft. For example, when approaching an airport, the rising air from any land devoid of trees or other vegetation produces a ballooning effect that can cause a pilot to overshoot the intended landing spot. On the other hand, if the pilot approaches over a large body of water or forest, he or she may suddenly sink

**Figure 3.5** Varying surface conditions generate convective currents.
*Reproduced from US Department of Transportation/ Federal Aviation Administration*

Labels in figure: Cool sinking air; Warm rising air; Intended flight path

and land short of the intended landing spot (Figure 3.5). This is why pilots must be aware of the effects of thermal turbulence. They must be ready to counteract thermal turbulence when it affects the safe operation of the aircraft.

## Mechanical Turbulence

Friction is another cause of turbulence. Scientists refer to turbulence generated by the resistance of one object moving over another as *mechanical turbulence*. As air moves over Earth's surface, the friction that develops between the air and surface modifies the air's movement. For example, within 2,000 feet of the ground, the friction between the surface and the atmosphere slows the moving air.

You read in the previous lesson how terrain features such as mountains can create weather. Large objects—from mountains to man-made structures such as buildings—also generate mechanical turbulence. They affect how wind flows due to friction between air and surface, and this can produce hazardous atmospheric conditions for aircraft.

Natural surface features and large buildings can break up the wind's flow and create *wind gusts* (or strong bursts of wind) that rapidly change direction and speed. Obstacles range from man-made structures such as hangars to large natural barriers such as mountains, bluffs, and canyons. Pilots need to be especially alert during takeoffs and landings at airports near large buildings or natural obstructions (Figure 3.6).

How great the turbulence a pilot faces depends on the obstacle's size and the wind's velocity. During a landing, for instance, such turbulence may cause an aircraft to suddenly "drop in." When this happens, the plane may be too low to clear any obstacles during approach to the runway.

**Figure 3.6** Man-made structures can cause turbulence.
*Reproduced from US Department of Transportation/Federal Aviation Administration*

Mountains can be even more treacherous. While the wind flows smoothly up a mountain's windward side and the upward currents help carry an aircraft over its peak, the wind on the leeward side acts differently. As air flows down the leeward side, it follows the terrain's *contours* (or outline) and is increasingly turbulent. This tends to push an aircraft into the side of the mountain. The stronger the wind, the greater the downward pressure and turbulence become (Figure 3.7).

Valleys and canyons can be equally dangerous because of severe downdrafts. Pilots should take test runs through unknown, hazardous territory with someone familiar and qualified to fly through the area before attempting these routes on their own.

**Figure 3.7** Turbulence on a mountain's leeward side
*Reproduced from US Department of Transportation/Federal Aviation Administration*

## Wind Shear

Wind shear is yet another form of turbulence that can be hazardous to aircraft. As you read in Chapter 2, Lesson 1, this type of turbulence is an abrupt, dramatic change in wind speed, direction, or both. It can also take place in the vertical or horizontal planes. Wind shears take place over very small areas (Figure 3.8).

This type of turbulence subjects aircraft to violent updrafts and downdrafts. It can also abruptly change an aircraft's horizontal, or forward, movement. While wind shear can occur at any altitude, low-level wind shear is especially hazardous because of how close an airplane is to the ground. It can cause a plane to rapidly lose airspeed and crash.

**Figure 3.8** Wind shear over an airport
*Reproduced from NASA*

### Low-Level Wind Shear

Characteristics associated with low-level wind shear are changes in wind direction of 180 degrees and speed changes of 50 knots or more. Weather systems linked with low-level wind shear are passing frontal systems, thunderstorms, and temperature inversions with strong upper-level winds (greater than 25 knots).

### The Flip Side: Inversion

Generally, as air rises and expands in the atmosphere, the temperature drops. However, an atmospheric anomaly (something that's not the norm) can occur that changes this typical pattern of atmospheric behavior. When the air's temperature rises with altitude, you have a temperature inversion, that is, *an increase in temperature with altitude.*

Inversion layers are commonly shallow layers of smooth, stable air close to the ground. The air's temperature increases with altitude up to a point, which is the top of the inversion. The air at the top of this layer acts as a lid and traps weather and pollutants below it. If the relative humidity of this air is high, it can contribute to the formation of clouds, fog, haze, or smoke. Pilots must contend with reduced visibility in the inversion layer.

Surface-based temperature inversions occur on clear, cool nights when the ground's dropping temperatures cool the air close to the surface. This air within a few hundred feet of the surface becomes cooler than the air above it. If the air cools enough to reach the dew point, fog may form, reducing visibility below that required for landing. Frontal inversions come about when warm air spreads over a layer of cooler air, or cooler air is forced under a layer of warmer air.

Wind shear is dangerous when flying for several reasons. The rapid changes in wind direction and velocity alter the wind's relation to the aircraft. They disrupt the aircraft's normal flight attitude and performance. If a pilot runs into a wind shear situation, he or she may encounter anything from slight to very dramatic effects depending on wind speed and direction. For example, a tailwind that quickly changes to a headwind causes an increase in airspeed and performance. But when a headwind changes to a tailwind, the airspeed rapidly falls along with a corresponding decrease in performance. In either case, a pilot must be ready to react immediately to changes to maintain control of his or her plane.

Thunderstorms can sometimes tip off a pilot that the most severe types of low-level wind shear may be present. In Chapter 2, Lesson 5, you'll read about a particularly dangerous form known as a microburst.

### High-Level Wind Shear

Wind shear at the atmosphere's upper levels, beginning around 18,000 feet, can produce *clear air turbulence*. Because passenger airliners generally cruise between 30,000 feet and 34,000 feet, they will sometimes encounter this type of turbulence. It makes for a bumpy ride and can even cause large aircraft to gain or lose thousands of feet of altitude very abruptly in just a few seconds.

As its name indicates, clear air turbulence often takes place in a cloud-free sky. But sometimes stratiform clouds accompany this type of instability. Vertical wind shear that overcomes the stable upper atmosphere creates this type of turbulence.

Most clear air turbulence takes place around—but not in—jet streams. Pilots should look for upper atmosphere frontal systems with strongly contrasting temperatures. Mountain ranges are another area where pilots should be alert to the possibility of clear air turbulence. Winds crossing these ranges may be affected by temperature conditions that result in vertical wind shear. A third region where pilots may run into clear air turbulence is around thunderstorms. If a strong wind meets the upper reaches of a thunderstorm cloud, wind shear results.

Remember that wind shear can affect any flight and any pilot at any altitude. While a pilot may get reports of wind shear, this form of turbulence often goes undetected. Therefore, a pilot should always be alert to the possibility of wind shear, especially when flying in and around thunderstorms and frontal systems.

## How Types of Severe Weather Affect Aviation

Severe weather is a sign of turbulence. For instance, wind shear is partly to blame for tornadoes. As you've read weather can pose serious hazards to flight. Thunderstorms can pack just about every weather hazard, including tornadoes, known to pilots. These dangers can strike alone or in combination.

Lift jump-starts thunderstorms. Instability and water vapor (moisture) keep them going.

### Thunderstorms

Convection triggers thunderstorms. For a thunderstorm to form, the air must have enough water vapor, an unstable lapse rate, and a lifting action that powers up the storm process. The rising and falling air creates just the right conditions for this weather event.

Water vapors rising with warm air condense into clouds. The convection process then continues within the clouds because as the water vapors condense, they give off heat. This heat creates more warm air, which then also rises, and so the pattern repeats until the clouds reach the tropopause.

At the tropopause the unstable atmosphere stabilizes. The clouds may spread horizontally at this point because they can no longer rise because of the tropopause, which traps them somewhere below 7.5 miles (12 km) or so. Once rain falls, however, the air cools and this slows convection until the storm and clouds break up (Figure 3.9).

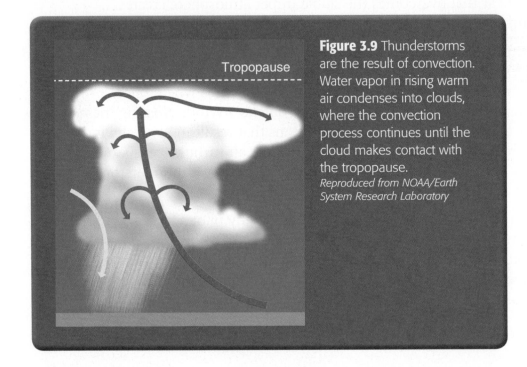

**Figure 3.9** Thunderstorms are the result of convection. Water vapor in rising warm air condenses into clouds, where the convection process continues until the cloud makes contact with the tropopause.
*Reproduced from NOAA/Earth System Research Laboratory*

## A Thunderstorm's Three Stages

A thunderstorm works through three phases before dissipating (Figure 3.10). They are:

1. *Cumulus stage*—A lifting action begins as warm air rises. Clouds grow vertically when there's enough moisture and instability. Steady, strong updrafts stop moisture from falling. The area where updrafts are taking place spreads beyond where the individual thermals— or rising currents of warm air that cause turbulence—were initially feeding the storm.

2. *Mature stage*—After about 15 minutes of the cumulus stage, a thunderstorm reaches its mature stage. This is the most violent period of a storm's life cycle. Moisture that's too heavy for the cloud to support falls at this stage as either rain or hail. The air now adopts a downward motion. What results is a meeting of three states: warm, rising air; cool, precipitation-created falling air; and violent turbulence all existing within and near the cloud. Below the cloud, the air that's rushing downward increases winds at Earth's surface and decreases the temperature. Once the vertical motion near the cloud's top slows down, the top of the cloud spreads out in an anvil shape.

**Wing TIPS**

As warm air rises in a thunderstorm's mature stage, it may move as much as 8,000 tons of water in its updraft every minute.

3. *Dissipating stage*—Now the storm enters its final phase. The downdrafts spread out and replace the updrafts that were feeding the storm.

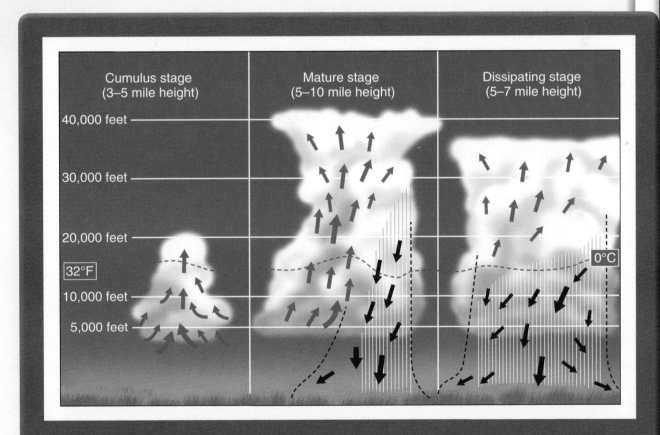

**Figure 3.10** A thunderstorm's life cycle
*Reproduced from US Department of Transportation/Federal Aviation Administration*

Some storms are random. They occur in unstable air, last for only an hour or two, and produce moderate wind gusts and rain. Meteorologists call these *air mass thunderstorms*. Generally warm surface temperatures cause this type of storm.

A more threatening type of weather is the *steady-state thunderstorm*. Fronts, big winds that run into one another, and *a long stretch of low pressure* called a trough spawn these storms. Steady-state thunderstorms often form into the narrow band of active thunderstorms called squall lines. When a steady-state thunderstorm reaches its mature stage, updrafts grow stronger and last much longer than in an air mass storm. This is how this more durable storm got the name of steady state.

## Weather Hazards Associated With Thunderstorms

Thunderstorms sometimes produce other types of severe weather and stir up hazardous flying conditions. When pilots find themselves flying in and around thunderstorms, they should be on the lookout for squall lines, tornadoes, turbulence, icing, hail, lightning, and poor visibility, among other weather hazards.

Light aircraft won't be able to fly over thunderstorms. Severe thunderstorms can even blow through the tropopause to as high as 72,000 feet. An aircraft flying under a thunderstorm may run into hail, damaging lightning, and violent turbulence. A good rule to follow when faced with a severe thunderstorm is to fly around it by a wide margin. Hail, for instance, can fall for miles outside of the clouds. The Air Force instructs pilots to fly 20 miles from the storm's edge. Sometimes the best answer is to remain on the ground until the storm passes.

### Turbine Engines and Updrafts

Updrafts fuel many thunderstorms. Sometimes an updraft's velocity is near or greater than the velocity of a storm's falling raindrops. When this occurs the updrafts can contain very high concentrations of water. This moist condition can overwhelm a turbine engine, which can handle some water but not to this great a degree. Therefore, severe thunderstorms with areas of high water concentration can damage a turbine engine or even cause it to fail.

## Squall Lines

Squall lines develop in moist, unstable air either on or in advance of a cold front. But they can also form in unstable air nowhere near a front. A squall line may be too long to fly around as well as too wide and severe to fly through. These elongated narrow storm bands often contain steady-state thunderstorms. They form quickly and are strongest in the late afternoon and early evening.

## Tornadoes

Tornadoes materialize out of the most violent thunderstorms. These thunderstorms draw air into their cloud bases with great vigor. If the rising air already has a rotating motion as it enters a storm, this swirling action can sharpen into a powerful vortex that reaches from the ground into the clouds. Air pressure is very low inside a vortex where wind speeds can reach 200 knots or higher. Dust and debris collect in the strong winds, and the low pressure spawns a funnel-shaped cloud that stretches earthward from a cumulonimbus base.

**Wing TIPS**

If a cloud with a vortex doesn't touch the ground, meteorologists refer to it as a *funnel cloud*. If it makes contact with land, it's a *tornado*. If it makes contact with water, it's a *waterspout*.

Not only squall lines but also isolated thunderstorms can produce tornadoes. In addition, groups of tornadoes can form—like fingers stretching out from the palm of your hand—several miles away from the main storm cloud where most of the lightning and precipitation are taking place. Pilots can't always detect these vortices, but they can easily damage an aircraft.

## Turbulence

Any thunderstorm may contain hazardous turbulence in the form of wind shear. A severe thunderstorm could actually destroy an aircraft. Wind shear produces turbulence in three areas: inside clouds, outside clouds, and at lower altitudes.

Inside a cloud, pilots encounter the strongest turbulence with wind shear located between the updrafts and downdrafts. Outside of clouds, pilots have run into shear-generated turbulence as much as several thousand feet above and 20 miles sideways from a severe storm.

Gust fronts accompany low-level turbulence. A gust front is *a rapid and sometimes drastic change in surface wind ahead of an approaching storm caused by cool air flowing out of a thunderstorm in a downdraft*. Gust fronts may stir up trouble as far as 15 miles ahead of any storm-related precipitation.

A dramatic shelf cloud forms in front of a severe thunderstorm moving over Rapid City, South Dakota.
*Courtesy of NOAA/National Weather Service*

## Wing TIPS

A shelf cloud with sharp, crisp edges tells you that strong winds are blowing beneath it. So the crisper a cloud's edges are, the more powerfully the winds are blowing.

Meteorologists look for shelf clouds and roll clouds, which appear on a storm's leading edge. Rising warm air builds these clouds, and the cool air flowing in the gust fronts carves out these wedge-shaped billows. A roll cloud is simply a shelf cloud that has broken away from the thunderstorm. Either can warn you that an extremely turbulent zone is coming.

### Icing

While gust fronts are an outcome of downdrafts, icing is a product of updrafts. As air flows up in a thunderstorm, it carries with it an abundant supply of liquid water in relatively large droplet sizes. Once the updraft lifts the water vapor above the freezing level, the water supercools with its temperature dropping to freezing even as the water remains in a liquid state. The temperature at the atmosphere's freezing level is 32 degrees F (0 degrees C).

As the updraft continues to rise, its temperature cools to around 5 degrees F (–15 degrees C). The water vapor in the upward current changes to ice crystals. Much of this remaining water vapor sublimates as ice crystals, that is, it changes to a gas without going back through the liquid state. As the air current continues climbing, the temperature drops even lower and the amount of supercooled water decreases.

Supercooled water freezes on impact with an aircraft. This can be extremely dangerous, as it affects lift and weight. You'll read more about icing's impact on aircraft in Chapter 2, Lesson 5.

### Hail

Hail is another form of supercooled water. It can be just as dangerous to aircraft as turbulence. Supercooled water lifted above the freezing level begins to freeze.

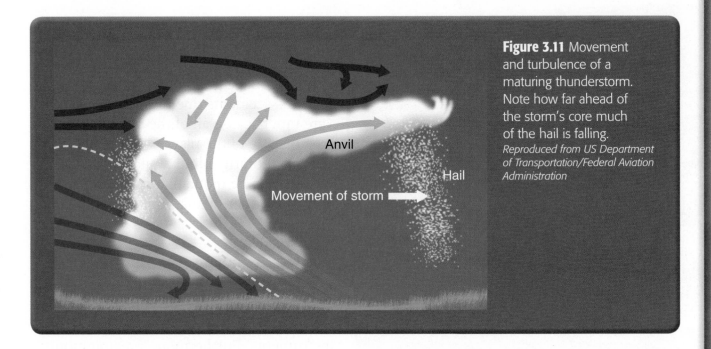

**Figure 3.11** Movement and turbulence of a maturing thunderstorm. Note how far ahead of the storm's core much of the hail is falling.
*Reproduced from US Department of Transportation/Federal Aviation Administration*

Once a drop freezes, other drops latch on and freeze to it. The hailstone can grow into a huge ice ball. Severe thunderstorms with strong updrafts that have grown to great heights can produce large hail softball size or larger. Eventually the hailstones fall. They may drop some distance from the storm's core. Therefore, a pilot could encounter hail in clear air several miles from thunderstorm clouds (Figure 3.11).

Hail starts to melt as it falls through air with a temperature above 32 degrees F (0 degrees C). It may hit the ground as either hail or rain. Look for hail when flying beneath the anvil of a large cumulonimbus cloud. Hailstones larger than one-half inch in diameter can cause significant damage to an airplane in a matter of seconds.

## Lightning

Lightning is a hazard that's most closely associated with thunderstorms, although other events such as volcanic eruptions and large wildfires can produce strikes. Rising and falling currents develop an electric charge when the water, ice particles, and hail the updrafts and downdrafts carry collide with one another.

Positively and negatively charged regions in the atmosphere try to connect with a moving positive charge on the ground. The ground beneath your feet generally has no charge—it is neutral because there are an equal amount of positive and negative charges. But the negatively charged clouds in a thunderstorm bring on a positive charge in the ground because the negative charges in the ground are repelled by those in the cloud and move away.

Wing TIPS

Earth experiences about 1,800 thunderstorms at any given moment. Lightning flashes 100 times per second, or 8 million times a day.

Lightning punched a small hole in the nose of this National Oceanic and Atmospheric Administration C-130 hurricane research aircraft.
*Courtesy of NOAA*

A lightning strike can puncture an aircraft's skin and damage its communications and electronic navigational equipment. However, serious accidents due to lightning strikes are rare because all of the charge stays on the outside of the plane. Nearby lightning can temporarily blind a pilot so that he or she may be momentarily unable to navigate either by instrument or by sight.

## Poor Visibility

Generally visibility is near zero within a thunderstorm cloud. A pilot won't be able to fly by sight (visual flight rules). Instead, the pilot will have to rely on instruments. Furthermore, precipitation and dust present between a cloud's base and the ground may severely reduce what the pilot can see and how well he or she can see.

### Wing TIPS

Cloud ceilings and cloud cover affect visibility. When clouds cover five-eighths to seven-eighths of the sky, meteorologists call the clouds *broken*. They call the sky *overcast* when clouds cover the entire sky.

Visibility refers to the greatest horizontal distance at which you can see big objects with the naked eye. Therefore, precipitation and dust can create the same types of visibility problems for pilots that low ceilings and cloud cover can make. But the hazards multiply when associated with other thunderstorm hazards such as turbulence, hail, and lightning.

In this lesson you've considered what causes bad weather, the different types of turbulence behind it, and the hazards a pilot should be alert to when in the air and before takeoff. It should be clear that flying in and around thunderstorms can be extremely dangerous. Pilots should avoid doing so in almost every circumstance. Now that you've read about everything from thunderstorms to hail and lightning, the next lesson will take you into a discussion of weather forecasting. Everyone's caught the weather report during the morning or evening news at some point or other. This next lesson will look at the history of weather forecasting, the equipment involved, and how the information is relayed to those who need to know—such as pilots.

# ✔ CHECK POINTS

## Lesson 3 Review

Using complete sentences, answer the following questions on a sheet of paper.

1. What does the atmosphere's stability depend on?

2. Where does the adiabatic process take place?

3. Which surfaces emit a large amount of heat, and which tend to absorb heat?

4. Why is low-level wind shear especially dangerous?

5. For a thunderstorm to form, what must air have?

6. Icing is a product of what?

# APPLYING YOUR LEARNING

7. A pilot of a large passenger airliner hears that a thunderstorm is in the area. What kinds of dangers might the thunderstorm pose to the aircraft, and what path should the pilot take in relation to the storm to keep aircraft, crew, and passengers safe?

### Quick Write

_____

_____

Why do you think the military needs its own weather service?

### Learn About

- the history of weather forecasting
- the types of instruments used in weather forecasting
- the various types of communication methods used to provide weather information

**W**eather forecasts can change the course of history. Several predictions made in the run-up to the D-day invasion on 6 June 1944 determined whether the mission to retake Europe from German hands had any chance at success.

The Allies under US Gen Dwight D. Eisenhower had to move more than 150,000 American, British, and Canadian soldiers by ship across the English Channel to the beaches at Normandy, France. Army Air Forces and Royal Air Force fighters would provide cover for Allied ships making the crossing. Some soldiers would also cross in air transports. In addition fighters and bombers did a lot of bombing and strafing ahead of the invasion to soften enemy defenses.

The 20-mile-wide English Channel often has dicey weather. Further, the coasts frequently see fog and other weather hazardous to flight. So the Allies turned to six forecasters with a seventh liaising between the weathermen and General Eisenhower. The liaison was James Martin Stagg, chief meteorological adviser to Eisenhower. The six forecasters worked in teams of two: American, Royal Navy, and British civilian. They looked ahead at what the weather might be like anywhere from 5 June to 7 June with the option of delaying the D-day invasion until 19 June. They had trouble agreeing on when the weather would be good.

In an interview with the two Royal Navy weathermen by the *Telegraph* of Britain in 2004, one of the forecasters named Lawrence Hogben said, "The outcome of D-Day, perhaps the whole future of the western world, rested on those forecasts, so I think you could say there was some pressure." According to Hogben, only the Americans voted to go forward with the invasion on 5 June. The two British teams determined that the weather would make air cover impossible and the crossings too dangerous on that date.

Early June had already produced plenty of poor weather, including rain clouds low in the sky and strong winds. Hogben told the *Telegraph* that the forecasts of bad weather for 5 June were accurate because the seas were high and the winds far too powerful for planes and ships to handle. "Things looked very bleak. We knew that, without a change, the invasion would have to be postponed until June 19, the next time the tides would be right. Luckily, the next day [6 June] there was a wholly unexpected break and we were able to change our forecast to favorable." Eisenhower readied the troops for the mission. While the weather wasn't perfect on 6 June, the mission turned the war in the Allies' favor.

It turned out to be more important than anyone realized that the operation wasn't put off until the 19th. Eleven days after the invasion, the three weather teams made forecasts for 19 June. All three agreed this time that the weather would have been perfect if they had waited. But all three forecasts were wrong. That day produced "the biggest storm of the twentieth century," said Hogben. If the mission had launched "that day, I doubt many landing craft would have even made it to the beaches. It does not bear thinking about."

## Vocabulary

- altimeter
- psychrometer
- anemometer
- Automated Surface Observing System
- radio waves
- reflectivity
- geostationary
- briefing

## The History of Weather Forecasting

Today when you board a plane, the pilots have access to some of the best weather forecasts in history. They use them to plot their routes. But the earliest pilots didn't fly with this advantage. Too often they crashed because of unexpected bad weather.

Since 1918 the National Weather Service has provided weather forecasts to pilots. Over time the quality and accuracy of these forecasts have greatly improved.

In the beginning weather forecasts mostly supported government flights for mail delivery and the military. Today millions of passengers and cargo shippers also benefit from advances in aviation forecasting.

### The First Aviation Weather Forecast

In flight's early years, meteorologists at the US Weather Bureau—now NOAA's National Weather Service—didn't have radar or other advanced technologies to spot bad weather in the upper atmosphere. So they used primitive means to issue the first aviation forecast on 1 December 1918. They attached instruments to kites and tethered balloons to find the temperature and wind direction. They combined these facts with measurements they took on the ground to produce their first forecast. The first report went to a government flight carrying mail from New York to Chicago.

An early weather forecaster tracks data.
*Courtesy of NOAA*

"In those early days, forecasters spent most of their time finding out what was happening at the moment with no real way of knowing what would happen in the immediate future," says Bob Maxson, director of NOAA's Aviation Weather Center in Kansas City, Missouri. "Meteorologists did not have the capability to collect complex atmospheric data, so they largely relied on what they observed on the ground coupled with post-flight reports from pilots."

The pilot flying the airmail route that December day took to the skies on a clear, windy day in New York and landed in Chicago amid clouds. Historical records from weather stations lining the route indicate that no precipitation was falling along the way.

## Wing TIPS

The *telegraph* was largely responsible for advances made in the study of weather in the nineteenth century. The telegraph delivers messages via electronic impulses through wires strung from one place to the next. With this invention, scientists could relatively quickly gather weather observations from distant points, then plot and analyze them.

## Weather Reports Address Safety Issues

By 1920 about half the airmail service pilots had died on the job. The focus up until this point had been to get mail delivered on time rather than get mail delivered safely. This was partly because no one really understood how much weather affected flight. Eventually, interest spread within the new aviation industry in putting safety first. The US government began to recognize the benefits of investing in an aviation-weather forecasting system.

On 20 May 1926 Congress passed the Air Commerce Act. It directed the US Weather Bureau to make air travel safer. This focus on safety also boosted the commercial passenger business. Today the United States has one of the world's most advanced aviation weather forecasting systems. It collects 76 billion weather observations each year. Scientists use the data to issue about four million aviation weather forecasts each year. This includes 94,000 warnings for 630 airports.

## Air Force Weather Agency

The US military also got into the weather game. US Army surgeons began recording the weather around 200 years ago. In 1870 Congress directed the secretary of war to start a weather service for the country. So, the Army set up its first military weather service in the US Army Signal Corps. However, after Congress authorized the launch of a civilian US Weather Bureau in 1890, the Army's weather arm waned.

Only with America's entry into World War I did the US military again organize its own weather team. Today's Air Force Weather Agency (AFWA) traces its history to the reemergence of a meteorological section within the US Army Signal Corps in 1917.

# Airmen Forecast the Weather

"Be prepared" may be the Boy Scouts' motto, but it's a very useful phrase in many fields. Being prepared for bad weather, for instance, can determine a mission's success. That's why the Air Force has its own weather team.

The Airmen of 3rd Operations Support Squadron (OSS) weather flight at Joint Base Elmendorf-Richardson, Alaska, check weather patterns for five airfields in the state: JB Elmendorf, Bryant, Juneau, Bethel, and Nome. The base supports Air Force, Army, and reserve members, so the weather flight team has to consider conditions for several different types of aircraft, including the F-22 Raptor, C-17 Globemaster III, C-12 Huron, E-3 Sentry, UC-35A, C-130 Hercules, HH-60 Pave Hawk, and C-23 Sherpa.

"A weather flight member's job is to provide the most accurate forecasts for the base, its residents, and the decision makers," says SSgt Raymond Polasky of the 3rd OSS weather flight. "The more accurate the forecast, the better they can do their mission planning, the better they can do their mission."

The weather flight crew uses a variety of instruments. These include radar, satellites, and weather sensors. Polasky adds that not only must the team members know how to read computer data, they must also understand the "physics of the atmosphere."

The sergeant admits that they don't always get the weather right, "But most of the time we're pretty accurate."

SSgt Raymond Polasky points to weather patterns from radar readings at Joint Base Elmendorf-Richardson, Alaska, in 2011. Polasky works with a team that forecasts weather for several airfields in the state.
*Courtesy of USAF/A1C Jack Sanders*

**CHAPTER 2** Working Through Flight Conditions

MSgt Stephen Hale (*left*) and SrA Ryan Unger of the 22nd Expeditionary Weather Squadron check a weather observation system and radar after setting them up at a control tower at Qayyarah West Airfield, Iraq, in 2011. The two Airmen were part of a three-man weather team that traveled from Baghdad to retrieve US Air Force equipment and install a newly purchased system for an Iraqi military weather service.
*Courtesy of USAF/SSgt Levi Riendeau*

AFWA partners with NOAA's National Weather Service to improve weather science, among other goals. AFWA provides worldwide coverage. It offers its services to Air Force and Army forces, and wherever else required. Its data helps the military engage targets around the world. This includes everything from safe passage for unmanned aerial vehicles during missions to troop safety in the air and on the ground.

In addition, AFWA personnel have conducted hurricane reconnaissance since World War II to protect life and property at home. In 1948 two AFWA officers issued the first tornado warning. AFWA also participated in the development of America's severe storm forecasting centers.

## The Types of Instruments Used in Weather Forecasting

In 1918 meteorologists used kites and balloons at 17 weather stations to produce forecasts. Today government branches such as the National Weather Service and private contractors create flight forecasts using data from thousands of sites and sources, including satellites and radar.

The data that observers gather from Earth's surface and upper altitudes form the basis of all weather forecasts. The four types of weather observations they make are surface, upper air, radar, and satellite. This section considers the types of tools scientists use to collect and study this information.

## Surface

When weather observers study surface weather conditions, they are generally looking at an area within a five-mile radius from a given airport. The factors they consider near Earth's surface include wind, visibility, pressure, temperature and dew point, sky conditions (such as cloud cover), and precipitation or other weather. While the data is local, pilots can still generate a good picture of the weather over a wide area by combining many of these reports.

Meteorologists use a range of tools to study surface weather. These include the thermometer for temperature, barometer for pressure, psychrometer for relative humidity, and anemometer for wind speed. Each works in unique ways.

There are several types of *thermometers*, which measure temperature. A common *bulb thermometer* has a rounded part filled with colored alcohol at the bottom of a long tube. The liquid expands and rises in the tube when temperatures increase. It condenses and falls when temperatures decrease.

As you read in Chapter 2, Lesson 1, a *barometer* measures atmospheric pressure. While mercury barometers are still in use, many of today's barometers use *aneroids*—flexible metal bellows that are tightly sealed after some air is removed. The bellows then respond to changes in atmospheric pressure. The average home weather station often contains an aneroid barometer (Figure 4.1).

**Figure 4.1** An aneroid barometer. The standard pressure exerted by the atmosphere at sea level is 29.92 Hg (or 29.92 inches of mercury). This is equal to 1,013.2 mb (millibars), which is another way to measure atmospheric pressure. At sea level the weight of the atmosphere exerts an average pressure of 14.7 pounds per square inch (or 14.7 lb/in$^2$).
© Hemera Technologies/PhotoObjects.net/Thinkstock

Sometimes the barometer is attached to an aneroid barograph, a cylinder with paper on it that records up to a week's worth of air pressure. A pen attached to the aneroid moves up and down to record the pressure changes. Generally when the air pressure falls, the weather is getting worse.

## The 29.92 Altimeter Setting—A Key to Aviation Safety

An altimeter is *an instrument that tells the pilot how high the plane is.*

Let's say you are at 35,000 feet in a Boeing 757 over the Atlantic headed to London from New York. Your aircraft is in the clouds and has been for an hour. There is an Airbus A-380 headed in the other direction from London to New York at 36,000 feet and on the same track as you. The two aircraft will pass each other in the clouds somewhere in the middle of the Atlantic Ocean and (hopefully) with 1,000 feet of altitude separation between them. But—will they actually have that separation? They will, but only if the pilots set their altimeters correctly.

When you took off from JFK in New York, there was an extremely strong high-pressure system sitting over the area—in other words, the barometer reading was very high. At the same time, on the other side of the Atlantic over England, there was a very deep low-pressure system with a very low barometer reading at the time the A-380 left Heathrow Airport. The altimeter settings of the two aircraft at the time of takeoff reflected the weather (barometric pressure) at their respective airports, but now each is 1,000 or more miles toward their destination in the middle of the Atlantic, in the clouds, and in a very different weather environment. If the pilots don't change their altimeter settings after takeoff (reflecting the barometric pressure in the area they are flying now) they will be flying hundreds or even more than a thousand feet off the altitude they have been assigned by air traffic control.

Pilots get the current local altimeter setting from air traffic control prior to takeoff, while en route, and at the destination airport. But when they are at cruising altitude away from the takeoff or landing airport, they use a standard, uniform altimeter setting—one that is used all over the world by all the world's aircraft—29.92. It is the standard air pressure at sea level and also the one used to keep aircraft traveling the world from running into each other. When you consider that airliners cruising at high altitude are traveling at 0.8 to 0.9 Mach (or eight to nine miles each minute), it is impractical to change altimeter settings every few minutes, and for that matter, over the oceans there isn't anybody on the "ground" to tell the pilots what the local altimeter setting is.

So as a matter of both flight safety and convenience, 29.92 has been adopted by air traffic organizations worldwide to make sure that all aircraft flying at high altitude (above 18,000 feet in the United States) use the same altimeter setting. This ensures they have the altitude separation needed to get you and everyone else safely from New York to London and back again and everywhere else in the world.

So 29.92 really is an important number—and now you know why.

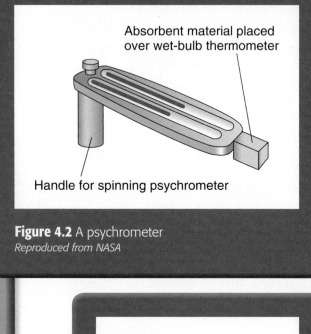

**Figure 4.2** A psychrometer
*Reproduced from NASA*

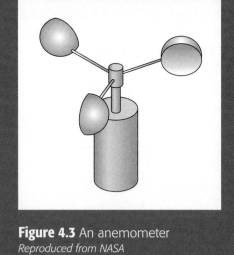

**Figure 4.3** An anemometer
*Reproduced from NASA*

A psychrometer is *a weather instrument that uses two thermometers to measure relative humidity* (Figure 4.2). One thermometer has a dry bulb and one has a wet bulb. The second bulb is wet because it has a cloth wound around it that has been dunked in water. You have different ways of taking a reading, depending on the type of psychrometer. If it's a *sling psychrometer*, then you spin the two thermometers round and round. If it's an *aspirated* or *ventilated psychrometer*, then you run air over the bulbs. In either case, the flow of air evaporates the moisture and cools the wet bulb.

If the air is damp, the wet cloth remains damp. But if the air is dry, it absorbs moisture from the wick. The evaporation cools the wet bulb. A weatherman reads the temperature recorded by each thermometer, and the difference between the two gives the relative humidity. If the thermometers show no difference in temperature, then the relative humidity stands at 100 percent. This would be because no moisture has evaporated from the wet cloth. If, in another example, the air is 86 degrees F (30 degrees C) and the thermometers differ by 27 degrees F (15 degrees C), then the relative humidity would be 17 percent (see Table 4.1). (The significance of the temperature difference between the two thermometers depends on the temperature of the surrounding air.)

| Difference Between Dry Bulb and Wet Bulb Temperatures | Relative Humidity |
|---|---|
| None | 100 percent |
| 0.5 degrees C | 96 percent |
| 1.0 degrees C | 93 percent |
| 1.5 degrees C | 89 percent |
| 9.0 degrees C | 44 percent |
| 9.5 degrees C | 42 percent |
| 14.5 degrees C | 19 percent |
| 15.0 degrees C | 17 percent |
| 18.0 degrees C | 5 percent |

**Table 4.1** Partial relative humidity for 30 degrees C
*Reproduced from NASA*

An anemometer is *an instrument that measures wind speed* (Figure 4.3). The most common anemometer has three or four small hollow metal cups set perpendicular to the ground. The cups catch the wind and revolve around a vertical rod. Their revolutions about the vertical rod help weather observers calculate the wind's velocity.

## Out of Thin Air

Mankind has used anemometers for more than 550 years to measure wind velocity and pressure. In 1450 an Italian inventor and architect named Leon Battista Alberti designed the earliest anemometer. He doesn't always get credit for being its original inventor.

Alberti's device was a disk placed perpendicular to the wind's direction. The wind would spin it, and the amount that the wind inclined the disk—that is, the angle at which the wind made the disk lean—showed how great the wind's force was.

In 1846 an Irishman named John Thomas Romney Robinson invented the hemispherical cup anemometer in wide use today. A combination of wheels recorded the number of revolutions in a given time.

Today's anemometers use many new technologies, including sound waves, laser and Doppler, and electric currents.

### Automated Surface Observing System

A widely used tool at airports across the country is the Automated Surface Observing System (ASOS). It's *a weather reporting system that provides surface observations every minute via digitized voice broadcasts to aircraft using ground-to-air radio*. When your local television weather anchor announces the day's high temperature, he or she got that information from an ASOS machine installed at your local airport. These machines sit at more than 900 US airports.

The system is a federally funded, joint program of the National Weather Service, Federal Aviation Administration (FAA), and the Department of Defense. It serves as the country's main surface weather observing network. It works nonstop, 24 hours a day, every day of the year. It's particularly helpful to pilots when they are about to land. Besides the computer-generated voice broadcasts, ASOS reports are also available in print and over the phone.

ASOS has a drawback, though. It can't report any weather that takes place over the horizon. The system's sensors can only detect what's directly overhead. It also isn't able to report on any event above 12,000 feet or phenomena such as tornadoes, freezing drizzle, or snow depth. People are still needed to add to and correct some ASOS reports.

## Upper Air

Upper air weather readings are more challenging to make than surface observations. Weather observers have two ways to learn about the upper air. They are radiosondes and pilot weather reports.

You first read about radiosondes in the Chapter 2, Lesson 1 Quick Write about weather balloons. These small instruments—suspended from inflated weather balloons—collect data such as air temperature, pressure, wind speed, and wind direction as they rise to tens of thousands of feet. They relay the data to ground stations via radio transmitters. When they reach more than 100,000 feet or so, the elastic balloons expand to their limits and pop. A parachute opens to safely return the radiosondes to the ground.

Pilots also provide vital details about the upper air to weather stations. In fact, they are the only real-time source of information regarding turbulence, icing, and cloud heights. Pilots gather and file what they see when flying. In addition, many airliners equip their aircraft with instruments that automatically send in-flight weather data to an airline dispatcher. The dispatcher then shares the data with weather forecasters.

## Radar

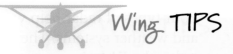

Weather observers and air traffic controllers rely on radar to get planes safely from airport to airport. Radar is an instrument that sends out and receives radio waves—*invisible electromagnetic waves in the portion of the electromagnetic spectrum longer than infrared light.* The radio waves sent off by the radar instrument reflect off of targets, such as airplanes, weather formations, and terrain, and bounce back to the radar. Radars can identify the distance, strength, direction, and speed of both moving and fixed objects such as aircraft, weather formations, and terrain.

A radar determines distance by measuring the time it takes—at the constant speed of light—for the radio wave to reach an object and then return to the radar antenna. It reads reflectivity—*the strength of a returned signal*—to gauge a storm's intensity. For example, the amount of reflectivity can depend on the size and number of raindrops. The bigger a raindrop is and the more of them there are, the stronger the reflected radio waves will be. Radar also can find out an object's velocity (its speed and direction). It reads whether wind is moving toward or away from its rotating antenna. People study radar readings for the presence of weather hazards. They also use them to pinpoint where planes are in relation to each other and to weather.

SSgt José Grimes monitors the radar on 8 March 2011 at the Air Force Rescue Coordination Center at Tyndall Air Force Base, Florida, as he coordinates with another agency during a search-and-rescue mission.
*Courtesy of USAF/Maj Steve Burke*

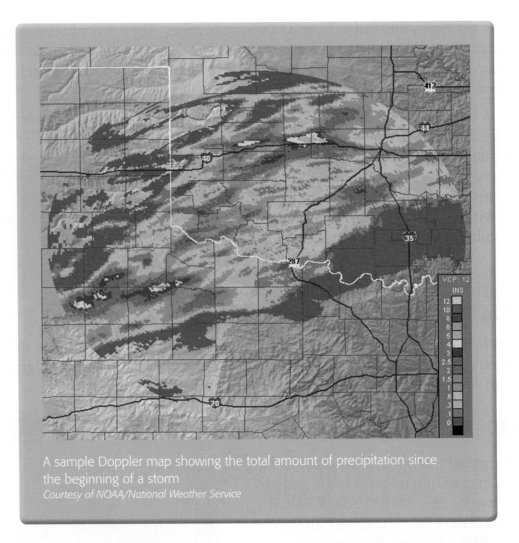

A sample Doppler map showing the total amount of precipitation since the beginning of a storm
*Courtesy of NOAA/National Weather Service*

The maximum range of each radar is 250 nautical miles. NEXRAD converts what the radars can read about a storm or other target—distance, strength, and velocity—into images. These are the brightly colored maps you might see during a TV weather report.

Doppler reads current weather. Pilots can access Doppler via electronic flight displays in their cockpit. Air traffic controllers use Doppler to compare where weather is with where aircraft are. The system has greatly improved the ability to issue warnings for severe weather and flash floods. It has also helped with air traffic and with protecting military resources. The average warning lead-time for a tornado is now up to 13 minutes, for example.

## Weather Outlets

Several government agencies, including the FAA, NOAA, and National Weather Service, work with private groups to get the latest weather data to pilots. They use NEXRAD and many other sources. One of the ways they get information to pilots is through the *Automated Flight Service Station (AFSS)*.

The Automated Flight Service Station is the main source for preflight weather reports. Pilots can get a preflight weather briefing—*the relay of information*—from one of these stations 24 hours a day by telephone. The service also offers in-flight weather data. The pilot will talk to a briefer at one of these stations, who then tailors weather and other data depending on the pilot's route, type of aircraft, type of flight (visual flight rules or instrument flight rules), destination, and altitude.

Another service available 24 hours a day by phone is the *Transcribed Information Briefing Service (TIBS)*. Some, but not all, Automated Flight Service Stations offer this shortened report. These briefings are prerecorded. They include special announcements. They do not replace the fuller preflight briefings. They are simply meant to be a quick rundown of weather and more.

Pilots may find the *En Route Flight Advisory Service (EFAS)* helpful once they're in the air. This service also goes by the name *Flight Watch*. Pilots must request this service to get timely in-flight weather data tailored to their route, type of flight, and cruising altitude. An En Route specialist can be one of the best sources for current weather along a route. This service is not available 24 hours a day, however.

The *Hazardous Inflight Weather Advisory Service (HIWAS)* is a national program that broadcasts bad weather alerts 24 hours a day. The broadcasts include reports such as SIGMETs and AIRMETs. These broadcasts are only a summary. For a more in-depth report, pilots should call an Automated Flight Service Station or En Route Flight Advisory Service specialist.

## Types of Briefings

Before every flight, pilots should gather all the data they can about their route. When a pilot speaks with a briefer at an Automated Flight Service Station, he or she may ask for one of three types of briefings: standard, abbreviated, or outlook.

A *standard briefing* is the most complete report. It provides an overall weather picture. Pilots should get this type of briefing before departure and use it to plan a flight. It supplies the following facts and more:

- Adverse conditions, including severe weather and airport closings
- A recommendation as to whether a pilot should fly under visual flight rules
- An overview of the larger weather picture, such as fronts and major weather systems
- Current conditions, such as ceilings, visibility, winds, and temperatures
- En route weather forecast for the proposed flight route
- Destination forecast for the estimated time of arrival
- Winds and temperatures aloft for specific altitudes along the flight route
- Notices to Airmen, called NOTAMs, which include such things as airport closings and restricted flight space (for example, Washington, DC); NOTAMs are published every 28 days.

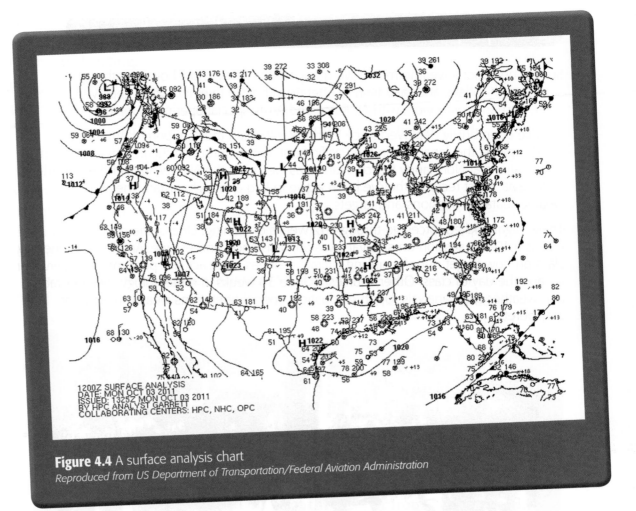

**Figure 4.4** A surface analysis chart
*Reproduced from US Department of Transportation/Federal Aviation Administration*

## Weather Charts

Weather charts show current weather or forecasts. Pilots use them when planning a flight. They show the movement of major weather systems and fronts.

A *surface analysis chart* assesses current surface weather (Figure 4.4). A computer prepares a new report every three hours. These charts show areas of high and low pressure, fronts, temperatures, dew points, wind directions and speeds, local weather, and visual obstructions. Symbols on the chart indicate everything from the type of cloud to who made a particular weather observation.

Another chart that lays out current surface conditions is the *weather depiction chart*. Every three hours, it updates the overall weather picture across the United States. It points out major fronts, areas of high and low pressure, and visual flight rules versus instrument flight rules weather. It does not include winds or pressure readings as surface analysis charts do. A pressure reading is not the same thing as a high- or low-pressure area. A pressure reading has to do with the actual pressure at sea level, for instance. Or it might note a pressure change over a three-hour period measured in millibars.

A *radar summary chart* also looks at current weather (but not surface conditions) (Figure 4.5). It's a collection of radar weather reports published every hour. It uses symbols to label type, strength, and movement of a precipitation cell (Figure 4.6). It marks areas that are on severe weather watch for tornadoes and severe thunderstorms with heavy dashed lines. If the radars aren't reading any kind of action, the chart will indicate this with an *NA* for *not available*. These charts are for precipitation only. They do not show such things as clouds or fog.

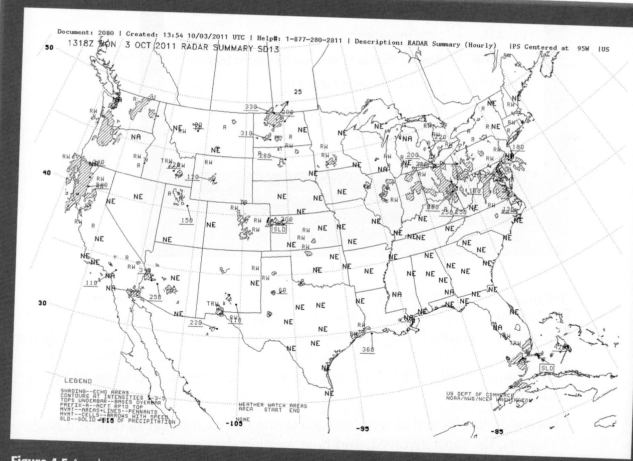

**Figure 4.5** A radar summary chart
*Reproduced from US Department of Transportation/Federal Aviation Administration*

| Symbol | Meaning |
|--------|---------|
| R | Rain |
| RW | Rain shower |
| S | Snow |
| SW | Snow shower |
| T | Thunderstorm |
| NA | Not available |
| NE | No echoes |
| OM | Out for maintenance |
| ↗ 35 | Cell movement to the northeast at 35 knots |
| LM | Little movement |
| WS999 | Severe thunderstorm watch number 999 |
| WT210 | Tornado watch number 210 |
| SLD | 8/10 or greater coverage in a line |
|  ╱ | Line of echoes |

**Figure 4.6** The symbols that you might find on a radar summary chart and their meanings
*Reproduced from US Department of Transportation/Federal Aviation Administration*

*Significant weather prognostic charts* are different from the three previous charts. These charts forecast weather. They don't report current weather. They predict turbulence, freezing levels, and instrument flight rules weather. They divide their forecasts into two readings. The first is for weather that could be found between Earth's surface and 24,000 feet, and the second for weather between 25,000 feet and 60,000 feet.

This lesson has looked at weather forecasting from a number of different angles: its history, the tools involved, and how weather data gets to pilots and others who need to know. Weather information is available to pilots around the clock. Computers have made constant improvements in forecast accuracy, methods, and timeliness. The next lesson will take you through several case studies of how weather affects flight, such as the impact of ice on safe flying operations.

**CHAPTER 2**   Working Through Flight Conditions

# ✔ CHECK POINTS

## Lesson 4 Review

Using complete sentences, answer the following questions on a sheet of paper.

1. Who has provided weather forecasts to aviators since 1918?

2. What was the name of the act that Congress passed on 20 May 1926 and what did it do?

3. What are the four types of weather observations that meteorologists make?

4. How does a radar determine distance?

5. What is one of the most important weather-tracking systems in the United States?

6. Which type of weather briefing is the most complete report?

# APPLYING YOUR LEARNING

7. A pilot is about six hours from departure. He or she needs to check the weather to plan the route. What are some of the preflight planning steps the pilot might take over the next six hours before taxiing down the runway for takeoff? Talk about the types of briefing the pilot might request, the outlets he or she might turn to, and the reports and charts the pilot might use.

# LESSON 5

## The Effects of Weather on Aircraft

### Quick Write

_____

_____

Do you think good flying instincts are something you're born with or a skill you can learn?

### Learn About

- how ice impacts flight
- how a microburst can induce wind shear on an aircraft
- how sandstorms can be hazardous to air flight
- how wake turbulence affects air flight

On 24 May 1988 a Boeing 737 from Belize was nearing New Orleans when crew members noticed signs of bad weather ahead. Their radar showed thunderstorm cells on either side of their intended flight path. Captain Carlos Dardano switched on the engine anti-ice system before his aircraft entered clouds at 30,000 feet. This is *a system designed to prevent ice buildup on an aircraft structure*.

Despite their precautions and their path between the red patches indicating severe weather on the radar, TACA Flight 110 ran into heavy rain, hail, and turbulence. Because the Salvadoran aircraft was coming in for a landing, the captain had decreased power for the descent. At about 16,500 feet both engines suffered a flameout, which is *a loss of power after too much water or some other substance such as volcanic ash has extinguished or drowned an engine's heat source*. In Flight 110's case, water ingestion from the surrounding precipitation caused the flameouts.

It's possible to recover in midair from a flameout. The crew restarted the engines. But the engines couldn't accelerate to idle, according to a US government report on the incident. The engines were operating far too slowly to generate thrust for lift. In addition, the engines began to overheat. The crew intentionally shut down the engines "to avoid catastrophic failure." Flight 110 was now in a glide.

The cockpit told air traffic control that they wouldn't be able to reach either of two nearby options: a small airport landing runway or a highway. They would head for some water.

Just then, the copilot spotted a levee, according to an interview with the pilot by *MSNBC* not long after the incident. A levee is a wall of earth that holds back water. They surround cities—like New Orleans—that sit near or below sea level. This levee was a long green strip of earth about a mile long. The pilot aimed his aircraft for the levee to attempt a deadstick landing. This is *a type of landing made without power*. The plane came down hard on the strip of land by the levee, according to *MSNBC*, but everyone survived. The pilot told the news station that his copilot laughed after the successful landing and cheered, "Hey, landing in a field, ain't we just like crop-dusters."

Dardano is a pilot who's faced other exciting flights. According to *MSNBC*, years before while flying a plane full of passengers in El Salvador, a guerilla shot from the ground at the plane and the bullet went through the captain's cheek and eye. Dardano kept his focus and flew his flight for another 25 minutes before landing, once again, safely.

## Vocabulary

- anti-ice
- flameout
- deadstick landing
- deice
- cockpit voice recorder
- pitot tube
- wind chill
- clear ice
- rime ice
- mixed ice
- buffet
- control column
- convective precipitation
- TOGA
- infrared
- arid
- radio silence
- rotate

## How Ice Impacts Flight

On 13 January 1982 Air Florida Flight 90 took off from Washington National Airport for Fort Lauderdale, Florida, with 74 passengers and five crew members. Snow was falling fast. The plane was barely airborne before crashing into the 14th Street Bridge that connects Virginia with Washington, DC. It fell nose first into the iced-over Potomac River and sank until only its tail was above water.

The aircraft—a Boeing 737-222—had sat on the ramp for nearly two hours before takeoff because of delays due to moderate to heavy snowfall. The captain had asked ground crews to deice the plane in the meantime. They did at 2:30 p.m. and again at 3:00 p.m. To deice is *to remove ice buildup from an aircraft structure, generally with a heated fluid.*

But too much time would pass between the final deicing and takeoff. The 737 headed toward the runway shortly after 3:00 p.m. However, it was the 17th plane in line for takeoff. Precipitation continued to fall in the interval. Not until 3:59 p.m. could the air traffic control tower clear the aircraft and its captain, Larry Wheaton, for takeoff. Visibility was only 0.25 miles.

According to government studies on Flight 90, the 737's nose pitched up abruptly right after liftoff. The cockpit voice recorder—*an instrument that records audio in the cockpit; also known as a "black box"*—revealed that the captain urged his copilot Roger Pettit with the instructions, "Forward, forward, easy," in an effort to reduce pitch. The aircraft continued forward but remained in its pitch-up attitude. The captain said again, "Come on. Forward, forward, just barely climb." Pettit answered, "Larry, we're going down, Larry." Wheaton said, "I know it." The plane hit the bridge. Only four passengers and one crew member survived. Several people crossing the bridge in cars also died.

The tail section of Air Florida Flight 90 is pulled from the Potomac River in Washington, DC.
*Courtesy of AP Photo*

## Causes of the Crash

The National Transportation Safety Board, which studies all aviation accidents, wrote a report on the crash later that year. It found several reasons for the accident. One reason was that the flight crew didn't turn on its engine anti-ice system. FAA and Air Florida regulations both said the system should be switched on when the weather is cold or wet. It runs hot air from the engine compressor over other engine parts to melt ice and prevent any more ice from building up. It also keeps probes that monitor engine performance clear of ice that might affect readings.

**Wing TIPS**

The National Transportation Safety Board is an independent federal agency appointed by Congress to investigate accidents related to transportation. The incidents can involve anything from aircraft to ships to railroads.

The board tested one of the main instruments that monitor thrust. According to its report, the *engine pressure ratio (EPR)* probe lets pilots know whether they are giving an aircraft enough thrust for takeoff and climb. With the engine anti-ice system turned off and the instrument iced over, the pilots wouldn't get an accurate reading of how much thrust they were applying. Instead the probe would have given them an incorrect reading. In the case of Flight 90, it indicated that they were producing more thrust than they really were. The investigators reasoned that had the crew applied more thrust, they might have avoided the crash altogether.

The board also found that the crew took off even though snow and ice covered the airfoils. Another captain who had seen Flight 90 before takeoff told the investigating board, "I commented to my crew, 'Look at the junk on that airplane,' … Almost the entire length of the fuselage had a mottled area of snow and what appeared to be ice … along the top and upper side of the fuselage above the passenger cabin windows. …"

The board further stated that the captain should have heeded irregular readings from the engine instruments. These readings were a warning not to take off in the plane's current state. The panel also cited the long delay on the ground and the heavy snowfall. The Boeing 737 has a tendency to pitch up when the leading edge of its airfoils have "even small amounts of snow or ice," the board reported. It also noted the pilot and copilot's limited winter flight experience.

The board heard from several witnesses to the crash. Most agreed that the 737's nose was flying with about a 30-degree to 40-degree attitude before it hit the bridge (Figure 5.1). This is far beyond the 15 degrees to reach a critical angle of attack leading to stall in most aircraft. A driver told the board, "I heard screaming jet engines. … The nose was up and the tail was down. It was [as if] the pilot [were] still trying to climb but the plane was sinking fast."

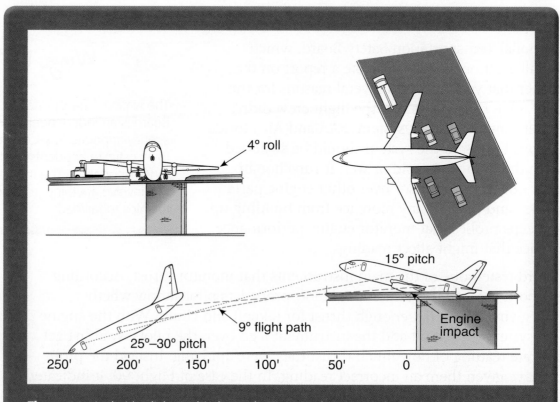

**Figure 5.1** Air Florida Flight 90 hit the 14th Street Bridge in a pitch-up attitude. It destroyed several cars as it swept across the top of the bridge. It then fell nose first into the Potomac River at a 25-degree to 30-degree downward pitch.
*Reproduced from National Transportation Safety Board*

## Air France Flight 447

In June 2009 Air France Flight 447 from Brazil to Paris, an Airbus A330, crashed into the Atlantic Ocean four hours after takeoff. All 228 passengers and crew on board died. It took two years before investigators could find the plane's black box on the ocean floor to begin figuring out what happened. Ice has been one of the main suspects.

The plane's autopilot disengaged while two copilots were on duty and the captain was taking a break (pilots and copilots often relieve each other on long flights). A crucial instrument called the pitot tube—or pitot probe—had most likely frozen over due to icing, so the copilots couldn't tell whether their cruising speed was too fast or too slow. A pitot tube is *a probe that senses temperature and relays airspeed information based on a combination of air pressures*. Bad weather made the situation even trickier.

Rather than dive to gain airspeed, the copilots climbed to 38,000 feet. Stall warnings went off in the cockpit but without accurate instrument readings the copilots couldn't be sure of the correct response to the alarm. The captain made it back into the cockpit a couple minutes before the crash. Even so, the three couldn't regain control. Flight 447 took three minutes 30 seconds to fall 38,000 feet into the ocean. France's Bureau of Investigation and Analysis continues to look into the incident. A large part of the aircraft was found on the ocean floor in April 2011.

**CHAPTER 2** Working Through Flight Conditions

## Ice Hazards

Ice can be hazardous for many reasons as Flight 90's crash proved. It affects lift, weight, and drag. It can cause stalls and affect the smooth operation of aircraft instruments and engines. Just as with a car, ice can also make it difficult to see out the windshield.

Ice, snow, and slush can affect aircraft on the ground and during flight. Icing takes place when a plane flies through visible water such as a cloud or fog—or sits on a runway surrounded by these elements—and the temperature where the moisture hits the aircraft is 32 degrees F (0 degrees C) or colder. Even if the air isn't at freezing, wind chill—*the cooling effect of wind on surfaces*—can cool an airplane's surface enough that icing can occur.

Supercooled water increases the rate of icing. As a supercooled water droplet strikes a plane's surface, a part of it freezes instantaneously. The way in which the remaining portion of the water droplet freezes determines what type of ice formation it is. It takes one of three forms (Figure 5.2):

- Clear ice is *glossy, see-through ice formed by the relatively slow freezing of large, supercooled water droplets.* When a droplet hits a surface, the liquid spreads out and gradually freezes. This ice freezes into a smooth, solid, hard, and heavy sheet. It is difficult to remove.

- Rime ice is *brittle and frostlike ice formed by the instantaneous freezing of small, supercooled water droplets.* It is rough, lighter in weight than clear ice, and easier to remove than clear ice. It also looks white because the rapid freezing traps air between droplets.

- Mixed ice is *a mixture of clear ice and rime ice formed when supercooled water droplets come in different sizes or the droplets mingle with snow or ice particles.* Because it freezes rapidly, it takes on a mushroom shape on a wing's leading edge. Ice particles embed in the clear ice and form a hard and rough mass.

### Wing TIPS

It is important to remember that temperature decreases with altitude. Even when temperatures are mild on the ground, a pilot will eventually reach below-freezing temperatures if he or she climbs high enough. This is called the *freezing level* and can be at the surface in winter or at 15,000 feet above mean sea level or higher in summer.

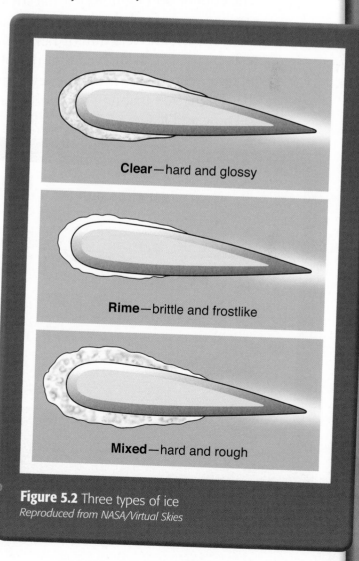

**Clear**—hard and glossy

**Rime**—brittle and frostlike

**Mixed**—hard and rough

**Figure 5.2** Three types of ice
*Reproduced from NASA/Virtual Skies*

The rest of this section discusses some icing dangers in relation to Flight 90's experience.

### Lift and Weight

The Air Florida Flight 90 crew fought stall from the moment they lifted off the runway. Temperatures were about 24 degrees F (−4.4 degrees C), just right for the accumulation of ice and snow. If ice and snow accumulate on an aircraft's surface, they disrupt the smooth airflow over the wings. This causes the boundary layer to separate at an angle of attack that's lower than that of the critical angle. The separation greatly reduces lift.

If ice builds up on an aircraft, the plane's weight increases while its ability to generate lift decreases. As little as 0.8 millimeters (0.032 inches) of ice on a wing's upper surface increases drag and reduces lift by 25 percent. The black box recording confirmed that the pilot and copilot saw anywhere from $1/4$ inch to $1/2$ inch of snow on the wings.

When an aircraft is about to stall, you can often feel it buffet, or *vibrate*. The copilot on Flight 90 felt his stickshaker act up from liftoff until the plane crashed 24 seconds later. (A stickshaker is a device that shakes the pilot's control stick and lets him or her know that the plane is about to stall.) Because tests by the NTSB showed that the plane should have been able to climb had all other conditions been fine, even at its lower rate of thrust, they concluded that Flight 90's inability to climb and fly at an airspeed above stall was due to snow and ice.

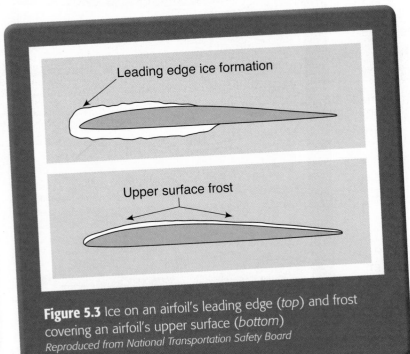

**Figure 5.3** Ice on an airfoil's leading edge (*top*) and frost covering an airfoil's upper surface (*bottom*)
*Reproduced from National Transportation Safety Board*

Snow and ice change an airfoil's shape and make it rough (Figure 5.3). This change in shape is what causes the boundary layer to separate at a lower-than-normal angle of attack and reduce lift. To overcome the airfoil's higher-than-normal angle of attack, Flight 90 needed more thrust for a higher airspeed. This way the aircraft could have produced enough lift to support its weight.

**CHAPTER 2** Working Through Flight Conditions

## Position of Ice on an Airfoil

The National Transportation Safety Board also looked at the position of snow, slush, and ice buildup on Flight 90's wings. The group said that while any accumulation on a wing will affect flight, the position of that buildup is just as important.

The worst place for buildup is at the wing's forward leading edge. This most affects its ability to produce lift. The investigators added that compounding the problem is the fact that the snow, slush, and ice won't be evenly spread over the wing.

A Boeing 737 has *swept wings*. These wings sweep or slant aft. They are critical to keeping a 737 balanced along its length. That is, when producing lift as they should, they prevent a plane from pitching up. When they can't produce the right amount of lift because of buildup or uneven buildup, this throws the longitudinal balance out of whack.

When a 737 readies for takeoff and climb, the trims on the tail are set to balance the wing lift and weight at the center of gravity. If, as the board suggested, more ice were on the wing's leading edge out toward the wingtips than toward the wing root, it changes where and how lift occurs along the wingspan. The aircraft "will be out of trim" and pitches up. While difficult, pilots can recover from this nose-up attitude with the use of extra force on the control column—*a device in the cockpit that a pilot uses to control pitch and roll; also known as a control yoke or a control or center stick*—and stabilizer trims.

Some of the ice on Flight 90's wings came from melted snow. While waiting for departure, Flight 90 intentionally got too close to another aircraft so its exhaust would melt snow and slush from Flight 90's wings. But this plan backfired. The safety board reported that the heat from the exhaust most likely turned any snow into slush, which would then have turned to ice on the wings' leading edge and the engine inlet nose cone. This may have increased the amount of ice that contributed to the accident.

## Drag

Snow and ice buildup also generates drag. The board's report said that snow and ice's most dangerous outcome is increased drag. It creates two kinds of drag, which you read about in Chapter 1, Lesson 2. The first is induced drag, which always accompanies lift. Induced drag and lift are always proportional. Each increases as the angle of attack increases. Therefore, the induced drag will be greater where an ice- or snow-covered wing is concerned because the angle of attack is greater.

The second kind of drag is parasite drag. While generally more of an issue at high airspeeds, it can be a problem for a plane contaminated with ice and snow flying at any speed. The board reported that Flight 90's increased angle of attack would have increased the amount of aircraft frontal area pushing air aside. The snow and ice would also have increased the amount of aircraft surface pushing air aside. This is one source of parasite drag. The other source would have come from airflow friction with the rough ice- and snow-covered surfaces.

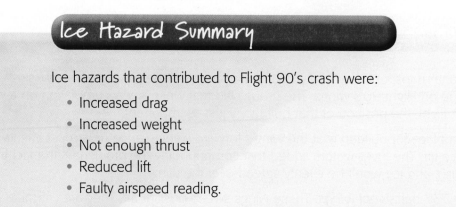

## Ice Hazard Summary

Ice hazards that contributed to Flight 90's crash were:

- Increased drag
- Increased weight
- Not enough thrust
- Reduced lift
- Faulty airspeed reading.

The safety board found many other factors also contributed to the accident. Still, the crash itself could have been averted. While the crew addressed the pitch issue right away, they didn't add more thrust in time to recover.

## How a Microburst Can Induce Wind Shear on an Aircraft

Wind shear can be an equally dangerous hazard to an aircraft. Wind shear accidents caused more than 500 deaths and 200 injuries in about 26 aircraft between 1964 and 1985. After a crash in Dallas, Texas, in 1985 that killed 134 out of 163 passengers and crew and one person on the ground, several organizations teamed up to prevent more shear accidents. These included NASA, the FAA, and private industry.

One of the most dangerous types of wind shear is the *microburst*, which is a violent downdraft. It takes place in a space of less than one mile horizontally and within 1,000 feet vertically. A microburst generally lasts about 15 minutes, and can create wind speeds greater than 100 knots and downdrafts as strong as 6,000 feet per minute. Radical changes in wind direction can occur in just seconds.

### Effects of a Microburst

When a pilot runs into a microburst while flying close to the ground, the powerful downdrafts and fast changes in wind direction can make it very difficult to control an airplane. For example, if a pilot takes off into a microburst, the plane experiences a quick series of events:

1. Performance-increasing headwinds
2. Performance-decreasing downdrafts
3. Wind rapidly shearing to a tailwind
4. Impact with the ground or aircraft pushed dangerously close to the ground.

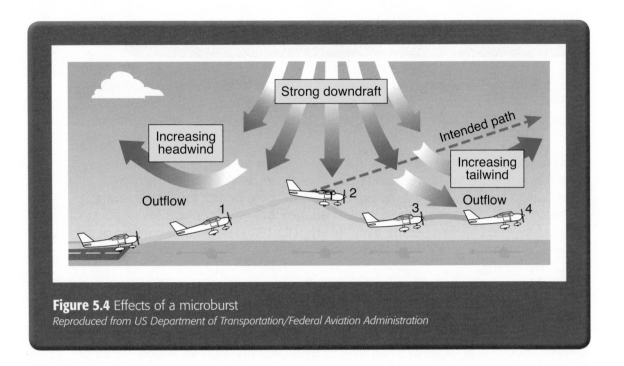

**Figure 5.4** Effects of a microburst
*Reproduced from US Department of Transportation/Federal Aviation Administration*

The downdrafts radiate outward as they rush toward the ground. As the downdraft spreads down and outward from a cloud, it creates an increasing headwind over an oncoming aircraft's wings (Figure 5.4). If a pilot is unaware that a microburst has caused the increased airspeed and lift, he or she may react by reducing engine power. But as the plane passes through the shear, the wind quickly changes to a downdraft and then a tailwind. This reduces the speed of air over the wings. The extra lift and speed vanish. The plane, now flying at reduced power because of the pilot's mistaken reaction, may suddenly lose airspeed and altitude. If the pilot adds power to the engines, he or she may be able to escape the microburst. But if the shear is strong enough, the plane may crash.

Planes are particularly vulnerable to wind shear and microbursts during takeoffs and landings. During landing, the pilot has already reduced engine power and may not have time to increase speed enough to escape a downdraft. During takeoff, an aircraft is near stall speed, which makes the plane very vulnerable to wind shear.

## Detecting Microbursts

Microbursts can be difficult to detect because they crop up in relatively small areas. But they are often associated with convective precipitation, *weather resulting from a vertical exchange of heat and moisture.* As cold, dense air sinks, it forces warm, moist air up. A thunderstorm is a type of convective precipitation.

Airports around the country have installed what is called a *low-level wind shear alert system.* A series of anemometers placed around an airport make up this system. The devices detect changes in wind speed. When wind speeds differ by more than 15 knots, the system issues a warning to pilots.

It was the accident that triggered seven years of research into microbursts. Delta Air Lines Flight 191 was coming in for a landing at Dallas-Fort Worth International Airport with 163 passengers and crew onboard on 2 August 1985 when a microburst slammed the plane into the ground.

A young Airman on leave witnessed the crash from his passenger seat in a nearby aircraft. Gary Creech, who eventually went to work for NASA's Dryden Flight Research Center, said, "I was looking out my window, sitting at the end of the runway aboard the second airplane lined up to take off. I had a window seat and was looking out the window when I noticed some really, really black thunderclouds at our end of the runway. Then I saw orange, extremely bright orange, light. My brain didn't register what I was seeing."

Flight 191 had just lost a battle with a microburst while trying to land.

"It was like a slow motion thing. There was the initial fireball, but then as the airplane rolled over, breaking apart and slowing down, the fire caught up with it and enveloped it," said Creech. "It all came to a halt directly even with my window, across the other side of the runway in the grass. As soon as the movement stopped, the rain hit. It was like a wall of rain and the fire quickly became a smoke ball, black and white smoke mixed."

Black box records retrieved after the accident revealed that the crew had tried to avoid bad weather by taking a more northerly approach to the Dallas-Fort Worth airport. Weather had started to dog them when they passed New Orleans. Air traffic controllers and the crew were in continual contact. But no one noticed that the weather was intensifying north of the airport because the meteorologist on staff was on his supper break as Flight 191 made its approach. The meteorologist had notified others he was stepping away from his desk and could be paged if needed.

Less than two minutes before Flight 191 crashed, the first officer alerted the captain of "Lightning coming out of that one [cloud]." The aircraft descended to 1,000 feet a little less than a minute later to prepare for landing. A few seconds after that the captain warned his first officer, "You're gonna lose it all of a sudden, there it is." Five seconds later he told the copilot to give the engines more power. The recording captured the engines at a higher power and the captain saying, "That's it." However, only 15 seconds later the captain gave his copilot the command, "TOGA." This is *a command that crews use to abort a landing; it stands for "takeoff/go around."*

Less than 15 seconds after the TOGA order, the plane crashed. A microburst had trapped the plane as it flew through a weather cell with a thunderstorm and heavy rain shower north of the runway.

The National Transportation Safety Board, which studied the accident, concluded that one of the causes of the accident was the crew's decision to fly through a cumulonimbus cloud containing lightning. In addition, instruments didn't exist on aircraft at that time that could detect wind shear. And finally, pilots didn't receive training on how to deal with wind shear. Today pilots and aircraft are much better equipped to handle these dangers.

Pilots hang on in the cockpit of NASA's Boeing 737 flying laboratory as it is about to enter a microburst wind shear cell in 1992.
*Courtesy of NASA*

When NASA, the FAA, and others joined forces to study wind shear from 1986 through 1993, they worked on developing a sensor for airlines and the military. They flew a Boeing 737 on more than 130 missions into severe weather to learn how to hunt for invisible shear anywhere from two to three miles ahead of the aircraft. NASA converted the 737 into a flying laboratory equipped with Doppler radar and infrared—*heat-seeking*—sensors. The crew entered around 70 microbursts.

The researchers' findings led to the development of a sensor that reads the speed and direction of invisible particles of water vapor and dust in the wind. It can give pilots enough advance warning that they can then anticipate how to react in the face of wind shear or a microburst.

## How Sandstorms Can Be Hazardous to Air Flight

Sandstorms and dust storms are another weather phenomenon that can be hazardous to flight. These strong, dry winds generally take place over arid— *hot and dry*—lands such as you find in the Middle East or the American Southwest. They can drastically reduce visibility, clog aircraft engines and instruments, and make it difficult to breathe.

These storms are walls of sand, dust, and debris. They usually form either in front of or behind a cold front. But more often they form behind a cold front where the winds are stronger. The bigger the particles, the stronger the wind must be to kick them up. The winds must also have vertical motion to move the sand, dust, and debris any real distance.

These types of storms may last only a few minutes or for several days, and can be many miles long and several thousand feet high. They typically die down at night as heat escapes Earth's surface, but not always. The strongest dust storms and sandstorms generally take place in the late morning or afternoon. If you find yourself in the middle of a sandstorm, visibility may be less than a half mile and wind speeds may be at least 30 miles per hour.

Sandstorms can damage aircraft and other machinery as particles work their way into their systems. NOAA research has found that the dust actually will solidify over time when it gets moist or mixes with lubricants in an engine.

A massive sandstorm cloud rolls over a military camp in Al Asad, Iraq, near nightfall in 2005. Storms like these can ground flights because they can reduce visibility to zero.
*Courtesy of USN/USMC/Cpl Alicia M. Garcia*

US Air Force crew chiefs assigned to the 332nd Expeditionary Aircraft Maintenance Squadron help guide an F-16 Fighting Falcon aircraft into a hardened shelter during a major dust storm on Balad Airbase in Iraq in 2008. Storms can damage an aircraft's engine and scratch its windows.
*Courtesy of USAF/SrA Julianne Showalter*

This can jam or clog machinery. Dust storms and sandstorms can also give off electrostatic charges that disrupt computer and electrical systems and fueling operations. Whipping sand and other particles can scratch windows as well, making it doubly difficult to see out a windshield.

## Operation Eagle Claw Runs Into Sandstorms

In 1980 two sandstorms in an Iranian desert shut down a US special operations mission. The mission was Operation Eagle Claw. Military troops were attempting to rescue 53 Americans who had been taken hostage on 4 November 1979 at the US embassy in Tehran, the Iranian capital. Earlier in 1979 the Islamic Revolution in Iran had ousted the country's longtime ruler, the Shah of Iran. In October 1979 President Jimmy Carter let the shah enter the United States for medical care. Supporters of the new Islamic rulers in Iran took the embassy personnel hostage in retaliation.

The rescue mission involved all four military branches: Air Force, Army, Navy, and Marines. It was to take place over two nights, 24 and 25 April 1980. It involved eight RH-53D Navy helicopters flown by Marines taking off from the aircraft carrier USS *Nimitz*. On the first night they would head to a meeting site in central Iran nicknamed Desert One. Three Air Force EC-130Es would be waiting at Desert One to refuel the helicopters. Three other Air Force MC-130E Combat Talons would drop off Army Delta Force troops for the helicopters to then ferry them 50 miles southeast of Tehran to prep for the next night's hostage rescue.

**Wing TIPS**

In the 1930s massive dust storms hit large swaths of the middle of the United States. Thousands of families headed west to find work or new land when they could no longer farm their own soil because of drought and erosion. Historians call this era and the affected region the *Dust Bowl*.

Weather played a large part in ending the mission prematurely. Before reaching Desert One, the MC-130s and EC-130s ran into two sandstorms. The first was mild, the second much more powerful, about 100 miles long and 5,000 feet high. The storms didn't have much impact on the big aircraft other than cutting down their visibility to less than one mile because the C-130s came equipped with radar and infrared sensors. But they had a far greater effect on the RH-53D Navy helicopters that eventually had to make their way through the dust clouds. The C-130s didn't warn the helicopters about the storms because the mission called for radio silence—*when no one is allowed to communicate by radio for security reasons.*

## Wing TIPS

In the Middle East, sandstorms and dust storms are called a *haboob*, which means "strong wind" in Arabic.

Because they couldn't see the ground and the view in front was down to less than a quarter of a mile, the helicopters had to spread out to avoid running into one another. This added confusion to an already difficult mission. One of the Marine helicopter pilots, Lt Col Jim Shaefer, later said the storms were the "hairiest conditions I have ever flown in." Helicopter after helicopter had to drop out of the formation. Before even encountering the storms, one crew had to abort the mission because of a damaged rotor blade. Once in the storm, another helicopter's navigation and flight instruments failed and it had to return to the aircraft carrier.

At the Desert One site yet another helicopter was having mechanical issues and it, too, was abandoned. Operation Eagle Claw now only had five helicopters to work with. Mission planners had agreed that the operation couldn't be carried out with fewer than six. The mission was also behind schedule because of how long it took the helicopters to fly through the storms. The ground assault commander in charge of the Delta Force, Col Charles A. Beckwith, recommended aborting the mission. Washington agreed and ordered the mission aborted.

The local conditions continued to plague the mission, however. As one of the helicopters lifted from the desert floor, the blades churned up dust which obscured the pilot's vision. It ran into a C-130. Eight Airmen and Marines died.

After the mission, the Joint Chiefs appointed officers from the Air Force, Army, Navy, and Marines to study the failed rescue. They named the team Special Operations Review Group. Admiral J. L. Holloway was its chairman. At the end of their study, he issued a statement listing all of the possible reasons behind the mission's collapse. The review group concluded that two factors in particular "directly" caused the mission to abort: "Unexpected helicopter failure rate and low-visibility flight conditions en route to Desert One." Both could be attributed in large part to the sandstorms.

A stained glass window at the Hurlburt Field Chapel honors the eight military members who died in Operation Eagle Claw with eight diamonds in the glass pattern.
*Courtesy of USAF/ Hurlburt Field*

Iran released the hostages on 20 January 1981, the day President Ronald Reagan was inaugurated into office. Reagan sent former President Carter to Ramstein Air Base in Germany to welcome the freed hostages.

## How Wake Turbulence Affects Air Flight

Aircraft produce some of their own hazards. One of these is *wake turbulence*. It is not a weather event. It's an invisible, atmospheric danger—much as microbursts and wind shear are—but it is comparatively easy to predict. This is because all aircraft generate wake turbulence while in flight.

### The First Lady's Flight

First Lady Michelle Obama and Dr. Jill Biden, the wife of Vice President Joe Biden, had a story to tell after an aborted landing at Andrews Air Force Base in Maryland in April 2011.

Air traffic controllers played it safe when they ordered a Boeing 737 with the first lady and Dr. Biden on board to circle Andrews' airfield because another larger aircraft also coming in for a landing was too close by. The National Transportation Safety Board reported that a C-17 cargo plane was less than three miles in front of the 737. Rules set down by the FAA say that planes must be five miles apart to avoid wake turbulence.

While no reports of wake turbulence came out of the incident, it was a public reminder of this particular aviation hazard.

## What It Is

Wake turbulence is a disturbance caused by a pair of vortices trailing from an aircraft's wingtips. Back in Chapter 1, Lesson 2, you read about wingtip vortices.

Vortices produced by larger aircraft can be particularly dangerous to other aircraft that get too close. Their wake can cause the aircraft that's following too close to roll violently left or right. This makes the aircraft difficult to control. In addition, turbulence generated within the vortices can damage aircraft parts and equipment if encountered at close range.

Wingtip vortices form when an airplane generates lift. Air spilling over the wingtips from high-pressure areas below the wings to low-pressure areas above them causes these rapidly rotating whirlpools of air. The swirling air masses trail behind the wings. Wake turbulence consists of two vortices rotating in opposite directions (Figure 5.5). Most of the energy is within a few feet of the center of each vortex.

## Intensity

The intensity of the turbulence depends on the weight, speed, and wing shape of the vortex-producing aircraft. If you change your speed, this affects vortex strength. If you deploy or retract your aircraft's flaps and other wing devices, this also affects vortex strength.

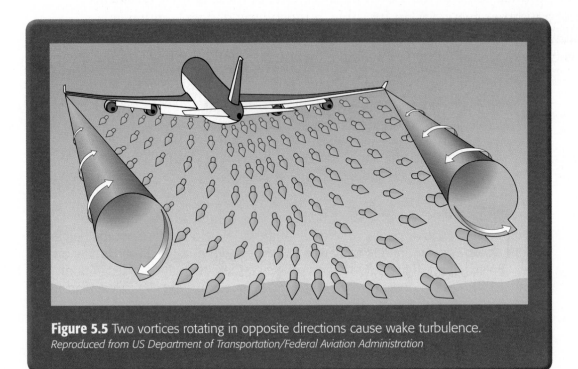

**Figure 5.5** Two vortices rotating in opposite directions cause wake turbulence.
*Reproduced from US Department of Transportation/Federal Aviation Administration*

Wingtip vortices are greatest when the generating aircraft is heavy, slow, and clean. The heavier and slower an aircraft is, the greater its angle of attack. This greater angle of attack gives the plane more lift, which in turn generates stronger wingtip vortices. So, an aircraft creates strong wingtip vortices during takeoff, climb, and landing when angle of attack is great. Furthermore, an aircraft with a *clean configuration* is one that's flying with its flaps and landing gear retracted to reduce drag. This also results in more lift and stronger vortices.

### Wing TIPS

Aircraft weight is directly proportional to the strength of vortices, while aircraft speed is inversely proportional to wingtip vortex strength. So a heavy plane results in strong vortices, while a light plane results in weaker vortices. And a slow plane results in strong vortices, while a fast plane results in weaker vortices.

## Behavior

Trailing vortices behave in predictable ways that can help pilots avoid wake turbulence. Aircraft produce vortices from the moment they leave the ground until they land because trailing vortices are a result of lift (Figure 5.6). They move outward, upward, and around the wingtips. However, vortices generally maintain a distance of a little less than a wingspan apart (see Figure 5.5 again). They also drift with the wind. They sink at a rate of several hundred feet per minute, with their descent slowing and decreasing in strength with time and distance behind the generating aircraft.

### Wing TIPS

After sinking to the ground, vortices tend to move sideways at two to three knots. Which way they move depends on the direction of the wind on the ground.

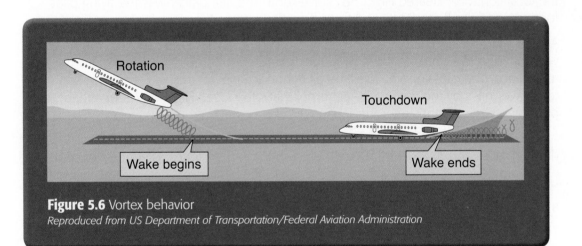

**Figure 5.6** Vortex behavior
*Reproduced from US Department of Transportation/Federal Aviation Administration*

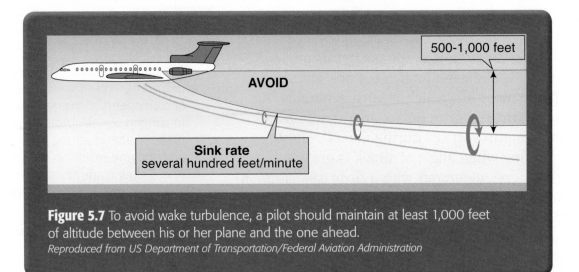

**Figure 5.7** To avoid wake turbulence, a pilot should maintain at least 1,000 feet of altitude between his or her plane and the one ahead.
*Reproduced from US Department of Transportation/Federal Aviation Administration*

## American Airlines Flight 587, Belle Harbor, New York

Wake turbulence played a part in the one of the deadliest aircraft crashes in US history, although according to investigators it was not the ultimate cause of the accident. A little after 9:00 a.m. on 12 November 2001, American Airlines Flight 587 taxied to runway 31L at John F. Kennedy International Airport in New York for a flight to the Dominican Republic. Visibility was more than six miles with only a few clouds scattered at 5,000 feet. Winds ranged from 12 knots to gusts of 22 knots. The Airbus A300-605R carried a pilot and first officer, seven flight attendants, and 251 passengers.

Just ahead of Flight 587 on the runway was Japan Air Lines Flight 47, a Boeing 747. The local air traffic controller cleared it for takeoff about 9:11 a.m. Within minutes, the local controller cleared Flight 587 for takeoff. The first officer was the flying pilot for this leg of the flight. Pilots and first officers take turns flying.

According to the cockpit voice recorder, the first officer asked the captain, "You happy with that distance" from Flight 47. His question referred to possible dangers from wake turbulence. The captain said, "We'll be all right once we get rolling. He's supposed to be five miles by the time we're airborne, that's the idea." Flight 587 took off about one minute 40 seconds after the Japan Air Lines plane.

Although American Airlines Flight 587 and Japan Air Lines Flight 47 had at least 4.3 nautical miles and 3,800 vertical feet between them at all times, Flight 587 encountered wake turbulence at 9:15:36 a.m. at about 1,700 feet. The captain said to his first officer, "Little wake turbulence, huh?" The plane was now at about 2,300 feet. At 9:15:51 a second wave of wake turbulence hit the plane. The situation quickly grew serious.

The airplane's flight data recorder recorded the rudder drastically deflecting five times in about seven seconds. This full swing of the rudder from one side to the other put lots of stress on the vertical stabilizer, which houses the rudder. At about 9:15:58 a.m. the vertical stabilizer and rudder ripped from the fuselage. The National Safety Transportation Board that studied Flight 587 concluded that the first officer overreacted to the second wave of wake turbulence by pushing hard on the rudder pedal five times.

## Avoiding Wake Turbulence

Pilots may run into turbulence on the runway and in the air. But because this disturbance is an everyday occurrence, they have many ways to avoid the danger. These include the following:

- *General rules*—
  - Avoid flying through another aircraft's flight path.
  - Avoid flying below or behind another aircraft.
  - Maintain at least 1,000 feet of difference in altitude between aircraft on similar flight paths (Figure 5.7).
- *During takeoff*—
  - Departing behind another aircraft, the pilot should rotate his or her aircraft before reaching the point on the runway at which the plane ahead rotates. To rotate is *to raise an aircraft's nose wheel off the runway*. Also, the pilot should climb above the other aircraft's climb path until clear of its wake.

The board reported that this rapid succession of inputs on the pedal overtaxed the control surfaces and tore the vertical stabilizer from the fuselage.

The board also deduced that the captain wouldn't have been able to see the first officer's overreaction on the pedal. Therefore, he must have "believed that the wake was causing the airplane motion, even after the vertical stabilizer had separated from the airplane (saying, at 09:16:12, 'get out of it, get out of it')." The plane crashed two seconds later in Belle Harbor, New York. No one survived the crash and five people on the ground died. Investigators found the tail section one mile from the crash site, and the airplane's engines—which subsequently broke from the plane—several blocks from the main wreckage.

After the accident, American Airlines changed its pilot training program, which may have contributed to the accident. The airline's simulator training did not accurately reflect the forces that would build up when the pilot overused the rudder pedal. In the new training, pilots were taught to react less aggressively to wake turbulence when flying an Airbus A300-600. American retired all its aircraft of that model in 2009.

A vertical stabilizer attachment point. It is a component that attaches a vertical stabilizer to the fuselage. When the NTSB studied Flight 587, it found that the attachments remained fastened to the body. Only the stabilizer and rudder had been torn away.

- *During landing—*
    - Approach the runway above the other aircraft's path when landing behind it, and touch down *after* the point at which the other aircraft's wheels contacted the runway.
    - But, when landing behind a departing aircraft, land *before* the departing aircraft's rotating point.

Wind plays an important role when avoiding wake turbulence. Wingtip vortices drift with the wind at the speed of the wind. For example, a wind speed of 10 knots will cause vortices to drift about 1,000 feet per minute in the wind's direction. So when following another aircraft, a pilot must consider wind speed and direction when picking his or her takeoff or landing point. Wake turbulence generally dies down in about three minutes.

This lesson has laid out some real-life cases in aviation through a series of accident reports, both civilian and military. Scientists continue to find ways to help pilots overcome these challenges. Training for pilots also continues to improve.

The next lesson you read will be the first in a chapter about flight's impact on the human body. It comprises two lessons. The first lesson will look at such issues as the effects of reduced pressure at higher altitudes on the body and motion sickness.

**CHAPTER 2** Working Through Flight Conditions

# ✔CHECK POINTS

## Lesson 5 Review

Using complete sentences, answer the following questions on a sheet of paper.

1. When does icing take place?

2. What happens when ice and snow accumulate on an aircraft's surface?

3. What is one of the most dangerous types of wind shear?

4. When are planes particularly vulnerable to wind shear and microbursts?

5. Where do sandstorms usually form?

6. During the Operation Eagle Claw rescue mission, why didn't the sandstorms have much impact on the big aircraft (the C-130s) other than cutting down their visibility to less than one mile?

7. What do all aircraft generate while in flight?

8. When are wingtip vortices greatest?

## APPLYING YOUR LEARNING

9. According to the National Transportation Safety Board's report on American Airlines Flight 587, the first officer overreacted to the wake turbulence by stressing the rudder. Based on all you've read about airplanes' physics, parts, and principles, what do you think would be the best way to handle an encounter with wake turbulence?

Air Force pararescuemen and Navy SEALs take part in free-fall parachute training on 21 January 2011 over Marine Corps Base Hawaii. The jumpers will pull the release cords on their parachutes after they lose some altitude.

*Courtesy of US Marine Corps/Lance Cpl Reece E. Lodder*

# the Human BODY

## Chapter Outline

"Man's flight through life is sustained by the power of his knowledge."

*Austin "Dusty" Miller, the quote on the Eagle and Fledgling statue at the US Air Force Academy*

# Human Physiology and Air Flight

## Quick Write

_____

_____

What devices and steps do you think could have prevented loss of consciousness?

## Learn About

- the four zones of the flight environment
- the physical laws of gases according to Boyle's law, Dalton's law, and Henry's law
- the respiration and circulation processes
- the effects on the human body of reduced pressure at high altitude
- the effects on the human body of acceleration and deceleration or increased g-forces
- spatial disorientation and motion sickness
- other stresses of flight operations

Payne Stewart was a well-known American golfer who won 11 tournaments hosted by the PGA Tour during an 18-year career. Besides his impressive wins, Stewart was also famous for the outfits he wore on the golf course—mostly old-fashioned knickers, often in bright patterns and colors. On 25 October 1999 he boarded a Learjet in Florida to head to Texas for yet another PGA Tour contest, the last of the season. The small business jet took off from Orlando International Airport with two crew and four passengers, including the golfer.

The flight was little more than 1,000 miles westward in clear weather. But the jet hit a technical snag just a few minutes into the flight. It crashed several hours later and hundreds of miles away in the wrong direction. From the time the jet departed Orlando International Airport at about 9:20 a.m. until 9:27 a.m., the crew was in repeated contact with air traffic control. All seemed well. Transmissions between the crew and controllers largely related to flight level, _the altitude at which an aircraft is flying_. The crew's flight plan requested an ultimate flight level of 39,000 feet, also written out as _FL 390_.

Around 9:21 a.m. the crew let the Jacksonville (Florida) Air Route Traffic Control Center know that they were climbing from 9,500 feet to 14,000 feet. A controller told them to next climb to FL 260 (26,000 feet) and maintain that flight level. At 9:27 a.m. the crew told another controller in Jacksonville that they had reached FL 230. The controller instructed them to next aim for FL 390. All transmissions from the Learjet ceased after the crew acknowledged the last flight level orders.

A little more than an hour later air traffic control still couldn't raise the crew on the radio. This was unusual. A US Air Force F-16 test pilot got orders to check on the Learjet. He flew within 2,000 feet of the small jet, which was flying at about 46,400 feet. A Learjet's maximum operating altitude is around 45,000 feet. He reported that while the plane looked undamaged, the cockpit windows seemed to be coated in ice or some other moisture on the inside. He also noted that he didn't see any flight control movements, that is, no adjustments to trims or other control surfaces.

Yet another hour later, two F-16s assigned to the Air National Guard in Oklahoma flew around the Learjet as it neared Minneapolis. About half an hour later two more F-16s from North Dakota's Air National Guard joined the survey. Barely 20 minutes later, the F-16s saw the Learjet turn right and descend. The lead pilot from North Dakota reported that "the target is descending and he is doing multiple aileron rolls, looks like he's out of control. …" One of the Oklahoma pilots said, "It's soon to impact the ground. He is in a descending spiral." The plane crashed in an open field in Aberdeen, South Dakota. No one survived.

As is usual, the National Transportation Safety Board investigated the crash. It found that the Learjet ran into trouble by at least 9:33 a.m. as the plane climbed above 36,400 feet. Not much later the jet swerved from its intended route and rose above its assigned 39,000 feet all the way to 48,900 feet. The board wrote that the evidence showed the crew lost consciousness.

As you read in Chapter 2, Lesson 1, air pressure drops at greater altitudes to levels that make it difficult for humans to breathe. If a plane doesn't maintain the right cabin pressure, crew and passengers pass out from lack of oxygen. Safety investigators concluded that because nothing else seemed wrong with the Learjet that it must have suddenly lost cabin pressure. With no one conscious at the controls, the plane just flew until it ran out of fuel and crashed.

## Vocabulary

- flight level
- physiology
- partial pressure
- respiration
- circulation
- trachea
- bronchi
- bronchioles
- alveoli
- diaphragm
- tissue
- pulmonary arteries
- pulmonary veins
- capillaries
- hypoxia
- tunnel vision
- hyperventilation
- Eustachian tube
- decompression sickness
- decompression
- g-force
- free fall
- terminal velocity
- linear acceleration
- radial acceleration
- angular acceleration
- g-suit
- rocket sled
- spatial disorientation
- carbon monoxide
- diuretic
- dehydration

## The Four Zones of the Flight Environment

Atmosphere can affect the human body just as readily as it impacts how an aircraft performs. Decreasing pressure with altitude makes it more difficult to breathe, for instance, because of the wider spread of air molecules. In fact pressure plays a large role in human health. This lesson looks at flight and how its environment interacts with physiology, which is *the study of how the body functions.*

As you read in Chapter 2, Lesson 1, the atmosphere is 78 percent nitrogen, 21 percent oxygen, and 1 percent trace gases. This composition remains the same even as you climb through the sky to areas of lower pressure. But the air density—the number of air molecules per cubic inch—and weight decrease as your aircraft ascends.

At lower altitudes where the air density and pressure are greater, the body can take in more air molecules with each breath, and each breath contains more oxygen even though oxygen always makes up about 21 percent of the atmosphere. This greater density and pressure is also why aircraft get more lift with less thrust at lower altitudes. At higher altitudes where the air contains fewer molecules, aircraft must exert more thrust to get enough lift.

The four zones of the atmosphere in which flight takes place are the troposphere, stratosphere, ionosphere, and exosphere. Chapter 2, Lesson 1 covered most of these terms. However, *ionosphere* is new. It is a combination

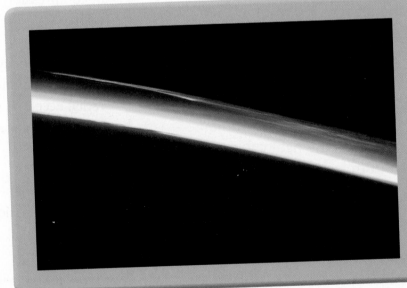

Clouds can form in the higher flight zones, including brilliant light blue polar mesospheric clouds over the Southern Hemisphere. An astronaut from Expedition 22 to the International Space Station took this photo in 2010.
*Courtesy of NASA/NASA's Earth Observatory*

of the mesosphere and thermosphere and is a zone of charged particles called ions. If you were to travel from Earth's surface through the exosphere to the edge of space, you would encounter lower and lower pressure the higher you traveled.

Most flight takes place in the troposphere and stratosphere. The troposphere is where you'll find the largest percentage of atmospheric mass.

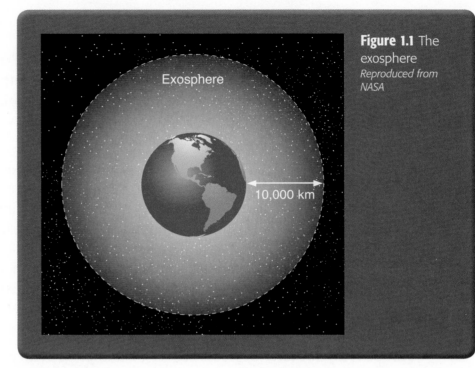

**Figure 1.1** The exosphere
*Reproduced from NASA*

The weight of this mass is why the pressure and density increase the closer you get to Earth's surface. It's where weather and life are. The troposphere's upper boundary differs depending on where you are flying, about four miles (eight km) above the poles and nine miles (14.5 km) above the equator. The pressure falls gradually in the lower and middle troposphere but then drops dramatically above that.

**Wing TIPS**

Ninety-nine percent of "air" is located in the troposphere and stratosphere.

Passenger airliners will fly above the clouds in the stratosphere to avoid weather. Pressure continues to fall in this zone. Ozone is the principal ingredient in the stratosphere. It forms when solar energy strikes and splits an oxygen molecule ($O_2$), and one of the lone oxygen atoms (O) drifts until it strikes and bonds to a complete oxygen molecule to form ozone ($O_3$). Ozone prevents harmful solar radiation from reaching Earth.

Nitrogen and oxygen ions make up the mesosphere and thermosphere (known collectively as the ionosphere). The space shuttle generally orbited in the thermosphere. Atmospheric pressure keeps falling in these zones, as the pull of Earth's gravity grows weaker with distance. The atoms in the exosphere are mostly hydrogen and helium. They have very low density, so collisions between atoms are rare. This zone eventually merges with space. To reach space, all launch vehicles and rockets must travel through all four zones, including the exosphere (Figure 1.1). For example, the astronauts traveled through all the regions to reach the moon.

## The Physical Laws of Gases According to Boyle's Law, Dalton's Law, and Henry's Law

Three scientists from centuries ago studied the relationship of gases and pressure, which today helps explain how the flight zones affect human health. The scientists were Robert Boyle, John Dalton, and William Henry. They each have a law named after them: Boyle's law, Dalton's law, and Henry's law.

In Chapter 1, Lesson 5, you read about Boyle's law, which looked at the relationship of pressure and volume of a gas at a constant temperature. Boyle maintained that when the pressure of a confined gas increases, its volume decreases, and when the pressure decreases, the volume increases. When an aircraft climbs, the drop in atmospheric pressure causes gases in the human body to expand. To correct this increase in volume, a pilot can descend to a lower altitude or pressurize the cabin.

About 150 years after Boyle's discovery, Dalton and Henry unearthed more facts about gases. In 1801 the British chemist Dalton looked at the example of a mixture of gases in a container, once again at a constant temperature. In this scenario the gases don't combine to form another type of gas; they remain separate even as they expand throughout the space inside the container. Dalton's law states that *the total pressure of a mixture of gases is equal to the sum of the partial pressure that each gas exerts individually.* Partial pressure is *the amount of pressure each gas applies individually.*

Gas escaping from a carbonated soft drink bottle illustrates Henry's law.
© iStockphoto/Thinkstock

According to Dalton's law, the total pressure that the human body experiences is the sum of the partial pressures exerted on it by oxygen, nitrogen, and trace gases. (The composition of air in the lungs is somewhat different from the atmosphere because of water vapor and carbon dioxide, but the theory is the same.) The total pressure decreases with gains in altitude. But whether measured at sea level or at 18,000 feet, oxygen will still make up 21 percent of the atmosphere, nitrogen will continue to compose 78 percent, and trace gases will make up the remaining 1 percent.

In 1803 Henry, another British chemist, learned that *the amount of gas dissolved in a volume of liquid is proportional to the pressure of the gas.* In other words as pressure increases, more gas dissolves in the liquid; and as pressure decreases, more gas escapes the liquid. Scientists often equate this process with what you see when you open a bottle filled with a carbonated soft drink. Once you open the bottle, this releases pressure that allows bubbles of gas to escape, or *undissolve.* In aviation this means that when an aircraft ascends, the pressure drops and gases in the blood and tissues can try to escape in the form of bubbles. This can lead to physical problems, which will be addressed later in this lesson.

## The Respiration and Circulation Processes

The gases present in the human body that expand and compress depending on pressure affect two physical processes. The first is respiration, which is another word for *breathing.* The second is circulation, which is *the process of moving blood about the body.*

Wing TIPS

Most people breathe more than 20,000 times each day.

The body manages respiration through the *respiratory system* (Figure 1.2). The respiratory system's main purpose is to take in oxygen and get rid of carbon dioxide. It does this with the help of several parts. These are:

- The trachea—*The windpipe, which moves air drawn in through the nose or mouth into the lungs*

- The bronchi—*Big passageways at the bottom of the windpipe that get air to the lungs*

- The bronchioles—*Small passageways that branch off the bronchi inside the lungs*

- The alveoli—*Air sacs that are at the end of all the other passageways in the lungs.*

In addition numerous muscles expand and contract the lungs. The diaphragm is *the principal muscle that helps the lungs draw in and expel air.* It's located right below the lungs. Lungs and blood vessels distribute oxygen to the body as well as retrieve carbon dioxide, a waste gas, from it.

The air you breathe in travels through the respiratory system's many parts until it reaches the alveoli, or air sacs. Oxygen in the air then enters the bloodstream, which circulates to the body's organs such as the heart. The air you breathe out flushes carbon dioxide from the body. The carbon dioxide moves from tissues in the body to blood vessels in the lungs and finally into the air sacs. When you exhale you are breathing out carbon dioxide from the air sacs.

The *cardiovascular system* controls the circulation of blood throughout the body. One of the cardiovascular system's main functions is to use blood as a way to carry oxygen from the lungs to organs or body tissue, that is, *the cells and any substance found between cells that make up muscles, skin, or other body parts.* It also uses blood to move carbon dioxide from the tissues to the lungs.

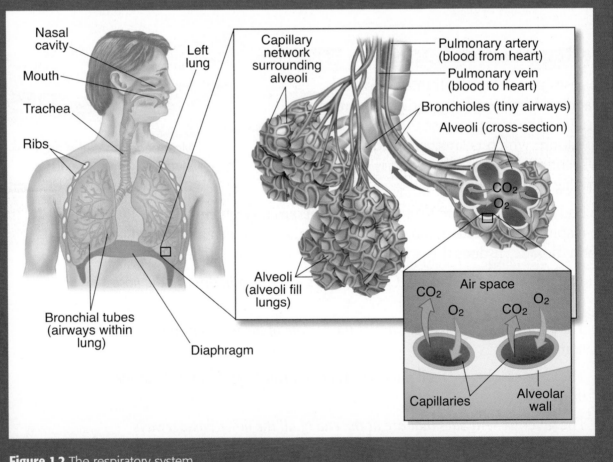

**Figure 1.2** The respiratory system
*Adapted from National Institutes of Health*

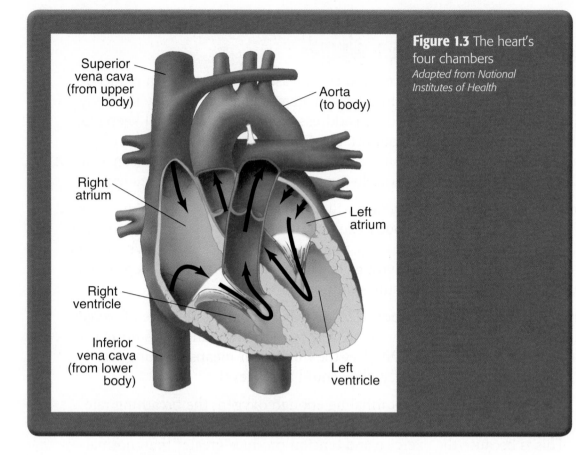

**Figure 1.3** The heart's four chambers
*Adapted from National Institutes of Health*

Labels: Superior vena cava (from upper body); Aorta (to body); Right atrium; Left atrium; Right ventricle; Inferior vena cava (from lower body); Left ventricle

The cardiovascular system's principal parts are:

- The heart

- The pulmonary arteries—*Blood vessels that carry blood from the heart to the organs and tissues*

- The pulmonary veins—*Blood vessels that carry blood from the organs and tissues to the heart*

- The capillaries—*Very small blood vessels branching off arteries and veins.*

The heart, which is a muscle that pumps blood throughout the body, has four chambers: right atrium, right ventricle, left atrium, and left ventricle (Figure 1.3). The body's biggest veins—the superior and inferior vena cavae—carry blood that's low in oxygen from different parts of your body to the right atrium. The heart then pumps it into the right ventricle. From there it flows to the lungs where the blood gathers oxygen from capillaries.

Now that the blood has plenty of oxygen, it empties into the left atrium. The left atrium pumps it into the left ventricle, and then the aorta, a giant artery, sends it back into circulation to feed oxygen to body organs and tissue.

## The Effects on the Human Body of Reduced Pressure at High Altitude

The interactions of the respiratory and cardiovascular systems with reduced pressure that you find at high altitudes can adversely affect the way the body operates. Knowing when and how to address reduced pressure will keep pilots and their passengers safe and comfortable. For instance, maintaining a cabin altitude of about 8,000 feet, which you read about in Chapter 2, Lesson 1, helps prevent some of the ill effects from flying thousands of feet above sea level.

But sometimes a system malfunction will cause an aircraft to lose cabin pressure or an oxygen tank to fail. The body can't take in enough oxygen in these cases and the organs suffer. An aircrew must be able to recognize the body's warning signs that something has gone wrong. They must also know what steps to take to counteract the reduced-pressure symptoms.

According to Dalton's law, by about 12,000 feet, the partial pressure of oxygen is reduced enough that it begins to interfere with the human body's normal activities and functions. By 18,000 feet the thinner air means the lungs can draw only half as much oxygen as they could at sea level.

When a body senses that it's not inhaling enough oxygen, the breathing rate picks up. For instance, people breathe faster when they exercise or are under stress. This is because the exercise and tension produce greater than normal amounts of carbon dioxide. The body's response is to breathe faster to flush out the excess carbon dioxide and draw in more oxygen.

At high altitudes where the air is thin, people also tend to breathe faster than they would at sea level. The following are some ill effects the body may experience at reduced pressure.

### Wing TIPS

With each breath you inhale one-half liter of air, 21 percent of which is oxygen.

## Hypoxia

One hazard of reduced pressure is hypoxia. Hypoxia is *a state of too little oxygen in the body*. It impairs how the brain and other organs function. The 1999 plane crash with the golfer Payne Stewart on board that you read about in the Quick Write was most likely the result of hypoxia.

Because federal regulations require aircraft to maintain safe cabin pressure and pilots of smaller aircraft to have masks that provide them with supplemental oxygen, hypoxia generally isn't a problem. But on occasion—as with the Stewart crash—systems fail. This is why aircrews need to be able to recognize if they are exhibiting any signs of hypoxia and then address the danger right away.

Most pilots won't face dangers from oxygen deprivation when flying below 12,000 feet. However, from 12,000 feet to 15,000 feet its effects begin to be felt when an aircraft's cabin altitude isn't maintained at 8,000 feet or the crew isn't getting enough additional oxygen through facemasks or other breathing devices.

Symptoms include increased breathing rate, headache, lightheadedness, dizziness, drowsiness, tingling in the fingers and toes, warm sensations, sweating, poor coordination, impaired judgment, and decreased alertness. They also include tunnel vision—*a condition in which the edges of your sight gray out to a point where you only have a narrow field of vision straight ahead.* Emotions such as euphoria (great happiness) or its opposite, belligerence (anger, a readiness to fight), may show up as well. The euphoria may give a pilot more confidence than he or she should have.

Oxygen starvation generally first affects the brain. This means that your judgment is the first thing to go, which makes it even trickier to recognize the onset of hypoxia. Pilots have only a limited amount of time—referred to as *time of useful consciousness*—to ward off the effects of hypoxia. The time grows shorter the greater the altitude (Figure 1.4).

Once you spot the symptoms, the answers to hypoxia are really quite simple. Fly at lower altitudes, and use supplemental oxygen. Pilots can also experience the beginnings of hypoxia in a safe environment at a training facility to learn firsthand which signs to watch for.

| Altitude | Time of Useful Consciousness |
|---|---|
| 45,000 feet MSL | 9 to 15 seconds |
| 40,000 feet MSL | 15 to 20 seconds |
| 35,000 feet MSL | 30 to 60 seconds |
| 30,000 feet MSL | 1 to 2 minutes |
| 28,000 feet MSL | 2½ to 3 minutes |
| 25,000 feet MSL | 3 to 5 minutes |
| 22,000 feet MSL | 5 to 10 minutes |
| 20,000 feet MSL | 30 minutes or more |

**Figure 1.4** Time of useful consciousness
*Reproduced from US Department of Transportation/Federal Aviation Administration*

## An Invention to Make Everyone Breathe Easier

After the Payne Stewart crash, an engineer at NASA decided something needed to be done to prevent more such accidents.

Over his Christmas holiday, Jan Zysko built a device that could monitor cabin pressure and altitude. He showed the prototype to managers at NASA, who handed Zysko and his team $100,000 to pursue the idea. It took them only six months to perfect the sensor.

Today everyone from pilots to skydivers and mountain climbers can buy a cabin pressure altitude monitor to warn them if the air pressure is too low. Now everyone can breathe easier at high altitudes.

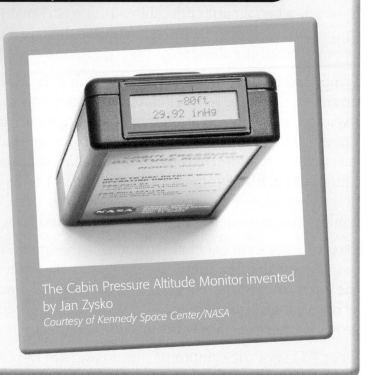

The Cabin Pressure Altitude Monitor invented by Jan Zysko
*Courtesy of Kennedy Space Center/NASA*

## Hyperventilation

A pilot can confuse the onset of hypoxia with hyperventilation because the two share many of the same initial symptoms. Hyperventilation is *an abnormal increase in the volume of air breathed in and out of the lungs.* It can be the result of stress or of flying at a higher altitude, either with or without oxygen. The body's natural reaction to both of these situations is to breathe faster.

Breathing too quickly rids the body of too much carbon dioxide. Too little carbon dioxide in the bloodstream reduces the necessary amounts of calcium in the body. The results are lightheadedness, suffocation, drowsiness, tingling in the hands and feet, and coolness. Very few people actually pass out from hyperventilating, however.

The solution to hyperventilation is to breathe more slowly, sometimes into a paper or cloth bag. Carbon dioxide builds up in the bag each time you exhale. Then when you inhale, you get more than the usual amount of carbon dioxide, which eventually returns you to the proper carbon dioxide level. Just talking out loud can also help overcome hyperventilation.

# Trapped Gas

If you've ever flown you've likely experienced the discomfort of gas trapped in your ear. This is because body cavities like the ear contain gas that expands during climbs and contracts during descents due to the difference between the air pressure outside the body and the air pressure inside the body. Trapped gas is therefore a good demonstration of Boyle's law, which explains increased volume with decreased pressure. If gas can't escape a cavity, it can be painful.

## Ear Block

The middle ear is a small cavity in the skull. The eardrum closes off the middle ear from the external ear canal. Normally the Eustachian tube—*a tube leading from inside each ear to the back of the throat*—equalizes the pressure differences between the middle ear and the outside world (Figure 1.5). These tubes are usually closed. But when you chew, yawn, or swallow, they open up and can equalize the pressure. Even a slight difference in pressure can cause discomfort.

During a climb air pressure in the middle ear may be greater than air pressure in the external ear canal. This causes the eardrum to bulge outward. As the air pressure in the middle ear cavity equalizes with the lower pressure at altitude during a climb, a person on an airplane may experience alternating sensations of fullness in the ear and then it will feel clear.

During descent the opposite happens. Air pressure in the external ear canal exceeds the air pressure in the middle ear. With a higher outside pressure, the eardrum now bulges inward. This is the more difficult condition to relieve.

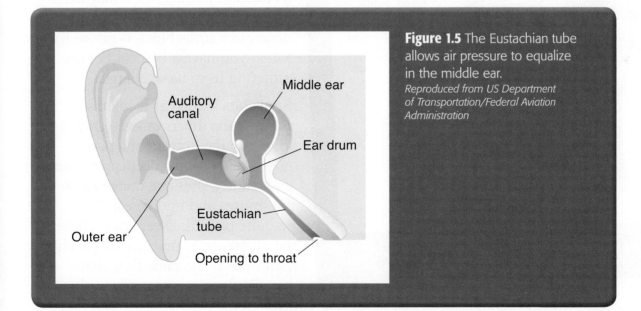

**Figure 1.5** The Eustachian tube allows air pressure to equalize in the middle ear.
*Reproduced from US Department of Transportation/Federal Aviation Administration*

Descent creates a partial vacuum that constricts the Eustachian tube walls. Remedies include yawning, chewing gum, and swallowing. If none of these works, pinch the nostrils shut, close the lips, and very gently blow in the mouth and nose. These procedures force air through the Eustachian tube into the middle ear. Sometimes colds can hinder the equalization of pressure and even damage the eardrums.

### Sinus Block

Another case of trapped gas is sinus block. The skull contains four sets of sinuses, all leading to the nasal passages. Two of the sets are the *frontal sinuses*, one above each eyebrow. The two other sets are the *maxillary sinuses* found in each upper cheek (Figure 1.6).

Air pressure in the sinuses equalizes with pressure in the flight deck during climbs and descents through small openings that connect the four sets of sinuses to the nasal passages. Colds, infections, and allergies can produce enough congestion around one of these openings to slow equalization of pressure.

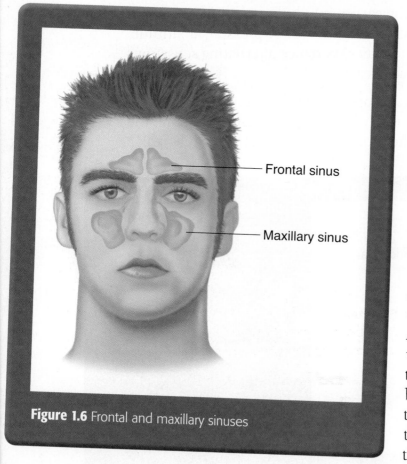

**Figure 1.6** Frontal and maxillary sinuses

Frontal sinus

Maxillary sinus

As the pressure difference increases between the sinuses and flight deck, congestion may plug the opening. This is a sinus block, and it occurs more frequently during descent than climb. A slow descent can reduce any pain associated with sinus block. A maxillary sinus block may even make a tooth ache. The best way to avoid a sinus block is not to fly when you have a bad cold or are suffering from nasal allergies.

Sinus block and ear block also affect scuba divers or even someone diving to the bottom of a pool. As the diver swims deeper and deeper underwater, pressure increases on the sinuses and middle ear. It can be quite painful. Divers can relieve the difference in pressure by pinching their nose at the same time that they lightly attempt to exhale also through the nose.

# Altitude-Induced Decompression Sickness

Decompression sickness is *a condition resulting from exposure to low pressure that causes dissolved gases in the body to form bubbles.* The type of exposure that could cause this sickness might be flying in an unpressurized aircraft to high altitude or in an aircraft that suddenly experiences decompression, *the loss of cabin pressure*, because of a malfunction or accident.

Nitrogen is the main gas that the body stores in its fluids and tissues. When the body is exposed to decreased pressures, the nitrogen dissolved in the body comes out of the solution (the fluids and tissues). If the low pressure forces the nitrogen to leave the solution too rapidly, bubbles form in different areas of the body and cause a number of signs and symptoms. The most common symptom is joint pain, which is known as *the bends*.

Decompression sickness illustrates Henry's law because as the pressure decreases the amount of gas in the solution decreases as well by escaping as bubbles. To address any symptoms (Figure 1.7), a pilot or crew member should do the following:

- Put on an oxygen mask right away and switch the regulator to 100 percent oxygen. The oxygen regulator is a device that controls the oxygen level in the aircraft, mixing cabin air with cylinder oxygen in a ratio that depends on the aircraft's altitude. For example, below 8,000 feet, the regulator supplies zero percent cylinder oxygen and 100 percent cabin air. By about 34,000 feet, it supplies zero percent cabin air and 100 percent cylinder oxygen.

- Begin an emergency descent and land as soon as possible. Even if the symptoms disappear by the time the aircraft reaches the runway, the individual should seek medical help and continue breathing the supplemental oxygen.

- Not move his or her joints to try to work out any pain. The person should keep the affected areas as still as possible.

Treatment once the individual has returned to land may include a hyperbaric chamber, a sealed chamber that helps people suffering from decompression sickness. Some symptoms may not appear until after landing. This is another reason to get advice from a doctor if one has gone through a sudden decompression at high altitudes.

| DCS Type | Bubble Location | Signs and Symptoms (Clinical Manifestations) |
| --- | --- | --- |
| **Bends** | Mostly large joints of the body (elbows, wrists, shoulders, hip, knees, ankles) | • Localized deep pain, ranging from mild (a "niggle") to excruciating—sometimes a dull ache, but rarely a sharp pain<br>• Active and passive motion of the joint aggravating the pain<br>• Pain occurring at altitude, during the descent, or many hours later |
| **Neurologic Manifestations** | Brain | • Confusion or memory loss<br>• Headache<br>• Spots in visual field (scotoma), tunnel vision, double vision (diplopia), or blurry vision<br>• Unexplained extreme fatigue or behavior changes<br>• Seizures, dizziness, vertigo, nausea, vomiting, and unconsciousness |
| | Spinal Cord | • Abnormal sensations such as burning, stinging, and tingling around the lower chest and back<br>• Symptoms spreading from the feet up and possibly accompanied by ascending weakness or paralysis<br>• Girdling abdominal or chest pain |
| | Peripheral Nerves | • Urinary and rectal incontinence<br>• Abnormal sensations, such as numbness, burning, stinging, and tingling (paresthesia)<br>• Muscle weakness or twitching |
| **Chokes** | Lungs | • Burning deep chest pain (under the sternum)<br>• Pain aggravated by breathing<br>• Shortness of breath (dyspnea)<br>• Dry constant cough |
| **Skin Bends** | Skin | • Itching usually around the ears, face, neck, arms, and upper torso<br>• Sensation of tiny insects crawling over the skin<br>• Mottled or marbled skin usually around the shoulders, upper chest, and abdomen, accompanied by itching<br>• Swelling of the skin, accompanied by tiny scarlike skin depressions (pitting edema) |

**Figure 1.7** Signs and symptoms of altitude-induced decompression sickness
*Reproduced from US Department of Transportation/Federal Aviation Administration*

# The Effects on the Human Body of Acceleration and Deceleration or Increased G-Forces

How quickly aircraft decelerate and accelerate and in what direction can also have an important and sometimes serious effect on the human body.

Humans are constantly exposed to the force of gravity. It's why people fall to the ground when they trip over an object and a pencil hits the floor after rolling off a desk. Because of gravity these objects accelerate toward Earth's surface at a constant 32 feet per second squared.

An F-16 Fighting Falcon pilot from the 79th Fighter Squadron at Shaw Air Force Base, South Carolina, banks while heading to a training exercise. A change in speed, direction, or both can induce g-forces.
*Courtesy of USAF/MSgt Kevin J. Gruenwald*

## What G-Force Is

G-force is *a measure of gravity's accelerative force*. At Earth's surface its strength is equal to a single g-force, or 1 G. Objects such as the falling pencil or the tripping person are in free fall, in other words *descending without a device to slow the descent*. Because they don't have very far to go, they hit the ground without gaining much accelerative force beyond 1 G, or 32 feet per second squared. But an object such as a plane way up in the sky gains more than 1 G as it plummets toward Earth. It gains speed until it either pulls out of a dive, hits the ground, or reaches terminal velocity, which is *the point at which drag on an object equals the force of gravity and the object stops accelerating*.

Steep turns in the air can generate a force of acceleration that's many times the force of gravity. Fighter pilots may experience forces as high as 9 Gs.

A pilot may experience a combination of linear, radial, and angular acceleration when working with flight controls. These accelerations induce g-forces on the body that scientists refer to as Gx, Gy, and Gz. The three types of g-forces act along three different axes that are perpendicular to one another. Think in terms of the x, y, and z axes you may have read about in your math classes.

## Wing TIPS

If you jumped out of an airplane without a parachute, you would accelerate until you reached a terminal velocity of about 120 mph. Experienced skydivers have reportedly been able to reach a terminal velocity of more than 200 mph, although they decelerate before opening their parachutes.

Gx is a force that acts on the body from the chest to the back, and from the back to the chest. For example, +Gx is the force from chest to back that occurs during takeoff when an aircraft is gaining speed on the runway. The force from the increase in speed pushes the pilot back into the seat. The force called −Gx is the force from back to chest during landing. This force pushes the pilot forward in the shoulder strap.

## Types of Acceleration

Acceleration comes in three forms. They are:

- Linear acceleration—Is *a change of speed in a straight line*. This type of acceleration takes place during takeoff, landing, and in level flight.
- Radial acceleration—Takes place with *a change in direction* such as in sharp turns, a dive, or pulling out of a dive.
- Angular acceleration—Results from *a simultaneous change in both speed and direction*, which happens in spins and climbing turns.

An F-22 Raptor performs a maximum climb takeoff during the 2010 air show at Nellis Air Force Base, Nevada, on 12 November 2010. Pilots experience linear acceleration during takeoffs.
*Courtesy of USAF/MSgt Kevin J. Gruenwald*

Gy is a lateral force that acts from shoulder to shoulder. Pilots encounter this force during aileron rolls about their longitudinal axis. Gz is a gravitational force that acts on the body's vertical axis. Applied from head to foot, such as when pulling out of a dive, it is +Gz. When it runs from foot to head, as in a dive, it is −Gz.

## The Effects

The most hazardous g-force is along the Gz axis. A pilot experiences +Gz when entering a high-speed turn or pulling out of a steep dive. The cardiovascular system has to act quickly to keep blood flowing to the brain to remain conscious. The body tries to counteract +Gz with a harder, faster heartbeat to keep the blood flowing up to the head.

If the body isn't able to respond quickly enough to +Gz, one of the first signs is a progressive loss of vision as the aircraft enters the maneuver. The eyes are extremely sensitive to low blood flow. The first thing a pilot may notice is the onset of tunnel vision. After that comes an even smaller field of sight referred to as *gun barrel vision*, then *gray out*, and finally *blackout* of all vision. If the accelerating g-forces don't let up, a pilot may finally pass out. An aircraft with enough altitude may be able to give a pilot enough time to recover and get out of the excessive Gs and recover sight and consciousness.

How quickly a pilot goes from reduced sight to unconsciousness depends on how fast the aircraft is accelerating. If the aircraft is gaining about 0.1 G per second, then a pilot will slowly lose vision but this should give him or her enough time to reduce Gs and handle the situation. But if the acceleration is as rapid as 1 G per second or more, unconsciousness can hit without any visual warning.

Even more dangerous than head-to-foot +Gz is the foot-to-head −Gz. When a pilot pushes over into a dive, for instance, blood can't flow back down through the veins into the heart. Yet the arteries are carrying more blood than ever to the head. Once again this force affects the eyes first. A pilot may experience *red out*, which clouds the vision with a field of red. The next phase is loss of consciousness because while lots of blood is flowing up to the brain, it's not able to flow through it and back down toward the heart.

By being alert to the dangers of excessive g-forces, a pilot can look for warning signs and try to avoid situations likely to induce dangerous levels of g-force. Being in good physical shape also helps pilots physiologically handle the stresses of g-forces. Furthermore fighter pilots and astronauts wear the g-suit, *a piece of clothing that protects pilots from the effects of g-forces*. The suit prevents blackouts by applying pressure to the legs and abdomen to keep blood pressure up and blood circulating to the brain. The first g-suits were worn during World War II for aerial combat. In the next lesson, you'll read about a different set of suits called *pressure suits*, which guard against the ill effects of decreased pressure at altitude.

Col John Paul Stapp is known as the "fastest human on Earth." In 1954 he also decelerated faster than any human on Earth. It was all because he took part in a number of military crash tests. Voluntarily!

Col John Paul Stapp rides a rocket sled in pursuit of greater aviation safety.
*Courtesy of USAF*

Stapp had earned his medical degree in 1944 and entered military service not long after that. Beginning in 1945 he studied aviation medicine, in other words, some of the topics you've been reading about in this lesson. Around this time the military decided to conduct some tests to see if it could increase the number of survivors in airplane crashes. Too many military pilots were dying in noncombat-related accidents that seemed as if they should have been preventable. Stapp volunteered to help with the tests.

Northrop Aircraft of California built a railroad track 2,000 feet long on a heavy concrete bed. The company placed a 1,500-pound rocket sled on the track, which had a 45-foot mechanical braking system. A rocket sled is *a rocket-powered vehicle that rides at ground level on tracks and is used for a variety of speed, acceleration, and deceleration tests*. Between the late 1940s and December 1954, Stapp volunteered for 29 rides on a series of rocket sleds. The *Sonic Wind I* was his last. Scientists built it for speed and an incredibly short braking distance.

With Stapp on board, *Sonic Wind I* sped him along at 632 miles per hour, a world-record land speed at that time. It took him only five seconds to go from a standstill to a speed faster than a .45 bullet, subjecting him to a force of about 20 Gs. This is a prime example of linear acceleration at takeoff. The rocket sled then stopped in fewer than 1.4 seconds, exerting more than 40 Gs on Stapp's body. That amount of g-forces is equal to hitting a brick wall in a car traveling 120 miles per hour. Stapp walked away from this final test with two wrist fractures (he set one of them himself on his walk back to their labs), rib fractures, and eye injuries. Everything healed up nicely, though. Stapp's offer to be the "human decelerator" helped the military and others find ways to improve pilot and even automobile safety. Because of his work scientists built safer helmets and aircraft seats, among many other discoveries.

According to historians Stapp also gets credit for coining a famous phrase. One of his assistants, Capt Edward A Murphy Jr., incorrectly rigged a harness and it failed to register strains Stapp was being subjected to during a test. After he learned what happened, Stapp said that "Whatever can go wrong, will go wrong." This saying has been called *Murphy's law* ever since.

## Spatial Disorientation and Motion Sickness

Flight can play tricks on the mind and make it uncertain of where exactly it is. Spatial disorientation refers to *the lack of knowing an aircraft's position, attitude, and movement.* The body includes three systems that work together to figure out these three facts. They are:

- *Visual system*—Eyes, which sense position based on what they see

- *Vestibular system*—Organs found in the inner ear that sense position by the way the body is balanced

- *Somatosensory system*—Nerves in the skin, muscles, and joints, which, along with hearing, sense position based on gravity, feeling, and sound.

The brain pieces all of this information together, and when the three systems agree the brain can accurately tell you where you are and in what direction you're heading. But flying conditions can sometimes confuse the three systems. This leads to spatial disorientation.

When a pilot flies by visual references, the eyes can correct any false readings coming from the other two systems. This is because the eyes can see the horizon and ground and use these references to balance out input from the other systems. But when a pilot must turn to instrument flight because of thick weather, incorrect feedback from the three systems can be disorienting.

The vestibular system in the inner ear is key to the senses of balance, motion, and position. It lets pilots determine movement and orientation while flying. The left and right inner ears each contain three *semicircular canals* that sit at right angles to one another. The canals sense the direction and speed of angular acceleration. Each canal is filled with fluid and has a section full of fine hairs embedded in the *cupola*. Acceleration causes the tiny hairs to deflect, which stimulates nerve impulses that transmit orientation through the *vestibular nerve* to the brain.

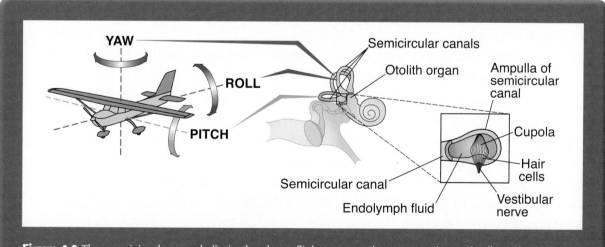

**Figure 1.8** The semicircular canals lie in the three flight axes and sense motions of roll, pitch, and yaw.
*Reproduced from US Department of Transportation/Federal Aviation Administration*

## Vestibular Illusions

Spatial disorientation can lead to any number of problems. Some issues relate to false readings from the inner ear.

*The leans* can occur when a pilot enters a banked attitude too slowly to set the fluid in motion in the semicircular canals. In this example the pilot has banked too slowly to the left. If he or she then abruptly corrects the aircraft to the right, this movement rolls the fluid around in the canal. It creates the illusion of a banked attitude in this opposite direction to such a degree that the disoriented pilot may make the error of rolling the aircraft back to the left.

Another illusion is the *graveyard spiral*. When a pilot has been in a turn long enough, the fluid in the ear canal will eventually move at the same speed as the canal. He or she will have the illusion that the aircraft is not turning at all. During recovery to level flight, the pilot will experience the sensation of turning in the opposite direction. As a result the disoriented pilot may return the aircraft to its original turn. An aircraft tends to lose altitude in turns unless the pilot compensates for the loss in lift. In this scenario the pilot may notice the loss of altitude, but the absence of any sensation of turning creates the illusion of being in a level descent. The pilot may pull back on the controls in an attempt to climb or stop the descent. This action actually tightens the spiral and increases the altitude loss. If carried on long enough, the plane can crash.

To counter these problems, pilots are taught that when they get spatially disoriented they should switch to instruments and ignore the faulty indications their bodies are giving them.

The vestibular nerve penetrates a large area at the base of each canal called the *ampulla*. The orientation of each canal lies along one of three flight axes that help pilots determine yaw, pitch, and roll (Figure 1.8).

Connected to the canals are two thin and tissuelike membranous sacs called the *saccule* and *utricle*. Together the saccule and utricle are often referred to as the *otolith organs*. The otolith organs sense the direction and speed of linear acceleration and the position, or tilt, of the head. The vestibular nerve sends impulses from the saccule, utricle, and semicircular canals to the brain to interpret motion.

The somatosensory system sends signals from the skin, joints, and muscles to the brain, which interprets the signals in relation to Earth's gravitational pull. The signals determine a pilot's posture, which is the attitude or position in which someone sits or stands. The brain receives constant updates based on the body's posture and movement in the pilot's seat. If a pilot flies based on these signals, he or she is said to be flying by the "seat of the pants." When used along with visual and vestibular clues, these signals can be fairly reliable. However the body can't tell the difference between acceleration forces due to gravity and those due to aircraft maneuvers.

## Motion Sickness

Motion sickness, like spatial disorientation, is a product of the brain receiving conflicting messages about the body's true position. For instance, a movie on a giant screen at a theater can sometimes induce motion sickness. This is because what's taking place on the screen has a different orientation than your own posture. You might lean along with the movement on the screen—such as a glider swooping down over a massive waterfall—while in reality your seat hasn't moved an inch. This visual effect can make you feel sick to your stomach. Your visual and vestibular systems don't agree. The answer at the theater is simply to shut your eyes to end the sensory conflict.

Motion sickness symptoms include nausea, dizziness, paleness, sweating, and vomiting. Student pilots sometimes suffer from airsickness because they aren't used to the motion and spatial orientation clues. But they generally get over the physical reaction within the first few lessons. Anxiety and stress can be part of what causes the sickness. The more you fly, the more readily you'll get over the likelihood of airsickness. In addition opening air vents, breathing supplemental oxygen, focusing on an object or point outside the aircraft, and not moving your head unnecessarily can help.

### Wing TIPS

Astronauts may also sometimes suffer from motion sickness referred to as *space sickness*. Like motion sickness it's due to a disagreement between what the eyes see and what the body feels. Without gravity's pull, it's hard to tell what's up and what's down.

## Other Stresses of Flight Operations

Pilots sometimes create their own stress. Cigarettes, alcohol, and drugs can seriously degrade pilot performance, as one string of examples. Smoking tobacco raises the concentration of carbon monoxide, *a colorless, odorless, tasteless, and toxic gas*, in the blood. At sea level, where normal conditions reign, its physiological effects are similar to flying at 8,000 feet. Just imagine what its effects are at high altitudes.

### The Pilot and the Cigarette

A pilot out West had a serious run-in with hypoxia, according to a true story told to the FAA. While flying his small aircraft over mountains at 13,500 feet, he lit up a cigarette. After one particularly deep drag, he passed out. He didn't know he'd lost consciousness until he woke up to find himself in a high-pitched dive.

With very little altitude left in which to recover, he pulled out of the nose-first plunge. The drag on the cigarette had replaced oxygen with carbon monoxide in his brain. This exchange of oxygen for a toxic gas knocked him out.

Alcohol diminishes mental capacity. A pilot's job calls for making hundreds of difficult decisions each flight. Even small amounts of alcohol can affect coordination, limit one's field of vision, impact memory, reduce reasoning power, slow reflexes, and lower one's attention span.

Like coffee, tea, and caffeinated soft drinks, alcohol is also a diuretic—*a drink or other substance that drains the body of needed water.* Dehydration is *the critical loss of water from the body.* Common signs of dehydration include headache, fatigue, cramps, sleepiness, and dizziness. These symptoms can also impact pilot performance. So it's best to avoid diuretic drinks. When the cockpit is hot or while flying at high altitudes where there's more water loss, pilots should be sure to drink plenty of water.

Besides illegal drugs, which should be avoided at all times, prescription and over-the-counter medications can also pose a problem when flying. Some prescribed medications have side effects such as loss of balance and nausea. Federal regulations actually prohibit pilots from performing crew duties while using medications that affect the body in any way contrary to safety. The safest rule is not to fly as a crew member while taking any medication unless approved to do so by the FAA. If a pilot has any questions, he or she should ask an aviation medical examiner, that is, someone designated by the FAA as having aviation medicine training.

This lesson has covered a range of health issues relevant to flight. The more pilots know about the dangers, the more alert they'll be to the warning signs and the safer they'll be. In addition, pilots can take lots of steps to improve their performance as a pilot right here at sea level—they can stay in shape, avoid alcohol and other diuretics, and avoid smoking. In addition, they should get plenty of sleep, because fatigue can also affect how well they function. Next up will be a discussion of protective equipment for pilots and astronauts and aircrew training.

# ✔CHECK POINTS

## Lesson 1 Review

Using complete sentences, answer the following questions on a sheet of paper.

1. Why does decreasing pressure with altitude make it more difficult to breathe?

2. What is physiology?

3. According to Dalton's law, what is the total pressure that the human body experiences?

4. What did Henry, another British chemist, learn in 1803?

5. What is the respiratory system's main purpose?

6. What are the cardiovascular system's principal parts and what do they do?

7. What does oxygen starvation generally affect first?

8. What type of exposure could cause decompression sickness?

9. Which is the most hazardous g-force?

10. How quickly a pilot goes from reduced sight to unconsciousness depends on what?

11. What is it that the eyes do when a pilot flies by visual flight rules?

12. How is motion sickness like spatial disorientation?

13. What are some stresses that can seriously degrade pilot performance?

14. What are some common signs of dehydration?

## APPLYING YOUR LEARNING

15. A plane is flying at 25,000 feet when suddenly the cabin loses pressure. What dangers do the crew and passengers face? What steps should the pilot take to address the situation and make sure everyone gets home safe?

## The Protective Equipment Used by Pilots and Astronauts

Because high altitudes are not the human body's normal habitat, people who fly above livable space need special equipment to survive. If they don't, they can suffer from many of the maladies you read about in the previous lesson: hypoxia, decompression sickness, and more. The protective equipment that pilots and astronauts rely on varies, depending on just how high they intend to go.

### Pressurized Cabins

Pilots fly at high altitudes to get above bad weather and turbulence and to increase their aircraft's fuel efficiency. To combat the physiological drawbacks that come with flying at high altitude, many modern aircraft have pressurized cabins, which you first read about in Chapter 2, Lesson 1. These systems make it possible to breathe naturally without any supplemental device such as a mask.

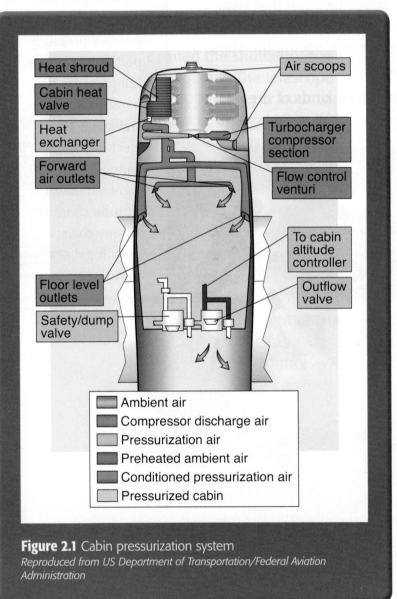

**Figure 2.1** Cabin pressurization system
*Reproduced from US Department of Transportation/Federal Aviation Administration*

A typical pressurization system creates a kind of protective bubble inside the cockpit and passenger cabin. It's a sealed unit that contains air under a pressure that's higher than the atmospheric pressure outside the aircraft. On aircraft powered by turbine engines, the system uses bleed air—*compressed air from the engine compressors*—to pressurize the cabin.

A cabin pressurization system typically maintains a cabin pressure altitude of about 8,000 feet while the plane is flying at its maximum cruising altitude (Figure 2.1). (*Cabin altitude* is another way of referring to cabin pressure.) This prevents rapid changes in cabin altitude that may be uncomfortable or injure crew and passengers. The system also exchanges air inside the cabin for fresh air outside it to get rid of odors and stale air.

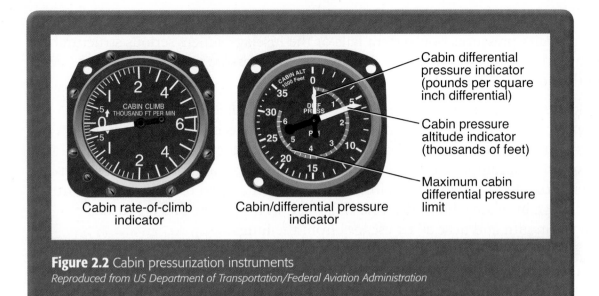

**Figure 2.2** Cabin pressurization instruments
*Reproduced from US Department of Transportation/Federal Aviation Administration*

## The System

Pressure systems include a regulator, an outflow valve, and a safety valve. The *regulator* maintains cabin pressure within a preset range. It also limits the differential pressure, which is *the difference in pressure between cabin pressure and atmospheric pressure outside the aircraft*. Aircraft fuselages generally have a maximum amount of differential pressure that the structure can withstand. Windows and doors that can rupture as well as cabin size play a role in determining structural strength.

Sometimes an aircraft will reach the altitude at which the difference between the pressure inside and outside the cabin is at the greatest spread allowed. In such a case, any further increases in altitude will result in similar increases in cabin altitude. Regulators prevent pressure systems from exceeding the maximum differential pressure.

The *outflow valve*, which you read about in Chapter 2, Lesson 1, is a device that releases air from the fuselage. By regulating the air exit, the outflow valve allows for a constant inflow of air to the pressurized area.

The system's *safety valve* includes three parts: a pressure relief valve, a vacuum relief valve, and a dump valve. The *pressure relief valve* keeps the cabin pressure from exceeding a chosen differential pressure above ambient pressure. Ambient pressure is *the pressure in the area immediately surrounding the aircraft*. On the other hand, the *vacuum relief valve* prevents ambient pressure from exceeding cabin pressure by letting outside air into the cabin. The *dump valve* releases cabin air to the outside.

Pilots also use several devices to keep tabs on the overall system. These instruments do everything from measuring the difference between inside and outside pressure to tracking an aircraft's rate of climb and descent in thousands of feet per minute (Figure 2.2).

### If the System Fails

If an aircraft's pressurization system fails or the fuselage suffers serious structural damage, the plane loses pressure. The differential pressure between the cabin and the outside evaporates.

Explosive decompression is *a change in cabin pressure that takes place faster than the lungs can decompress.* This can damage the lungs. Decompression that occurs in less than 0.5 seconds is explosive and potentially dangerous. In the case of an explosive decompression, you may hear a loud noise, and then see the cabin fill with fog, dust, or debris. Fog forms because of the rapid drop in temperature and change in relative humidity. The ears clear on their own, and air rushes from the mouth and nose as air escapes from the lungs.

Rapid decompression, on the other hand, likely won't hurt the lungs. It's *a change in cabin pressure in which the lungs decompress faster than the cabin.* However, it cuts down on a pilot's time of useful consciousness as the lungs rapidly exhale oxygen, which reduces pressure on the body. (The time of useful consciousness is the amount of time a pilot can perform flying duties while having an insufficient oxygen supply.) This decreases the partial pressure of oxygen in the blood, and so reduces pilot performance by one-fourth to one-third of normal.

The main danger from decompression is hypoxia, although decompression sickness is another hazard. Supplemental oxygen and a rapid descent are two solutions. Furthermore, crew and passengers should always wear their seat belts and safety harnesses in flight. The winds and forces at work during decompression can toss individuals around the cabin or even pull them out of openings in the aircraft.

## Oxygen Masks

Aircrews can protect themselves from the risks of decompression with oxygen equipment. Most high-altitude aircraft come equipped with some type of fixed oxygen equipment. Aircraft without built-in gear should carry portable oxygen equipment.

### Pressure Drop

If the *ambient temperature*—or temperature immediately surrounding an object—around an oxygen cylinder falls, you may notice that the pressure in the container also decreases. This agrees with Gay-Lussac's law, which you read about in Chapter 1, Lesson 5. Pressure varies directly with temperature if the volume of a gas remains constant.

So if you've stored your supplemental oxygen container in an unheated area of the aircraft, you may notice the pressure dropping in the cylinder. This drop won't be due to any decrease in the oxygen supply. It simply means that the lower temperatures in the surrounding air have compacted the oxygen supply in the cylinders.

Maj Eric Nodland with the Michigan Air National Guard's 107th Fighter Squadron conducts preflight checks in the cockpit of an A-10 Thunderbolt II before a training mission in 2011 at Nellis Air Force Base, Nevada. His oxygen mask covers his nose and mouth.
*Courtesy of USAF/TSgt Michael R. Holzworth*

## The Equipment

Portable equipment, as one example, includes a container, regulator, and mask. Containers hold oxygen under high pressure—usually 1,800–2,200 pounds per square inch (psi). They should be filled only with aviation oxygen, which is 100 percent pure oxygen. Medical oxygen contains water vapor, which could freeze at high altitudes, and industrial oxygen can contain impurities.

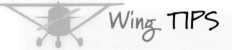

**Wing TIPS**

Oxygen is highly combustible. Materials that are nearly fireproof in ordinary air may catch fire easily in oxygen. Oils and greases may ignite if exposed to oxygen, so mechanics don't use them to seal oxygen equipment valves and seals.

The regulator supplies the flow of oxygen. How much it supplies depends on the design intended for different altitudes. For example, regulators for use up to 40,000 feet provide no oxygen from the cylinders and 100 percent cabin air at cabin altitudes of 8,000 feet or less. But as cabin altitude increases, that ratio will eventually change to 100 percent oxygen and no cabin air at around 34,000 feet cabin altitude. Regulators for use up to 45,000 feet, however, provide 40 percent cylinder oxygen and 60 percent cabin air at lower altitudes. The ratio changes to 100 percent oxygen at higher altitudes.

Not every mask works with every oxygen system. The two must be compatible. Masks should fit securely enough over the face to prevent oxygen from leaking out. Most masks cover only the nose and mouth. Some masks include microphones so the crew can speak to one another.

**Wing TIPS**

Pilots with beards or mustaches should keep them trimmed so they don't interfere with an oxygen mask's fit. If outside air can sneak in where the mask isn't well sealed to the face, it dilutes the oxygen.

Another device that administers oxygen is the cannula, *a piece of plastic tubing that runs under the nose*. Cannulas are more comfortable than masks, and crews can use them up to 18,000 feet. Before every flight pilots should inspect their oxygen system, including checking for cracks in masks and cannula tubing, and verifying oxygen quantity and pressure.

### Types of Oxygen Delivery Systems

How an oxygen system delivers air depends again on the intended altitude. It also depends on who will be using it.

Users of the *continuous-flow oxygen system* generally include passengers because it's less expensive than other systems. It works up to 28,000 feet. It provides a steady supply of air through a cup that fits easily over the nose and mouth and a bag that collects oxygen from the system until the user is ready to inhale. It is an inefficient system, however, because the oxygen flows whether the person is inhaling or exhaling. This uses up the oxygen supply more quickly than other systems.

The *diluter-demand oxygen system* operates up to 40,000 feet. It supplies oxygen only when the user inhales through the mask. That is, it supplies oxygen only when there is a *demand* for it, as the system's name implies. It stops the flow while the user is exhaling. This preserves the oxygen supply in the container. Depending on the altitude, the system also *dilutes*, or weakens, the oxygen flowing into the mask by mixing it with cabin air.

The *pressure-demand oxygen system* functions above 40,000 feet. It works very much like the diluter system, except that it supplies oxygen at positive pressure at cabin altitudes above 34,000 feet. Positive pressure is *a forceful flow of oxygen that slightly overinflates the lungs*. It creates a pressure in the lungs that is similar to the pressure found at lower altitudes. Without positive pressure, 100 percent oxygen isn't enough to protect pilots at great altitudes. Positive pressure allows pilots to fly above 40,000 feet.

## Pressure Suits

Oxygen systems alone, however, can't protect crew and passengers in every instance. Very high altitudes call for an additional piece of equipment called the pressure suit. While atmospheric pressure at sea level is 14.7 psi, the human body can actually survive with pure oxygen delivered to the body at a pressure slightly less than 3 psi. Pilots and astronauts find these conditions at very high altitudes and in the vacuum of space, which contains a near absence of gases and pressure.

The partial-pressure suit is *a close-fitting garment that covers most—but not all— of the body and creates pressure to support life*. Fighter, bomber, and test pilots were some of the first to wear these suits. Test pilot Capt Charles E. "Chuck" Yeager was one of these pilots. In 1947 he flew the experimental aircraft X-1 at 43,000 feet and broke the sound barrier at Mach 1.06.

## Life Under Pressure: A Summary

The effects of altitude are as follows:

- Above 10,000 feet some people grow short of breath and dizzy.
- Above 20,000 feet they need supplemental oxygen.
- At 34,000 feet pilots need 100 percent oxygen to equal oxygen's partial pressure at sea level.
- Above 40,000 feet they need 100 percent oxygen under positive pressure.
- Above 50,000 feet, a near-space environment, the body requires a pressure suit.
- At 55,000 feet atmospheric pressure is so low that water vapor in the body appears to boil. It makes the skin inflate like a balloon.
- At 63,000 feet blood at normal body temperature (98 degrees F, or 36.6 degrees C) appears to boil. Just as your ears pop while traveling on a plane, reduced atmospheric pressure at these high altitudes allows the gases in your body to expand to the point that they appear to boil off.
- Above 65,000 feet the atmospheric pressure nears that of space. Pressure suits must offer enough protection for a human to survive in a near vacuum.

Suit designs included inflatable tubes called *capstans*. When filled with air, they tightened the suit around the body. The purpose was to exert enough pressure to balance the breathing pressure and prevent hypoxia. However, the suits generally restricted pilot movement, provided no way to release heat and sweat, and were too heavy.

Test pilot Joseph Walker stands beside the X-1E at the NASA High-Speed Flight Station at Edwards Air Force Base in 1958. He's wearing an early Air Force partial-pressure suit meant to protect him if the cockpit lost pressure above 50,000 feet.
*Courtesy of NASA*

In addition pilots couldn't use them at very high altitudes for long stretches or to explore space because the suits didn't adequately protect the body. They needed something more. This led to the invention of the full-pressure suit. This is *a type of suit that covers the entire body from head to toe, surrounding the wearer in a protective, contained environment similar to a pressurized cabin.* It consists of a gas bladder that maintains a pressure of about 3 psi on the body when inflated. It has a built-in ventilation system to help with heat. A full-pressure helmet tops off the ensemble.

Astronauts with Project Mercury, America's first manned space program, which lasted from 1958 to 1963, wore two-layer full-pressure suits. The inner layer surrounded the astronaut with pure oxygen. The outer layer exerted pressure on the inner layer to keep it from blowing up like a balloon. Scientists designed the suits to inflate only if a spacecraft lost cabin pressure. During Project Mercury, all the spacecraft maintained their pressure, so the suits never inflated.

Astronauts in today's space program also wear full-pressure suits. They need them when they go on space walks outside the International Space Station to make repairs. Modern-day pilots also make frequent use of full-pressure suits. For instance, pilots who fly the U-2 spy plane soar more than 70,000 feet into the air where full-pressure suits are necessary. When pilots first started testing the U-2 in the 1950s, they only wore partial-pressure suits.

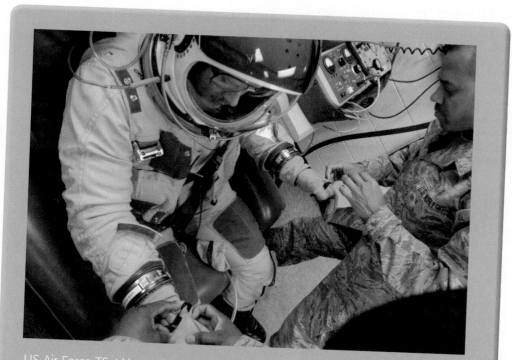

US Air Force TSgt Vontez Morrow preps US Air Force U-2 pilot Capt Beau Block for a mission on 13 March 2011. Block took off from Osan Air Base, South Korea, to capture imagery of the earthquake- and tsunami-affected areas of Japan. His full-pressure suit protects him during high-altitude flight.
*Courtesy of USAF/SMSgt Paul Holcomb*

Navy SEALs free-fall jump from the ramp of a C-17 Globemaster III at 12,500 feet on 15 April 2010 over Fort Pickett Maneuver Training Center, Virginia.
*Courtesy of USAF/SSgt Brian Ferguson*

## Emergency Equipment

In rare cases pilots may have to escape from their aircraft. Engines may lose power because of mechanical failure or a large bird strike. Enemy fire may damage engines or the aircraft body beyond repair. Pilots have options when they aren't able to safely glide their plane back to the ground.

### Parachutes

A parachute is *a large cloth device that slows descent*. Parachute parts include a large cloth canopy, lines, risers, a harness fastened around the body, and a container. Lines attach the canopy to the risers. Risers are bands that connect the lines to the container and harness. The parachute itself is packed away in the container that the pilot wears either on his or her back or in front. The harness hugs the container and risers to the body so the pilot is secure during descent.

After jumping from an aircraft, a pilot enters free fall. Gravity draws the pilot toward Earth. The parachute opens when the pilot yanks a ripcord. Some parachutes include a pilot chute, which is a very small parachute that hauls the big canopy out of its pack. Some also include a reserve parachute, which the user can deploy if everything else malfunctions.

Airmen from the 720th Special Tactics Group at Hurlburt Field, Florida, use a static line as they jump out of a C-130 Hercules over the Florida coastline during a water rescue exercise.
*Courtesy of USAF/Amn Matthew R. Loken*

A parachute works by creating drag. The canopy encounters air molecules, and this slows the pilot's descent. The pilot can guide the parachute either left or right with toggles, which are lines attached to the canopy.

Starting in 1922 the US Army ordered all its pilots to wear parachutes. It also switched to manually deployed parachutes. Up to this point most parachutes had automatically deployed as the user jumped from the aircraft. Paratroopers today still sometimes use the static line—*a cord attached to the aircraft and pack that automatically deploys the parachute as a jumper exits the aircraft*—to engage their parachutes when jumping in nonemergency situations. But in an emergency pilots nowadays generally use a parachute that doesn't use a static line to deploy. This gives them time to drop away from a damaged aircraft, which may be on fire or spinning out of control, before releasing the canopy from its pack. That way they don't get tangled up with the aircraft.

### Wing TIPS

In 1922 Lt Harold Harris with the Army Air Service became the first pilot to deploy a manual parachute before his plane crashed.

**CHAPTER 3**   Flight and the Human Body

## Ejection Seats

Modern jet fighters fly so fast that escape by parachute alone isn't an option. This is why fighter pilots use an ejection seat, *a seat that thrusts a pilot out of an aircraft in an emergency.* The military refers to these seats as a pilot's "last chance for life."

If a pilot tried to escape a damaged fighter with only a parachute, he or she would be battling g-forces and windblast. Windblast is *the effect of air friction on a pilot who ejects from a high-speed aircraft.* It can injure or even kill a pilot in part because the windblast can slam the pilot into the fuselage before he or she can jump away from the plane. G-forces can pin a pilot in place so that movement, such as jumping from a plane, is next to impossible.

An ejection seat helps pilots overcome these hazards. It has many parts. First is the pilot's seat, which is padded. Second is a system built into the seat made up of explosives or electrical circuitry to trigger the ejection process. Third is a rocket on the back of the seat to propel the seat with the pilot in it from the cockpit.

Generally the pilot starts the ejection process by tugging on a firing handle. Before the seat can leave the aircraft, the ejection system must kick off the overhead canopy. Otherwise the pilot will crash into it. Once the canopy is out of the way, the seat rides out of the cockpit on rails that steer it up and away from the aircraft.

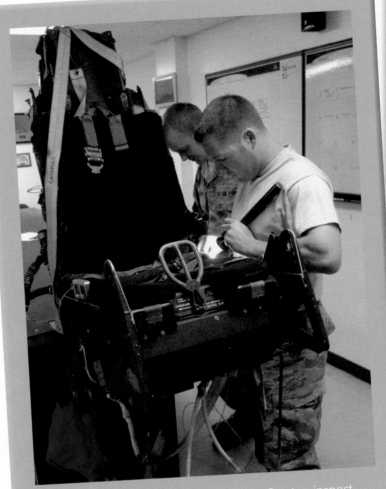

SSgt Ronald Kavanaugh and A1C William Stanton inspect an ejection seat from an F-22 Raptor on 12 May 2010 at Langley Air Force Base, Virginia.
*Courtesy of USAF/A1C Jason J. Brown*

### Wing TIPS

A pilot is subject to the head-to-toe +G force when first ejected out of an aircraft.

Capt Christopher Stricklin ejects from the USAF Thunderbirds number six aircraft less than a second before it hits the ground at an air show at Mountain Home Air Force Base, Idaho, on 14 September 2003. Stricklin, who was not injured, ejected after both guiding the jet away from the crowd of more than 60,000 people and ensuring he couldn't save the aircraft. This was only the second crash since the Air Force began using F-16 Falcons for its demonstration team in 1982. The ACES II ejection seat performed flawlessly.
*Courtesy of USAF/SSgt Bennie J. Davis III*

Seats used to be less stable than they are today. Engineers now build them with devices called gyroscopes that prevent the seat from weaving and rolling in midair. Seats may deploy small parachutes called *drogue chutes* or *drag chutes*. These also steady the ejector seat's motion through the air and help with deceleration. Modern ejection systems work not only at high speeds and altitudes, but also when an aircraft is moving slowly or even when stationary on the ground.

The ejection seat has a system of harnesses that secures the pilot to the seat. Once the pilot clears the aircraft and has decelerated to a safe speed, the harnesses automatically unlatch. At the same time a parachute called the *reserve parachute* deploys and yanks the pilot away from the seat. The pilot can then drift to the ground. Pilots ejecting at high altitudes also need an oxygen mask or some other device to protect them from hypoxia. The right clothes are also important because the temperatures can be dangerously low at high altitude.

# The Function and Use of Flight Simulators

Besides all of the protective gear available, pilots and astronauts have a range of equipment they can train on to prepare for flight hazards. A flight simulator is *a machine that imitates real-life situations and dangers that pilots may face.* Military and civilian pilots use flight simulators to train for many types of in-flight situations. Crew members can do things with a simulator that can't be safely replicated with a normal flight. In addition, the National Transportation Safety Board uses simulators when investigating accidents to try to re-create conditions in the aircraft cockpit before the crash.

When pilots train on a simulator, they sit in a room that looks like a cockpit with realistic seats, harnesses, headsets, and flight controls. They have all the blinking lights before them that a real cockpit might have. Finally, screens surround the mock cockpit with images that the pilots might see if they were looking out of the windows of a real cockpit. A simulator is a safe place to make mistakes so a pilot doesn't make them out in the real world. It also saves wear and tear on real aircraft and on fuel.

Two pilots and a flight engineer fly inside an HC-130 Hercules simulator during training at Moody Air Force Base, Georgia. Pilots, navigators, flight engineers, and other crew members train on simulators as if on a real mission.
*Courtesy of USAF/A1C Benjamin Wiseman*

A1C Brendan Davis gives instruction to Coast Guard aviation maintenance technician students from Elizabeth City, North Carolina, Coast Guard Station, in an altitude chamber during training at Langley Air Force Base, Virginia.
*Courtesy of USAF/SSgt Samuel Rogers*

## Altitude Chamber

In addition to using flight simulators, military pilots and astronauts undergo special training for the unusual flight conditions they will encounter. Training for high-altitude flight may include in-class instruction. The class covers many of the topics you've already read about, from hyperventilation to trapped gas. It may also include time in an altitude chamber. This is *a sealed room that reproduces a high-altitude environment*. Because not everyone reacts the same way to reduced pressure, a monitored session in an altitude chamber helps a pilot get familiar with the warning signs of hypoxia.

Pilots have to meet certain requirements to take part in an altitude chamber exercise. They can't have a cold because colds can make it more difficult to equalize pressure in the ears. Beards aren't allowed because they can interfere with the oxygen mask's fit. And pilots must present proof that they've had a medical exam, usually within the past 12 months.

Before starting the mock flight in the chamber, participants breathe pure oxygen for about 30 minutes. An instructor outside the chamber gives instructions and controls the pressure. Other instructors join the students inside the altitude chamber to make sure everyone is safe throughout the exercise. Students wear their oxygen masks as the chamber loses pressure. Once the chamber gets to the equivalent of about 25,000 feet, the students take turns removing their masks to test their reactions to low pressure. Instructors then put them through some easy exercises that can seem very difficult when one is deprived of oxygen. Those still wearing their masks can watch how others respond.

A West Virginia University student named Ryan Coder who had the chance to try an altitude chamber flight through NASA said, "After about 10 seconds you get used to it. For me it wasn't too bad. At the end I got kind of lightheaded and by the end I was just staring at a fixed point. One kid in my chamber passed out." Another college student who also took part in the same exercise—Duncan Miller of the University of Michigan—said, "When you breathe in the air, it feels like normal air. You start to feel a little tingly."

After the participants have spent enough time at low pressure, they put their masks back on. The instructor outside the chamber slowly normalizes the pressure. For several hours after a mock flight, participants are not allowed to fly as a crew member, exercise, or drink alcohol.

## Barany Chair

Another training device is called a Barany chair—*a spinning device that demonstrates the hazards of spatial disorientation.* Pilots not only grapple with spatial disorientation when they train in a Barany chair—they might also fight motion sickness, just as when flying. Instructors use the chair to help students overcome both problems. The Barany chair simulates aircraft maneuvers using the centrifugal effect, *the inertia that appears to fling a body away from the center of rotation.*

When the eyes and the inner ear don't agree, a pilot might not know which way is up, down, or sideways. People are land animals. When on the ground, the human senses are very accurate at reading direction and speed. But up in the air where a pilot can't put his or her feet on solid, level ground, the senses can get confused.

College student Duncan Miller from the University of Michigan shows off his Superman moves during a zero-gravity flight aboard a modified Boeing 727 in 2010. Miller conducted experiments in electric propulsion.
*Courtesy of NASA*

The Barany chair, which rotates smoothly, isolates the vestibular system to train pilots to accurately read signals from the inner ear.

The chair has a metal hoop around it about waist high. Once seated and strapped in, the student puts on ear covers to block out noise and a blindfold to prevent any visual clues. Now the pilot only has the inner ear to sense direction and speed. The student holds his or her head upright and rests his or her fists on the hoop. The instructor then spins the pilot, who indicates which way he or she thinks the chair is spinning by pointing with the thumbs. After several rotations, the instructor slowly stops the chair. Spatial disorientation sets in pretty quickly in this exercise because the fluid in the inner ear will either lag behind the chair's motion or get ahead of it once the chair has slowed or stopped. In addition, the instructor may ask the pilot to tilt his or her head to the left, right, down to the chin, or back, all of which confuse the signals from the inner ear fluid.

If a pilot is getting instruction for airsickness, the blinders and ear covers aren't needed. Training to end airsickness takes about three days. It includes trying to pinpoint what causes the motion sickness. A session in the chair may include getting spun for 10 minutes three times with 10-minute breaks between spins. The student spins at a rate of about 25 revolutions per minute. Usually by the third day, no one gets sick.

One way to fight airsickness is to inhale deeply through the nose and exhale through the mouth. This exercise is meant to relax the body. Another is to clench and release different muscle groups, which is also supposed to get rid of tension.

A1C Brendan Davis (*left*) takes Coast Guard Amn Paul Jaquish through a spatial disorientation exercise in a Barany chair at Langley Air Force Base, Virginia.
*Courtesy of USAF/SSgt Samuel Rogers*

**CHAPTER 3**   Flight and the Human Body

## The "Vomit Comet"

Astronauts train in simulated weightlessness in an aircraft dubbed the *Vomit Comet* because people sometimes get airsick during their sessions in it. Scientists and college students also get to join in on NASA's Reduced Gravity Program to conduct experiments on the aircraft. The space agency uses twin-jet C-9 and modified Boeing 727 aircraft that let astronaut pilots experience weightlessness before they head into space.

The *Vomit Comet* ascends to a cruising altitude of about 24,000 feet. Then it soars to more than 32,000 feet at a 45-degree angle, only to descend back to 24,000 feet also at a 45-degree angle (Figure 2.3). Each *parabola*—the roller coaster ride up and down— gives the astronauts-in-training about 25 seconds of near-weightlessness. Each flight runs through anywhere from 40 to 60 parabolas, and takes two to three hours to complete.

On the leg up, 2 Gs pin passengers to the aircraft floor. But at the parabola peak, it's low gravity. During these moments of apparent weightlessness, the students can float, spin, and tumble harmlessly about the aircraft. Aircraft walls have straps students can grip if they need to orient themselves.

### Wing TIPS

The Air Force started the Reduced Gravity Program in 1957. NASA has operated it since 1973.

### Wing TIPS

Before the C-9 and the Boeing 727, NASA used a KC-135A model aircraft. The space agency retired its last KC-135A in 2004. That final aircraft—referred to as *NASA 931*—flew more than 18,000 passengers, more than 34,000 parabolas, and used more than three million gallons of fuel.

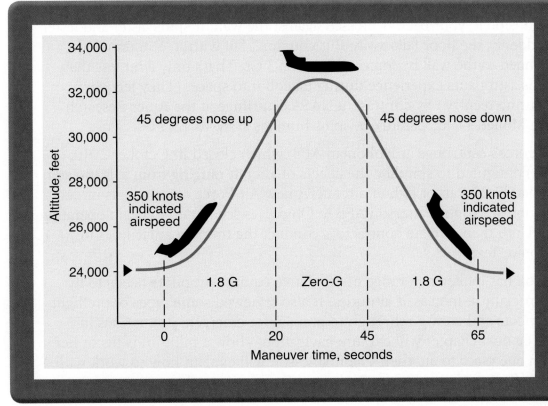

**Figure 2.3** A typical zero-G maneuver by NASA's Reduced Gravity Program aircraft
*Reproduced from NASA/JSC*

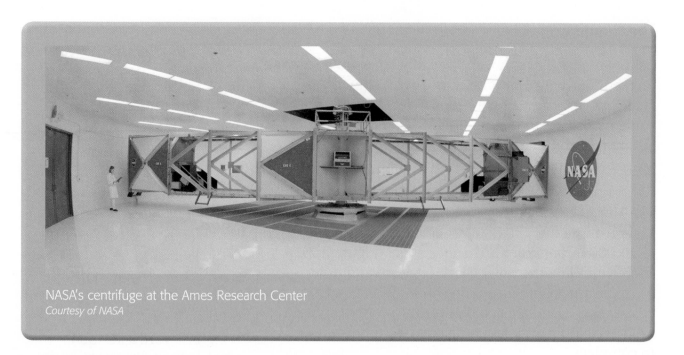

NASA's centrifuge at the Ames Research Center
*Courtesy of NASA*

## Centrifuges

Pilots and astronauts may also train on a machine called a centrifuge—*a device that creates artificial gravity forces by spinning around a central point.* The main reason for centrifuge training is to prevent gravity-induced loss of consciousness, or G-LOC. This can occur when pilots are exposed to high Gz, which you read about in the previous lesson.

Perhaps you've ridden on a circular carnival ride that spins dizzyingly fast. Standing inside it, your back is pressed against the wall. It spins faster and faster until, suddenly, the floor falls away. But you don't fall with it. You remain in place, pinned to the wall by forces as great as 3 Gs. That's only a bit less than the 3.2 Gs astronauts experience during takeoff into space. (They feel about 1.4 Gs during reentry.) By contrast, a NASA centrifuge at the Ames Research Center at Moffett Field, California, spins humans at up to 12.5 Gs.

The Air Force's centrifuge at Holloman AFB, which closed in October 2010, could be configured to simulate the effects of aircraft ranging from a T-38 Talon trainer to an F-22 Raptor fighter aircraft. A new Air Force centrifuge is under construction at Wright-Patterson AFB in Ohio. It is scheduled to begin operating in 2012. Until then, private contractors conduct the training at facilities in San Antonio, Texas.

This lesson has looked at a range of protective equipment pilots needs to fly safely, especially at increased altitudes. It also reviewed some types of preflight training pilots and crew go through to prepare for emergency situations in the air. The next chapter will explore navigation—how a pilot finds his or her way from one place to another. The first lesson will explain how to work with navigation basics—latitude, longitude, and Earth's size and shape.

**CHAPTER 3** Flight and the Human Body

# ✔ CHECK POINTS

## Lesson 2 Review

Using complete sentences, answer the following questions on a sheet of paper.

1. What is positive pressure, and what does it create?

2. What would happen if a pilot tried to escape a damaged fighter with only a parachute?

3. What do military and civilian pilots use flight simulators for?

4. Why do instructors join the students inside the altitude chamber?

5. What is a Barany chair?

6. What happens on the leg up in the *Vomit Comet*?

# APPLYING YOUR LEARNING

7. If you're a test pilot flying above 40,000 feet, what kinds of protective equipment will you need and why?

A protractor, ruler, flight computer, and a chart are important tools for developing a flight plan.
© Anthony DiChello/ShutterStock, Inc.

# Here to THERE

"There are no signposts in the sky to show a man has passed that way before. There are no channels marked. The flier breaks each second into new uncharted seas."

*Anne Morrow Lindbergh*

# LESSON 1

## ✈ Navigational Elements

### ✏ Quick Write

_____

_____

What other steps do you think the American pilots could have taken to navigate to friendly territory?

### ✦ Learn About

- the history of navigation and navigation instruments
- the relationship of Earth's size and shape to navigation
- the correlation of latitude and longitude to flight position
- how to determine navigational direction
- chart projection characteristics
- how chart projections are used in navigation
- the problems associated with projections

**D**uring World War I American pilots faced German aces, antiaircraft fire, bad weather, and troubles due to low-tech navigation. Because the United States couldn't produce anywhere close to enough aircraft in time to affect the outcome, American aviators flew French and British combat planes throughout the war.

On 10 July 1918 a group of US pilots took off from a French airfield to attack German-controlled rail lines in France. They were with the Air Service of the American Expeditionary Forces, the name for the US presence in Europe during the war. The pilots flew Breguets, French-built biplanes. But instead of finding their targets, clouds blew in and blocked the crewmen's view of the ground below. The wind picked up. Rain fell. Now the pilots couldn't see the railroad, and they weren't quite sure where they were.

The pilots pulled out their compasses to navigate to a friendly airfield. When their commander spotted a likely landing spot, the group followed him in. However they were off course by more than they realized because of having to navigate through the bad weather. They ended up on a French airfield that was in enemy hands. The Germans took the two dozen Americans prisoner and seized their planes.

According to US Air Force historians, Col William "Billy" Mitchell—who oversaw air combat operations—blew up when he learned about losing six aircraft because of poor navigation. It would take at least three weeks to get new planes. Navigation was proving to be as important as worthy aircraft to hang on to the skies over Europe.

## The History of Navigation and Navigation Instruments

When pilots like the Wright brothers first flew, they didn't have a cockpit filled with electronic gadgets to help them find where they were going. In fact they didn't have a cockpit to speak of, and they couldn't travel more than a few hundred feet anyway. But as planes evolved into mature machines that could journey hundreds and eventually thousands of miles, pilots—with the help of engineers—invented ways to keep track of where they were. They learned to *navigate*.

Air navigation is the act of flying an airplane from one place to another while keeping tabs on your position as the flight progresses. It's what travelers on land and sea have done for millennia. Pilots have two ways to navigate: by sight or with instruments. For some flights a pilot can simply peer through the window for familiar landmarks to check that he or she is sticking to the route. At other times a pilot must rely on equipment that monitors his or her progress.

## Of Bonfires and Other Bright Lights

Just as airmail flights in the early part of the twentieth century prompted the development of better weather forecasts—which you read about in Chapter 2, Lesson 4—so airmail flights spurred advances in air navigation. Airmail pilots flew day and night to finish their routes. During the day they could follow railroads, paved roads, and familiar buildings and landscapes. The night passages were what gave them trouble. During the night pilots couldn't find landmarks, nor could they see the airfields.

### Vocabulary

- airway
- attitude indicator
- great circle
- great circle navigation
- arc
- longitude
- prime meridian
- course
- true north
- compass rose
- heading
- track
- vector
- groundspeed
- chart projection
- cartographers
- tangent
- secant
- apex
- rhumb line
- developable surface

### Wing TIPS

Even though they were a couple hundred feet in the air, early pilots sometimes referred to the same road maps that car and truck drivers used to get around.

A beacon sits on a 51-foot tower. Below it is a shed with the beacon number painted on it big enough to be seen from the sky during the day.

*Courtesy of US Department of Transportation/Federal Aviation Administration*

The answers were crude in the beginning. The US Post Office teamed up with private citizens to light an airway—*the path a plane follows in the sky.* Their first really big event took place at night in 1921. Post Office staff and the public lit bonfires to help Jack Knight, an airmail pilot, find his way from North Platte, Nebraska, to Chicago, Illinois. Although darkness swallowed the landscape below him, Knight could follow the bonfires from point to point along his route. Around the same time, the US Army illuminated a 72-mile trial airway from Dayton to Columbus, Ohio, with beacons, floodlights, and more. They also outlined airfields in lights and shone spotlights on windsocks.

Between 1923 and 1933 the Post Office and the Department of Commerce's Aeronautics Branch set up 18,000 miles (29,000 km) of airways running back and forth across the country. They placed beacons every 15 to 25 miles (24 to 40 km). In good weather the light from each beacon could reach up to 40 miles. Airmail pilots could work their way across the country by flying from light to light. The system had 1,500 beacons. Pilots could also use these beacons for navigation by day because each one was numbered in big letters that could be seen from the sky.

## Navigational Instruments

Next came navigational aids that reached into the cockpit. During the 1920s several branches of government—the Post Office, Navy, Army, and Bureau of Standards—developed a radio navigation beacon system along already established airways. They placed low-frequency radio transmitters on the ground that sent audio signals to aircraft receivers. Pilots could listen to the signals to make sure they were sticking to the right airway. Suddenly they weren't so dependent on visual markers on the ground, whether bright beacons at night or landmarks by day. The new system was referred to as a *four-course radio range* because of the four towers sending out four signals in Morse code at each station.

### Wing TIPS

Morse code is a way to transmit text using a series of off and on lights, tones, or clicks called *dots* and *dashes*. A different combination of dots and dashes represents each letter of the alphabet and the numerals zero through nine. It's used with telegraph equipment, and a trained listener can understand it simply by watching or listening.

By the mid- to late 1920s engineers had also developed two-way radios. Pilots could finally talk to people on the ground to gather weather and navigation data. The Commerce Department's Aeronautics Branch had 68 radio stations in place by 1934.

Other instruments useful in navigation began to show up in the cockpit around this time as well. These included the altimeter, an aircraft instrument that displays altitude, and the directional gyroscope, now known as an attitude indicator (Figure 1.1), *an instrument that measures aircraft attitude in relation to the horizon.* Four other instruments you might find in the cockpit starting in the 1930s were the *vertical speed indicator* for climb and descent rates, an *airspeed indicator*, a *turn-and-bank coordinator*, and a *heading indicator* (that is, a kind of compass). They gave pilots greater freedom to fly through total darkness and at greater heights because pilots no longer had to constantly reference Earth below for direction. You'll still find these six instruments in aircraft today.

The 1940s brought along improved radio navigation signals. The first, in 1941, was the ultrahigh-frequency (UHF) radio range. It covered 35,000 miles (56,000 km) of airways. Then in 1944 came the very high frequency (VHF) omnidirectional radio range (VOR). Invention rates were speeding up because of World War II when Allies and Axis Powers were racing to improve all kinds of technology so each could win the war. With VHF-VOR, pilots no longer had to tune in to a radio signal to hear code to find out where they were. They could just check a dial on their cockpit instrument panel to navigate. By the 1950s VHF-VOR began replacing the low- and medium-frequency four-course radio range.

**Figure 1.1** Attitude indicator

A very high frequency omnidirectional radio range (VHF-VOR) system
*Courtesy of the Minnesota Department of Transportation*

Radar also popped up during the 1940s with the war. It kept track of plane position by bouncing radio waves off passing aircraft. Radar is critical to air traffic control because it helps that operation maintain safe distances between aircraft. In 1967 satellites soared onto the scene. They tracked aircraft and broadcast their locations to ground stations. The country's navigation system until this point required pilots to fly from checkpoint to checkpoint. But as computers and satellite technology evolved, pilots could begin to follow more flexible courses within designated *area navigation routes*. This made for more efficient, shorter flights.

Newer technologies continue to emerge, such as the global positioning system (GPS) that transmits by satellite. You will read more about developments in navigation technology in Chapter 4, Lesson 5.

**Wing TIPS**

In 1973 the federal government extinguished its last lighted airway beacon, a system that had been in place since the 1920s. The state of Montana, however, still uses and maintains part of the network in the western mountains.

## The Relationship of Earth's Size and Shape to Navigation

Earth's size and shape affect how you get from here to there. Maps may make the world look flat, but Earth is an imperfect sphere. Its radius at the equator is 3,986 miles (6,378 km), while at the poles it's 3,972 miles (6,356 km).

When you take off from an airfield and climb high in the sky, you can see that the Earth is round. It's possible to know this because of the horizon. At the horizon Earth's surface appears to fall away from you—as any round surface would—in every direction you look.

Scientists define Earth's shape using a concept referred to as the great circle. A great circle is *an imaginary circle on a sphere's surface whose center is the sphere's center.* The equator is a great circle because its center is also Earth's center. A circle that passes through both the North and South Poles is a great circle because its center is Earth's center.

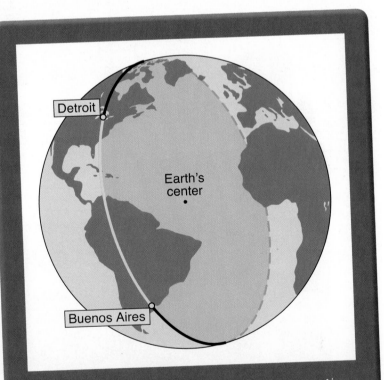

**Figure 1.2** A great circle route from Detroit to Buenos Aires is along a plane that cuts through the center of Earth. It is the shortest path a pilot can take between two points.

Great circle navigation is *the shortest distance across a sphere's surface between two points on the surface*. It follows great circles across Earth's surface. Pilots plot great circle routes when flying long distances (Figure 1.2). To help understand great circles, get a globe and insert a push pin with a piece of string attached at your starting point. Then stretch the string to your destination location and push a pin in there. The route the string indicates between the two pins is the great circle route.

## The Correlation of Latitude and Longitude to Flight Position

Pilots use a combination of great circles and imaginary lines to pinpoint exactly where they are on Earth and where they're headed. They calculate locations in degrees, minutes, and seconds of arc. An arc is *a unit of measurement along a circle*.

The equator is the starting point for measuring north-south locations on Earth. This great circle is equidistant from Earth's poles. Circles parallel to the equator are called *parallels*, or lines of *latitude*, which you first read about in Chapter 2, Lesson 1. Except for the equator, the planes of parallels do not run through Earth's center so they are not great circles. These lines of latitude measure degrees north and south of the equator. The angular distance from the equator to either of the poles is one-fourth of a circle, or 90 degrees. As another example, downtown Pasadena, California's latitude is 34 degrees, 08 minutes, 44 seconds of arc north of the equator, or 34° 08′ 44″ N.

**Wing TIPS**

Lines of latitude run east and west but measure arcs north and south of the equator.

Great circles that pass through both the North and South Poles are called *meridians*, or lines of longitude. Longitude is *a measurement east or west of the prime meridian in degrees, minutes, and seconds of arc from 0 to 180 degrees*. Meridians are at right angles to the equator.

The prime meridian is *0 degrees longitude and is a line of longitude that runs through Greenwich, England*. It is the starting point from which to make measurements in degrees east and west. The longitude for downtown Pasadena is 118 degrees, 8 minutes, 41 seconds of arc west of the prime meridian, or 118° 8′ 41″ W.

**Wing TIPS**

Lines of longitude run north and south but measure arcs east and west of the prime meridian.

Pilots can find the position of any point on Earth's surface by referring to both its latitude and longitude (Figure 1.3). For example, Washington, DC, is approximately 39 degrees N latitude and 77 degrees W longitude. One degree of latitude is equal to about 111 km or exactly 60 nautical miles. Because lines of longitude meet at the poles, the length of a degree of longitude varies from 111 km at the equator to 0 km at the poles.

## Time Zones

Meridians are useful for designating time zones. A day is the time required for Earth to make one complete rotation of 360 degrees. Because a day has 24 hours, Earth revolves at a rate of 15 degrees per hour. Noon is when the sun is directly above a meridian. Time zones west of that meridian are experiencing morning, and those east of that meridian are experiencing afternoon.

Most people observe this standard of time zones, one for each 15 degrees of longitude. The United States has four times zones (Figure 1.4). They are Eastern (75 degrees), Central (90 degrees), Mountain (105 degrees), and Pacific (120 degrees). The dividing lines between the zones aren't ruler straight because of the irregular shapes of town and state borders. But they're pretty close to being 15 degrees apart.

When the sun is directly above the 90th meridian, it is noon Central Standard Time. At the same time it is 1:00 p.m. Eastern Standard Time, 11:00 a.m. Mountain Standard Time, and 10:00 a.m. Pacific Standard Time.

Pilots particularly have to take time zones into account when they travel east because they may be traveling into the dark. They also have to be familiar with time expressed in terms of a 24-hour clock, which the military also uses. Air traffic instructions, weather reports and broadcasts, and estimated times of arrival are all based on this system. For example, a weather report would express 9:00 a.m. as 0900, 1:00 p.m. as 1300, and 10:00 p.m. as 2200.

Because a pilot may cross several time zones during a flight, the aviation field has adopted a standard time system. It is called Universal Coordinated Time (UTC). Many refer to it as Zulu time. UTC is the time at the 0 degree line of longitude running through Greenwich, England. All time zones around the world are based on this reference. To convert your time zone to the time in Greenwich, do the following:

- For Eastern Standard Time, add five hours.
- For Central Standard Time, add six hours.
- For Mountain Standard Time, add seven hours.
- For Pacific Standard Time, add eight hours.

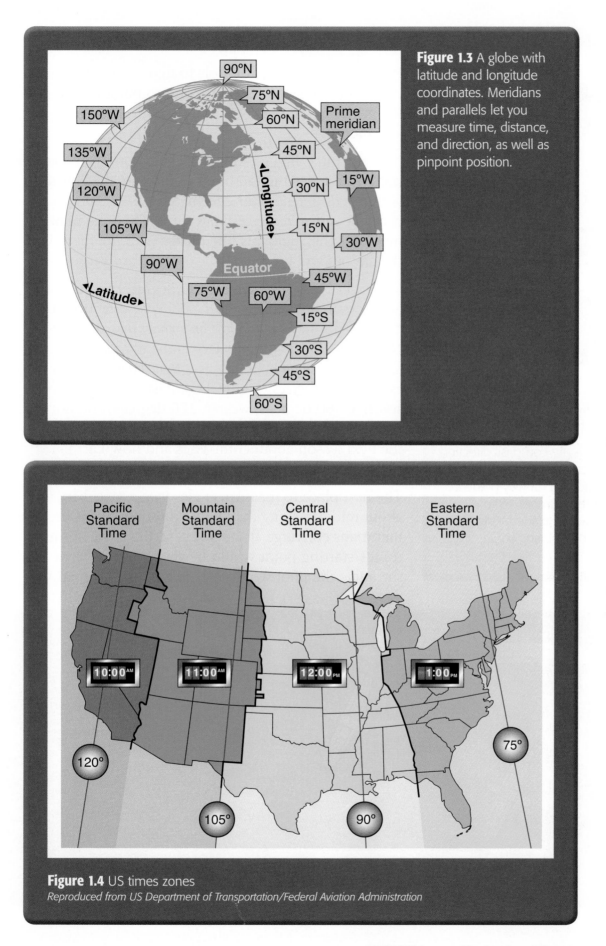

**Figure 1.3** A globe with latitude and longitude coordinates. Meridians and parallels let you measure time, distance, and direction, as well as pinpoint position.

**Figure 1.4** US times zones
*Reproduced from US Department of Transportation/Federal Aviation Administration*

LESSON 1 ■ Navigational Elements

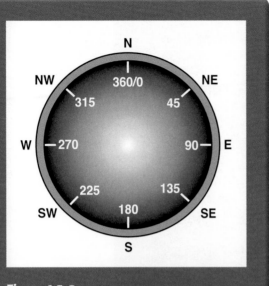

**Figure 1.5** Compass rose
*Reproduced from NASA*

*Wing* TIPS

Airport runways indicate compass direction with degrees written in shorthand. For instance, if one end of a runway points south, it will bear the number *18* to indicate 180 degrees. A plane taking off from this end will be pointing south, or 180 degrees.

# How to Determine Navigational Direction

Pilots need a number of tools to figure out which direction to take to get from point to point along the great circles. One mix of tools includes charts and meridians.

First the pilot draws a line on a chart from the departure point to the destination point to indicate the course, which is *the intended direction of flight in the horizontal plane measured in degrees from north.* Then the pilot measures the angle this line makes with the meridians the aircraft will cross.

Direction is always expressed in degrees, and degrees are measured on a chart in a clockwise direction from true north—*a meridian line on a chart that runs to the North Pole.* On a compass face is the compass rose, *a circle marked off in 360 degrees that shows direction expressed in degrees* (Figure 1.5). East is at 90 degrees, south at 180 degrees, west at 270 degrees, and north at 360 degrees or 0 degrees. You will read more in the next lesson about compasses and how Earth's magnetic field affects them.

The best place to measure angles is at a meridian about midway along the intended route. This is because meridians converge at the poles, so the angle at the route's starting point would be very different from the angle the pilot would measure at the destination.

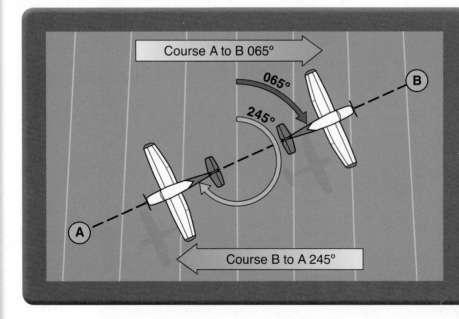

**Figure 1.6** A true course from point A to point B would be 065 degrees. The return trip would have a true course of 245 degrees.
*Reproduced from US Department of Transportation/Federal Aviation Administration*

The angle at the journey's halfway point will split that difference. The course the pilot measures on a chart is known as the *true course* (Figure 1.6) because he or she is measuring the angle with reference to a meridian, or true north.

A plane's heading is *the direction in which an aircraft nose is pointing during flight.* The *true heading* is its direction when measured in degrees clockwise from true north. An aircraft's track is *the actual path taken over the ground in flight.*

Wind can change an aircraft's direction. When pilots plot a course, they determine their vector, *an aircraft's direction and speed.* An example is a plane flying at 45 degrees (a northeast direction) at 100 knots per hour. If a wind is blowing south at 20 knots, the pilot can do some calculations to figure out that his or her aircraft's new vector is about 55 degrees at a groundspeed of 87 knots. (Groundspeed is *an aircraft's rate of progress in relation to the ground.*) You will read more about the effect of wind on flight in Chapter 4, Lesson 3.

## Distance and Time

Distance and time are two more ingredients needed to navigate. How far pilots fly, how fast, and in how much time all determine how much fuel the aircraft will need.

When measuring distance, pilots use the nautical mile, which you read about in Chapter 1, Lesson 1. It is equal to about 6,080 feet, or 1.15 statute miles. It is also equal to one minute of a great circle's arc. If a pilot flew 120 nautical miles in one hour, his or her speed would be referred to as 120 knots per hour.

How much time a pilot spends in the air depends on what winds the aircraft encounters. Wind can increase or decrease an aircraft's speed. Pilots measure two types of speed when flying—groundspeed and airspeed.

An aircraft flying eastward at an airspeed of 100 knots in still air has a groundspeed that's exactly the same—100 knots. If the winds are moving east at 20 knots, this doesn't affect the aircraft's airspeed. But the aircraft's progress over the ground is increased to 100 plus 20 for a groundspeed of 120 knots. Conversely if the mass of air is moving west at 20 knots, the aircraft's airspeed again remains the same. But the groundspeed drops to 100 minus 20, or 80 knots (Figure 1.7).

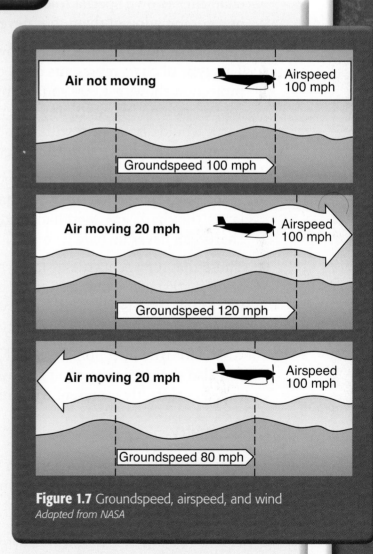

**Figure 1.7** Groundspeed, airspeed, and wind
*Adapted from NASA*

# Chart Projection Characteristics

Charts help pilots find and maintain the best route to a destination—but these visual aids have their drawbacks. The truest representation of Earth is a globe. The charts that pilots use are flat. A chart projection is *the portrayal of Earth's spherical surface on a flat surface*. Anytime you try to represent something that's three-dimensional in two dimensions—as on a chart or map—you get a distorted image.

Cartographers—*people who make charts and maps*—have created many kinds of charts to try to address this distortion. But they can never entirely escape warping some aspect of Earth's image when converting it to a flat surface.

## Area, Shape, Distance, and Direction

Charts portray four main characteristics: area, shape, distance, and direction. When a chart accurately portrays one of these characteristics, then that characteristic is referred to as *true area*, *true shape*, *true distance*, or *true direction*. Each type of chart fills its own special role by depicting at least one of these true characteristics or a combination of them, but never all.

Charts that illustrate true area go by the name *equal-area* or *equivalent* projections. This type of chart represents areas of the same size equally. So if an equal-area chart projection shows a 1,000 square-mile patch of ground in Wyoming within a single square-inch on the chart, any other 1,000-square-mile patch of land elsewhere on Earth would also appear within a single square-inch on the chart. This type of chart is particularly known for distorting characteristics such as shapes.

Charts showing true shape are *conformal* chart projections. Shapes tend to be fairly accurately represented on these charts although they may grow more distorted the larger the charts are. Meridians and parallels intersect at right angles. Conformal charts can never include true area.

Those that represent true distance are *equidistant* projections. They convey accurate distances, but only along certain lines on a chart or from the center of projection (that is, the point from which the image is being projected). They are never true for distance over an entire chart.

Charts for true direction are called *azimuthal* or *zenithal*. Conformal charts can also show true direction but over far smaller areas. Azimuthal projections can also correctly represent true area, true shape, or true distance.

## Geometry

Cartographers base their charts on three geometric projections, which are the cylinder, the cone, and the plane. A plane is also referred to as azimuthal. Some geometric forms are better at projecting area, shape, distance, or direction than others.

## Cylinders

The cylinder forms an image of Earth in one of two ways. It is tangent (*touching*) if it wraps around Earth along a single line such as the equator or a meridian. It is secant (*intersecting*) if it encircles Earth along two lines (Figure 1.8). The cylinder chart's most accurate regions are those closest to the tangent or secants. The most inaccurate are those farthest from the tangent or secants. If you slice the cylinder open along a meridian and spread out the chart, you can see that all lines of longitude and latitude are straight and parallel.

The *Mercator projection* is the most common example of a cylinder chart. It is tangent to Earth along the equator, so its greatest accuracy is at the equator. Its lines of longitude and latitude are straight and parallel, although the parallels lie farther and farther apart the closer they get to the poles. Parallels and meridians meet at right angles. Direction is true and represented by straight lines. Distances are true at the equator and fairly true within 15 degrees of it. The Mercator maintains shapes in small areas but distorts shapes, distances, and areas as it nears the poles.

## Cones

A cone projection's peak can line up with Earth's polar axis. The cone touches Earth along one or more parallels. Like the cylinder it is truest along the tangent or secants. Some projections use two or more cones to increase accuracy.

To read a conic chart projection, you roll it out after cutting it open along a meridian. The meridians are straight, equidistant, and converge toward the chart's apex (or *peak*). The parallels curve around the apex. They are also closer to one another the closer they get to the peak. Meridians and parallels meet

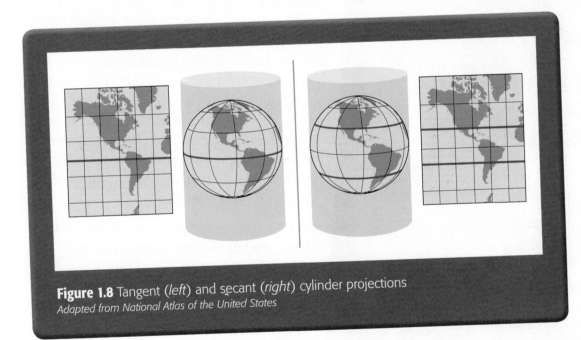

**Figure 1.8** Tangent (*left*) and secant (*right*) cylinder projections
*Adapted from National Atlas of the United States*

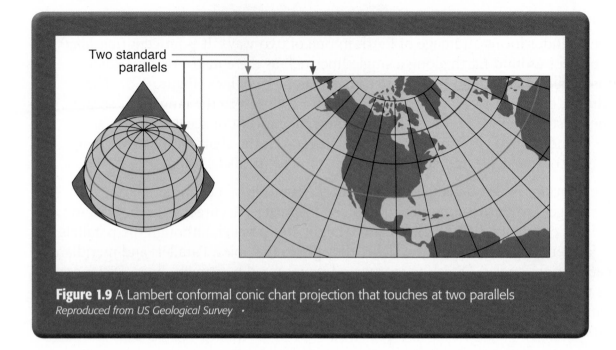

**Figure 1.9** A Lambert conformal conic chart projection that touches at two parallels
*Reproduced from US Geological Survey* •

at right angles. A *Lambert conformal conic* is one type of conic chart. It is secant along two parallels (Figure 1.9). Distances are true only along the secants, and directions are fairly true. Because it is conformal, shapes and areas also fair well near the secants but begin to warp the farther they get from the two parallels.

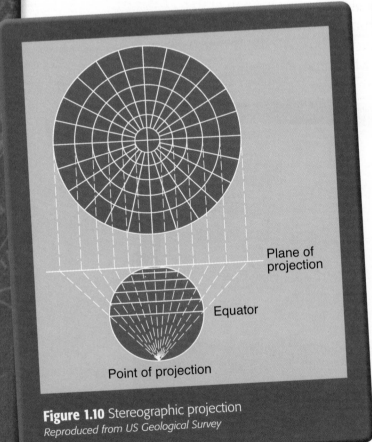

**Figure 1.10** Stereographic projection
*Reproduced from US Geological Survey*

## Planes

A plane is tangent to a point rather than a line. Scientists often use a plane projection called the *stereographic* projection to chart polar regions, although these projections can be used to chart large areas anywhere on the globe. If using a stereographic projection to chart the South Pole, the point of tangency is at the South Pole and the point of projection is at the North Pole. Therefore the point from which the image is projecting is always on the opposite side of the globe from the point of tangency in a stereographic projection (Figure 1.10).

In this stereographic chart of the South Pole, the meridians meet at the pole and extend from it in straight lines. The parallels are circles around the pole.

True direction is only from the center point of projection. Because a stereographic chart is also conformal, shapes are fairly true at the center. But distortion increases for shapes, areas, and distances the farther you get away from the pole.

Another common plane projection is the *gnomonic* chart. Its point of projection is from the center of the globe (Figure 1.11). All great circles show up as straight lines. True direction is only from the center point of projection. A gnomonic chart is not conformal, equal area, or equidistant.

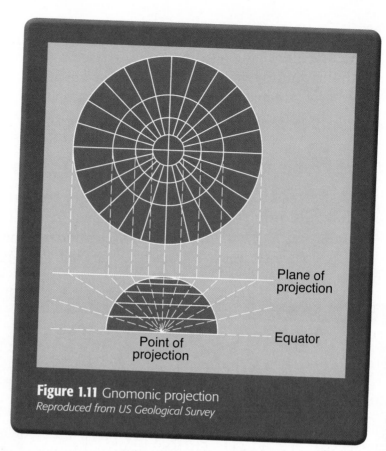

**Figure 1.11** Gnomonic projection
*Reproduced from US Geological Survey*

## How Chart Projections Are Used in Navigation

Pilots use different chart projections when trying to reach different destinations. Some are good for navigating around the poles and some around the equator. Others specialize in limited areas while a few are helpful when finding your way over a long, thin route.

The gnomonic chart projection shows the shortest route between two points—an important feature for long distances. The point of projection creates great circles that appear as straight lines on the *plane of projection*, that is, the flat surface on which the projection appears. A pilot can find true direction from this type of chart but can't rely on any of the other characteristics such as shape and area.

The stereographic plane projection works best for navigating around the poles where it is most accurate. At the poles it offers true direction. If a pilot needs to take a long-distance journey that stretches much beyond the center of this type of chart, he or she will run into a lot of distortion.

The Lambert conformal conic projection, which is secant along two parallels, is most accurate along these parallels. Therefore it is best for trips following an east-west route. It is also good for charting continents and countries that mostly stretch east and west between two parallels. This conformal projection offers pretty accurate shapes and areas near the parallels but not away from them. Directions are also fairly accurate. Distance is true but only along the secants.

The Mercator projection is best for navigating along the equator and highly inaccurate for flying over the poles. These projections show true direction with the rhumb line, which is *a line that runs across meridians at the same angle*. This is why Mercators are good for navigating. If a pilot follows a rhumb line, he or she can stick to a single compass reading.

## The Problems Associated With Projections

As you've read in this lesson, it is difficult to convert a sphere to a flat surface without distorting some part of the original image. This is why cartographers use a developable surface—such as the cylinder, cone, or plane—to make charts. A developable surface is *a geometric shape that doesn't stretch when flattened*.

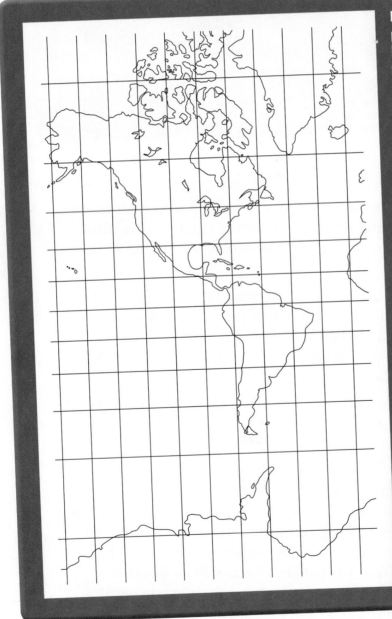

**Figure 1.12** In this Mercator projection, South America looks smaller than Greenland, while in actuality South America is eight times the size of Greenland. Projections tend to distort some characteristics such as area, shape, distance, and/or direction.
*Reproduced from US Geological Survey*

Even so charts present problems to navigators. They can't accurately represent all of the chief characteristics. For instance, distortions tend to be far greater when a chart tries to cover large regions. A specific example of area distortion is in the Mercator projection, which makes South America look smaller than Greenland (Figure 1.12). But in actuality South America is eight times the size of Greenland.

Shapes change in a Lambert conformal conic projection as another example. This is because the parallels are increasingly distant from one another as they get farther from the central parallels. While shapes are fairly true around the secants, they are less so as they move away from those parallels.

Distances beyond 15 degrees aren't very reliable on a Mercator. They appear increasingly great beyond those 15 degrees because the parallels grow farther apart as they near the poles. Other chart projections also have this problem.

Direction is at least partly true for most navigation chart projections. Without decent direction, a pilot won't get where he or she intends to go.

## Wing TIPS

Even the paper that charts are printed on can cause distortion. This is because humidity causes paper to expand. For example, if the atmospheric humidity jumps by 60 percent, the chart may swell by 1 percent.

This lesson presented you with some of the most basic elements of navigation, from Earth's size and shape to latitude and longitude. It also introduced you to the different characteristics and geometric forms of the chart projections pilots use to navigate. The next lesson will look at navigational aids, including the clock and compass. It will also explain how to use navigation charts, how to pull together a preflight plan, and what to do if you get lost.

## Lesson 1 Review

Using complete sentences, answer the following questions on a sheet of paper.

1. What is air navigation?

2. How did Jack Knight, an airmail pilot, find his way from North Platte, Nebraska, to Chicago, Illinois?

3. Why is the equator a great circle?

4. When do pilots plot great circle routes?

5. What is the starting point for measuring north-south locations on Earth?

6. How can pilots find the position of any point on Earth's surface?

7. Where is the best place to measure angles and why?

8. What is a plane's heading?

9. What is the truest representation of Earth?

10. On which three geometric projections do cartographers base their charts?

11. Which chart projection shows the shortest route between two points?

12. What is the Mercator projection best at doing?

13. What is a developable surface?

14. Which characteristic is at least partly true for most navigation chart projections?

## APPLYING YOUR LEARNING

15. If a pilot flew from Seattle, Washington, to Helsinki, Finland, what route would he or she take, and which chart projection(s) would the pilot use?

# LESSON **2**

## ✈ Navigational Aids

### ✏ Quick Write

_____

_____

What would you do if you lost sight of land and your compass didn't work? What clues might help you go in the right direction?

### ✦ Learn About

- the functions of the clock and compass
- the elements of a map
- how to use air navigation charts
- the purpose of flight planning
- how to draft a preflight plan
- the procedures to perform when lost

**N**avigational errors can have huge consequences. Some navigation stories are well known among pilots. The account of the Lost Patrol is one of them.

On 5 December 1945 14 crewmen in five US Navy TBM Avenger torpedo bombers took off on a training exercise from Naval Air Station Fort Lauderdale, Florida. The crews of Flight 19 consisted of the flight instructor, four pilots, and nine enlisted members. The outing was meant to test the pilots' navigation skills and the enlisted crewmen's bombing skills.

The name of their particular assignment was *navigation problem No. 1*. On the first leg of the journey, they were to fly a course of 91 degrees for 56 miles to Hen and Chickens Shoals off the Florida coast to practice bombing. Once they had completed their bombing exercise at the shoals, they were to continue flying at 91 degrees for another 67 miles. On the second leg they were to set a course of 346 degrees for 73 miles. Then on the final leg they were to travel at 241 degrees for 120 miles to return to the air station in Fort Lauderdale.

The flight instructor, Navy Lt Charles Taylor, was an experienced war veteran with about 2,500 hours in the air. Each of the 13 students had around 300 hours. Taylor let the students take the lead so they could practice navigating. Little more than an hour into the flight, an instructor back in Fort Lauderdale caught snatches of conversation over the radio among the crew indicating that they were lost. He identified himself as FT-74, the name of his aircraft, and offered his help. A few minutes later Taylor, identifying himself as FT-28, said, "Both my compasses are out and I am trying to find Fort Lauderdale, Florida."

By this time FT-28 had retaken the lead to guide his students home. But because Taylor didn't know his position, he didn't know which way to fly. If the five aircraft were east over the Atlantic Ocean, they could simply fly west until they hit land. But Taylor mentioned to FT-74 that he had seen some islands, which he assumed were the Florida Keys just off Florida's southern tip. That would have meant they were way off course, maybe even in the Gulf of Mexico, and any course correction would need to be north and possibly east depending upon how far they had strayed. Radio contact continued for about another three hours with no progress. The lost crews couldn't find land.

Experts who have studied the incident believe that FT-28 was exactly where they should have been, near the Bahamas. But Taylor's compass malfunctioned and they became disoriented. Thinking they were near the Florida Keys—southwest of the mainland—they flew east. Given that they were east of Florida, not west of it, continuing to fly east would have taken them out over the open ocean far enough that they didn't have enough fuel to reach land.

Many planes searched for the lost aircraft, including a seaplane with 13 men on board that exploded shortly after takeoff. For another five days hundreds of aircraft scoured 250,000 square miles of sea and land. No one found survivors or even wreckage. One of the last messages from FT-28 mentioned ditching in the ocean as a team when they ran out of fuel. Most likely this is exactly what happened to the crews of Flight 19.

**Vocabulary**

- chronometer
- celestial navigation
- magnetic North Pole
- magnetic South Pole
- variation
- isogonic lines
- agonic line
- deviation
- legend
- relief features
- contour lines
- airspace
- estimated time of arrival
- flight log
- megahertz

TBM Avengers
Copyright © NAS Fort Lauderdale Museum

## The Functions of the Clock and Compass

Navigational aids are the tools pilots use to find their way from point A to point B. They come in two forms. The instruments a pilot uses in the cockpit—such as the altimeter and attitude indicator—are one type of aid. Landmarks visible from an aircraft—including railroads, highways, mountains, and towers—are another type.

Some navigational tools have been around for centuries. Two of them are the clock and compass. For example, sailors measured the angle that the sun makes with the horizon at noon to find their latitude. They could also find their latitude (north/south position) by the stars at night. But to find longitude (east/west position) calls for a very precise clock.

## The Search for an Accurate Clock

Scientists learned as early as the 1500s that a clock could help calculate distances sailed east or west. As you read in the previous lesson, 15 degrees of longitude represents one hour. If you can figure out the time difference between where you are and a clock set to the time at the prime meridian, you can work out your longitude. For example, if your position is two hours behind the time at the prime meridian—it's noon where you are but 2:00 p.m. in Greenwich, England—then you are at 30 degrees west longitude.

But early clocks couldn't keep accurate time, especially at sea, because the oceans rocked ships too violently. This rocking motion upset a clock's smooth functioning. Not until the 1700s did an English inventor named John Harrison devise a clock dependable enough at sea that sailors could use it for navigation. His creation was called a chronometer, *an instrument that keeps exceptionally accurate time.*

In the early decades before satellites, sophisticated radar, and computers, pilots turned to sailors' time-tested techniques. They resorted to celestial navigation— *a method of finding your way using the stars and planets.* These older methods were actually quite accurate. In 2001 *Air & Space* magazine ran an article in which the writer put some old methods to use, including a World War II-era bubble sextant (a navigation tool) to measure the angle of celestial bodies to the horizon. By the end of an overseas flight, his coordinates were only five nautical miles off from what the modern instruments were reading.

When a pilot flies a great circle route, he or she follows a curve from point A to point B. Keeping track of time spent flying along this path is important for a pilot to monitor where he or she is. It is also important for keeping track of how much fuel is left. Pilots of smaller planes compute fuel use in *gallons per hour.* Pilots of large aircraft compute fuel use in nautical air miles per 1,000 pounds of fuel. Once a pilot knows how far he or she is going to fly as well as the amount of time required to cover that distance, the pilot can calculate how much fuel he or she will need.

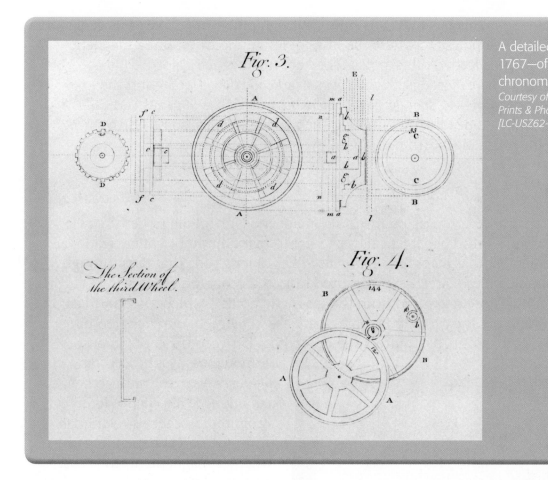

A detailed sketch—dated 1767—of John Harrison's chronometer
*Courtesy of Library of Congress, Prints & Photographs Division [LC-USZ62-110390]*

Wind is another factor that can affect flight time. If a pilot has a sturdy clock or watch that lets him or her know how much time has lapsed on a particular trip, the pilot will know how much fuel he or she has left. The type of clock used in both of these calculations—point-to-point travel and fuel use—doesn't have to be as perfect as what is needed to determine longitude. You'll read more about using clocks in navigation in the next lesson.

## The Magnetic Compass

The compass is another key navigation tool. Its main function is determining direction. One of the oldest and simplest instruments for indicating direction is the *magnetic compass*.

### Wing TIPS

Federal regulations require a compass for both visual flight rules and instrument flight rules flight.

Magnets are one of the moving parts of a magnetic compass. They work because Earth itself is a huge magnet, spinning in space, surrounded by a magnetic field. All magnets have a north pole and a south pole. Like poles repel (north and north, south and south), and unlike poles attract (north and south).

Earth rotates about an axis that runs from its geographic North Pole to its geographic South Pole. These poles are where the lines of longitude meet. But a magnetic compass responds to two other poles. They are the magnetic North Pole—*a spot on Earth where the magnetic field points vertically downward*—and the magnetic South Pole—*a spot on Earth where the magnetic field points vertically upward*. A compass needle lines up with the magnetic poles, so its north tip points to the magnetic North Pole.

## Variation

Because a magnetic compass points to the magnetic North Pole rather than the geographic North Pole, it can cause a pilot to make a number of navigational errors. Remember that maps and charts use lines of longitude that pass through the geographic poles. Directions measured from the geographic poles are *true directions*. Those measured from the magnetic poles—which are hundreds of miles away from the geographic poles—are called *magnetic directions*. So pilots must make corrections that address the differences between these two types of directions. Variation is *the difference between true direction and magnetic direction* (Figure 2.1). Variation is either *east variation* or *west variation* depending on whether magnetic north is east or west of true north.

Aeronautical charts include isogonic lines, which are *lines drawn across charts to connect points having the same magnetic variation* (Figure 2.2). On the US West Coast the magnetic compass needle points to the east of true north. On the East Coast the compass needle points to the west of true north.

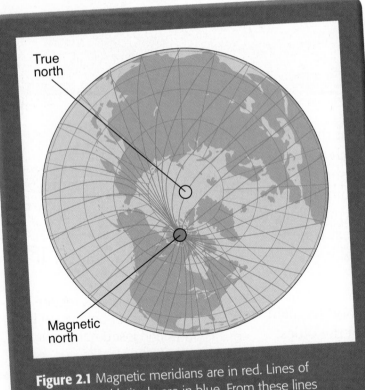

True north

Magnetic north

**Figure 2.1** Magnetic meridians are in red. Lines of longitude and latitude are in blue. From these lines of variation, you can find the effect of variation on a magnetic compass.
*Reproduced from US Department of Transportation/Federal Aviation Administration*

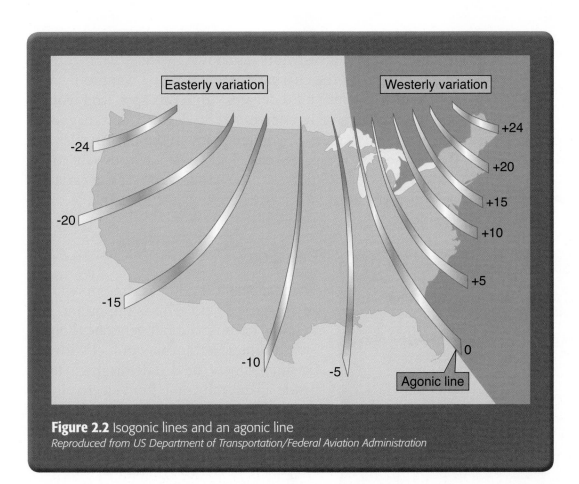

**Figure 2.2** Isogonic lines and an agonic line
*Reproduced from US Department of Transportation/Federal Aviation Administration*

In contrast an agonic line is *a line where magnetic variation is zero.* It is where magnetic north and true north agree. In Figure 2.2 the red line that passes near Chicago is an agonic line. A location east of the agonic line has a westerly variation that requires you to add to your true course. A location west of this line has an easterly variation that calls for you to subtract from your true course (Figure 2.3).

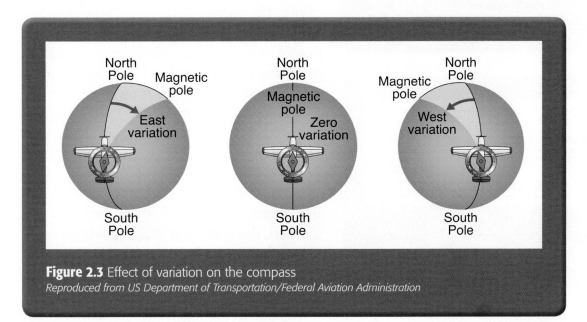

**Figure 2.3** Effect of variation on the compass
*Reproduced from US Department of Transportation/Federal Aviation Administration*

For example, if you are flying near Washington, DC, the variation is 10 degrees west—as shown by the +10 isogonic line passing over the city in Figure 2.2. If you want to fly a course of true south (180 degrees) from this spot, you would add the variation of 10 degrees to 180 degrees for a magnetic course of 190 degrees. If you are flying in the Los Angeles area, the variation is 14 degrees east. Once again, if you want to fly a true course of 180 degrees, you would subtract the variation of 14 degrees from 180 degrees for a magnetic course of 166 degrees.

## Deviation

The magnets in a compass line up with any magnetic field. This is important to know because aircraft generate their own magnetic fields that conflict with Earth's magnetic field. The magnetic fields inside an aircraft come from electrical circuits, radios, lights, tools, the engine, and magnetized metal parts. The fields in an aircraft cause deviation—*a magnetic compass error caused by local magnetic fields within aircraft.* Deviation deflects a compass needle from its normal reading (Figure 2.4).

To correct for deviation, pilots use a compass deviation card (Figure 2.5). Aircraft technicians create these cards for pilots. The card tells the pilot how much to add or subtract to correct for deviation. For any numbers not listed on the card he or she can do some quick math. For example, say a pilot needs to figure the correction for a heading of 195 degrees (using the sample card in Figure 2.5). The correction card shows only how much to correct for 180 degrees (0 degrees) and 210 degrees (+2 degrees). Because 195 degrees is exactly halfway between the two, the corrected heading would be 195 degrees plus 1 degree. A magnetic heading corrected for deviation is known as a *compass heading.*

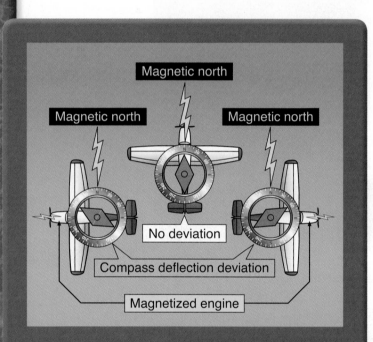

**Figure 2.4** Magnetized portions of an airplane cause a compass to deviate from its normal indications.
Reproduced from US Department of Transportation/Federal Aviation Administration

| For (magnetic) | N | 30 | 60 | E | 120 | 150 |
|---|---|---|---|---|---|---|
| Steer (compass) | 0 | 28 | 57 | 86 | 117 | 148 |

| For (magnetic) | S | 210 | 240 | W | 300 | 330 |
|---|---|---|---|---|---|---|
| Steer (compass) | 180 | 212 | 243 | 274 | 303 | 332 |

**Figure 2.5** A sample compass deviation card
Reproduced from US Department of Transportation/Federal Aviation Administration

## The Compass Rose

Many airports have a *compass rose*, which is a series of lines marked out on a ramp or maintenance area with no magnetic interference. Lines oriented to magnetic north are painted every 30 degrees.

The aircraft technician lines up the aircraft on each magnetic heading and adjusts the compass to minimize deviation. While tweaking the compass, he or she will turn on the aircraft radios and any other instruments, such as lights for night flying, that could cause deviation. The technician will then record corrections on a compass correction card. This card sits near the compass in the cockpit so it's handy for the pilot.

## The Elements of a Map

Being able to read a map is also important to navigation. Note that a map is not the same thing as a chart. A map isn't as detailed as a chart. But it's still a good place to start researching your route. It includes details not likely to change quickly on the ground below, such as roads, railroads, and coastlines.

The spot to begin research on a map or chart is at its legend, which is *a list of all symbols and their meanings on a given map* (Figure 2.6). Legends can include a variety of information, including:

- *Map scale*—This tells you how much a certain distance on the map is equal to on the ground. For example, it might say 1:3,000,000, where 1 inch equals 47 miles and 1 centimeter equals 30 km.

- *Physical, or topographic, features*—These include mountains, valleys, tree lines, sand, and more. Mountains, valleys, and hills are often represented in brown. Relief features are *symbols on a map that represent the elevation of physical features such as mountains, valleys, and cities.* Maps specify elevations with contour lines, which are *lines on a map drawn between points of the same elevation* (Figure 2.7). The closer the contour lines are to one another, the steeper the rise or drop in elevation.

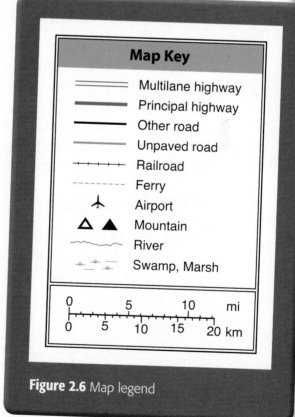

**Figure 2.6** Map legend

**Figure 2.7** Contour lines

- **Water, or hydrographic, features**—These include everything from shorelines to oceans, lakes, swamps, and even irrigation canals. A map may also indicate water depth. Blue usually represents water on a map.

- **Transportation**—Roads, highways, railroads, bridges, and tunnels are man-made physical features, which get their own special category. You can generally identify roads, highways, and even highways under construction by the red lines on a map. Railroad lines are usually black. Bridges and tunnels may be a mix of red and black with other special markings.

- **Cultural features**—This category highlights more man-made objects, such as dams, parks, oil pipelines, and battlefields. It may also include such features as roads and railroads. The symbols are often in black but may be small drawings such as crossed swords for a battlefield.

## How to Use Air Navigation Charts

An air navigation chart—also referred to as an *aeronautical chart*—is a road map for pilots. Pilots use charts to plot their routes. Charts and their legends include very detailed information that aids navigation. A pilot reviews the symbols in a legend when plotting a course on a navigation chart for features that might affect a flight. These features include airport locations, obstructions, and the map scale for distances. Charts go through frequent revisions to take into account changing conditions at airports and elsewhere on the ground.

## VFR Navigation Charts

Which type of air navigation chart you use depends upon the type of trip you plan to take. *VFR navigation charts* are for flying by visual flight rules. *IFR navigation charts* are for flying according to instrument flight rules. Three VFR charts that pilots use include:

- Sectional
- VFR terminal area
- World aeronautical.

## Sectional Charts

These charts are for pilots flying slow- to medium-speed aircraft (Figure 2.8). They have a scale of 1:500,000, where 1 inch equals 6.86 nautical miles or about eight statute miles. A chart drawn to this scale can include lots of detailed information (as opposed to, for example, a map with a scale of 1:12,000,000, which is more typical of what you'd see in an atlas covering an entire continent in a single spread). Its many details include airport data, navigational aids, types of airspace (*that space which lies above a country and is subject to a nation's laws*), and cultural and physical features. Most sectional charts undergo revisions every six months.

### Wing TIPS

The question of who controls the airspace above a country is governed by the Chicago Convention, originally signed in 1944. The convention, or treaty, created the International Civil Aviation Organization (ICAO). ICAO is a United Nations agency headquartered in Montreal, Quebec, Canada. It is governed by an assembly of those nations that have signed the convention. The assembly meets every three years to set policy on international aviation. For example, ICAO has decided that pilots and copilots of international flights, and the air traffic controllers who guide them, must speak a certain level of English.

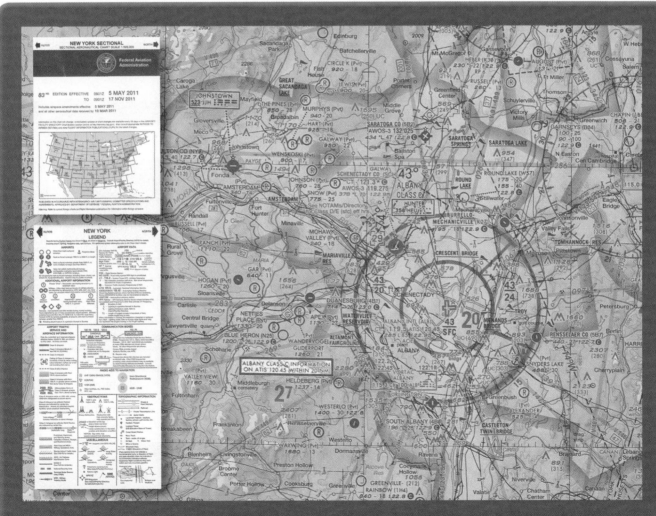

**Figure 2.8** Sectional chart and legend
*Courtesy of US Department of Transportation/Federal Aviation Administration*

Charts include airspace information. You can define airspace vertically and horizontally as well as in relation to time. The time factor relates to how efficiently airspace is managed.

The concept of airspace is a way to control traffic in the skies. Airspace is divided into two *categories*—regulatory and nonregulatory. Within these categories are four *types* of airspace—controlled, uncontrolled, special use, and other airspace. How airspace is divided into categories and types depends on:

- The complexity or density of aircraft movements
- The kinds of operations taking place within the airspace
- The level of safety required
- National and public interest.

### Controlled Airspace

This includes Class A, Class B, Class C, Class D, and Class E (Figure 2.9). Each of these classes covers specific dimensions in the air. Pilots must follow different rules to enter different classes, meet different qualifications, have the right equipment aboard their aircraft, and abide by visual flight rules (VFR). These rules are shaped by weather conditions (e.g., the weather may need to be clear with visibility up to three miles to enter a certain class of airspace). Air traffic control provides service to these classes according to the special rules of each. They break down as follows:

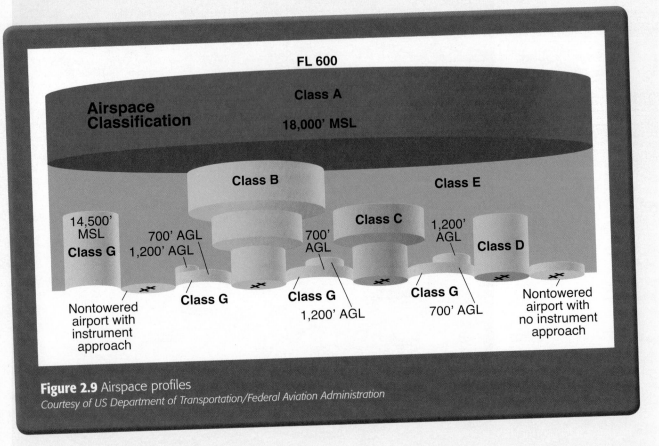

**Figure 2.9** Airspace profiles
*Courtesy of US Department of Transportation/Federal Aviation Administration*

*Class A airspace* is the most restricted class. It generally stretches from 18,000 feet up to and including 60,000 feet. Most aircraft flying in Class A airspace are moving at high speeds. Aircraft fly under instrument flight rules. They must have two-way radios and an instrument that transmits their altitude called an *altitude-encoding transponder*.

*Class B airspace* surrounds the nation's busiest airports. It generally extends from the surface to 10,000 feet, and takes the shape of an upside-down funnel. Air traffic control must provide clearance for an aircraft to enter Class B airspace. All aircraft must have two-way radios and an altitude-encoding transponder. VFR weather minimums require a pilot to have at least three miles of visibility and to fly clear of any clouds. Sectional charts note Class B airspace with a thick blue line.

*Class C airspace* surrounds busy—but not the busiest—airports. It generally extends from the surface to 4,000 feet at airports with an operational control tower and radar approach control. All aircraft must establish two-way radio contact with air traffic control before entering the airspace and have an altitude-encoding transponder on board. VFR weather minimums call for three miles' visibility. VFR flights must fly at least 500 feet below and 1,000 feet above clouds, and if flying level with clouds to travel at least 2,000 feet away from them. Sectional charts outline Class C airspace with a solid magenta line.

*Class D airspace* has less traffic than Class C. This airspace generally stretches from the surface to 2,500 feet surrounding airports with an operational control tower. Aircraft must establish two-way radio communication with air traffic control before entering and while within the airspace. VFR weather minimums are the same as in Class C. Sectional charts outline Class D airspace with a segmented blue line.

*Class E airspace* has the fewest restrictions. It is any controlled airspace not designated Class A, B, C, or D. IFR flights need clearance to enter Class E. VFR flights don't need clearance or special equipment.

## Uncontrolled Airspace

*Class G airspace* is uncontrolled airspace. Air traffic control has no authority in this airspace but VFR weather minimums still apply.

## Special Use Airspace

*Special use airspace* refers to areas with limits placed on flight because of activities taking place in that space, such as military training or missile firing. This type of airspace includes areas that go by the names of *prohibited*, *restricted*, *alert*, *warning*, and *military operation areas*. Examples of prohibited areas include the airspace over the White House and Capitol in Washington, DC, or certain military bases.

## Other Airspace Areas

*Other airspace areas* include nearly everything not covered under any of the other classes. This includes *national security areas* and *terminal radar service areas* where pilots can get additional help from radar services to maintain safe distances between IFR and VFR aircraft.

### VFR Terminal Area Charts

These charts are useful when flying in or near Class B or Class C airspace. Their scale is 1:250,000 (1 inch equals 3.43 nautical miles or about four statute miles). This larger scale lets them show topographic features in more detail. Most are revised every six months.

### World Aeronautical Charts

Pilots flying moderate-speed aircraft at high altitude use world aeronautical charts. The scale is 1:1,000,000 (1 inch equals 13.7 nautical miles or about 16 statute miles). Because of this small scale, world aeronautical charts don't show as much detail as sectional or terminal area charts. But they are useful when flying long distances. Most world aeronautical charts are revised once a year.

## IFR Navigation Charts

When a particular flight doesn't allow flight by VFR, such as bad weather, pilots use an IFR navigation chart. Two types of IFR navigation charts are:

- IFR enroute low altitude
- IFR enroute high altitude.

### IFR Enroute Low Altitude Charts

Pilots flying below 18,000 feet in conditions requiring IFR procedures use IFR enroute low altitude charts. The scale varies, so one inch may represent anywhere from five to 20 nautical miles. The charts include navigation aids, airports, controlled airspace details, minimum altitudes to clear obstructions, distances, and magnetic courses. The charts go through a revision every 56 days. Pilots flying small propeller planes use these charts.

### IFR Enroute High Altitude Charts

Jet pilots use the IFR enroute high altitude chart. It is for navigating at or above 18,000 feet. The scale again varies, where one inch may equal anywhere from 18 to 45 nautical miles. These charts show jet routes, navigation aids, airports, time zones, distances, special use airspace, and more. They are also revised every 56 days.

## The Purpose of Flight Planning

Federal regulations require pilots to collect a variety of material and information before a flight. The purpose of flight planning is to make flight as safe as possible. It should be done carefully and thoroughly in a reasonable amount of time before takeoff.

A sectional chart and charts for areas along a flight route should be among a pilot's supplies. Depending on the type of aircraft, additional tools may include a flight computer or electronic calculator. Other examples of equipment might include a flashlight if a pilot is flying at night. If flying over a desert he or she might pack some extra water.

**CHAPTER 4** Flying From Here to There

The data the pilot must gather includes weather reports and forecasts, fuel requirements, alternate routes and airports, and traffic delays. Much of this will be available from air traffic control. Checking the weather is one of the very first steps, because it determines whether flying is a wise idea to begin with and which route the pilot should take.

## Airport Information

The pilot should study the destination airport and any other airports where he or she might land if diverted because of violent weather or problems with the aircraft. Two FAA publications contain information about airports. They are Notices to Airmen (NOTAM)—a report issued every 28 days—and the Airport/Facility Directory (A/FD)—revised every 56 days (Figure 2.10). The information they contain includes airport locations, elevation, runway and lighting details, available services, and control tower and ground control frequencies.

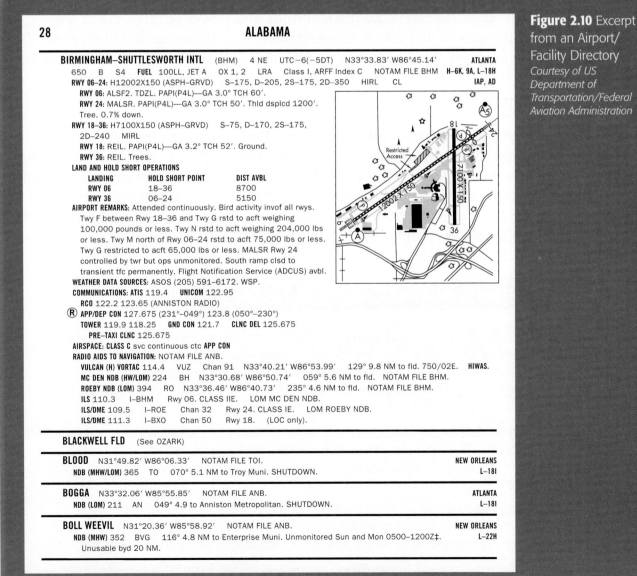

**Figure 2.10** Excerpt from an Airport/ Facility Directory
*Courtesy of US Department of Transportation/Federal Aviation Administration*

## Aircraft Information

Other flight factors directly involve the aircraft. The pilot should check the Aircraft Flight Manual or Pilot's Operating Handbook for weight and balance information. He or she must add up the weight of fuel, oil, passengers, and baggage, as well as how much the aircraft weighs when empty. An aircraft can only carry so much weight so the total can't exceed the particular aircraft's maximum. The pilot must correctly distribute the load as well to keep the center of gravity within limits.

The next step is to figure out how much takeoff and landing distance the plane needs based on the load, the airport's elevation, and the temperature. The pilot uses charts to find out elevation and weather reports for temperature. He or she then compares the distances with the amount of runway available at the airport. The heavier the load and the higher the elevation, temperature, or humidity, the longer the takeoff roll and landing roll will be and the lower the rate of climb.

The pilot looks at fuel consumption charts as well to find how much fuel the aircraft will use at the estimated flight altitude and power settings. He or she calculates this rate of fuel use, then compares it with the estimated flight time. For longer journeys small aircraft in particular may have to stop to refuel en route.

## Flight Plans

A *flight plan* is a form a pilot fills out before a flight and turns in to air traffic control. It contains numerous details about the trip (Figure 2.11). The details include whether the pilot intends to fly according to visual flight rules or instrument flight rules, the type of aircraft, departure point and time, cruising altitude, destination, estimated length of flight, how much fuel the aircraft has, alternate airports where it can stop in case of trouble, and the number of people on board.

### Defense Visual Flight Rules (DVFR)

In addition to flying by visual and instrument flight rules, pilots sometimes fly according to *defense visual flight rules*. The defense rules apply when operating civilian aircraft in zones that are sensitive for national security reasons. Examples include the airspace over Washington, DC. This type of zone is referred to as an *air defense identification zone (ADIZ)*.

To fly into an air defense identification zone, a pilot must have a two-way radio and constantly monitor the radio. The flight plan must contain the time the pilot intends to enter the zone and the point where he or she aims to cross into it. In addition the pilot must depart the zone within five minutes of the time he or she estimated leaving it on the flight plan.

If the two-way radio fails while flying over one of these secure zones, a pilot has two choices—continue flying according to the original flight plan or land as soon as possible. The pilot then reports the radio failure as soon as possible to an air traffic controller.

Form Approved OMB No. 2120-0026

| U.S. DEPARTMENT OF TRANSPORTATION FEDERAL AVIATION ADMINISTRATION **FLIGHT PLAN** | (FAA USE ONLY) ☐ PILOT BRIEFING ☐ VNR ☐ STOPOVER | | | | TIME STARTED | SPECIALIST INITIALS |

| 1. TYPE | 2. AIRCRAFT IDENTIFICATION | 3. AIRCRAFT TYPE/ SPECIAL EQUIPMENT | 4. TRUE AIRSPEED | 5. DEPARTURE POINT | 6. DEPARTURE TIME | | 7. CRUISING ALTITUDE |
|---|---|---|---|---|---|---|---|
| X VFR  IFR  DVFR | N123DB | C150/X | 115  KTS | CHK, CHICKASHA AIRPORT | PROPOSED (Z) 1400 | ACTUAL (Z) | 5500 |

**8. ROUTE OF FLIGHT**

Chickasha direct Guthrie

| 9. DESTINATION (Name of airport and city) | 10. EST. TIME ENROUTE | | 11. REMARKS |
|---|---|---|---|
| | HOURS | MINUTES | |
| GOK, Guthrie Airport Guthrie, OK | | 35 | |

| 12. FUEL ON BOARD | | 13. ALTERNATE AIRPORT(S) | 14. PILOT'S NAME, ADDRESS & TELEPHONE NUMBER & AIRCRAFT HOME BASE | 15. NUMBER ABOARD |
|---|---|---|---|---|
| HOURS | MINUTES | | Jane Smith | |
| 4 | 45 | | Aero Air, Oklahoma City, OK (405) 555-4149 | 1 |
| | | | 17. DESTINATION CONTACT/TELEPHONE (OPTIONAL) | |

| 16. COLOR OF AIRCRAFT | CIVIL AIRCRAFT PILOTS. 14 CFR Part 91 requires you file an IFR flight plan to operate under instrument flight rules in controlled airspace. Failure to file could result in a civil penalty not to exceed $1000 for each violation (Section 901 of the Federal Aviation Act of 1958, as amended). Filing of a VFR flight plan is recommended as a good operating practice. See also Part 99 for requirements concerning DVFR flight plans. |
|---|---|
| Red/White | |

FAA Form 7233-1 (8-82)    CLOSE VFR FLIGHT PLAN WITH ____McAlester____ FSS ON ARRIVAL

**Figure 2.11** Flight plan form
*Reproduced from US Department of Transportation/Federal Aviation Administration*

The FAA requires pilots to file a flight plan for two reasons. First, it helps keep tabs on traffic in the air. Air traffic controllers can then safely direct aircraft to unobstructed routes, altitudes, and runways. This prevents midair collisions and other problems. Second, a flight plan gathers enough information about an aircraft and its route for the FAA to conduct search and rescue if needed. If an aircraft doesn't show up at its estimated time of arrival—*the time someone intends to arrive at a destination*—this alerts air traffic controllers that a pilot may have run into trouble. They can then attempt to find the plane using information from the flight plan.

**Wing TIPS**

Although the FAA requires pilots flying by instrument flight rules to file a flight plan, it encourages but does not demand that those flying by visual flight rules turn in a plan.

## How to Choose Cruising Altitude

Pilots flying above 3,000 feet but below 18,000 feet using VFR choose their cruising altitude based on direction. Those on a magnetic course between 0 degrees and 179 degrees fly at altitudes that are an odd number of thousands of feet plus 500 feet (e.g., 3,500 feet, 5,500 feet, 7,500 feet). Those on a magnetic course between 180 degrees and 359 degrees fly at altitudes at an even number of thousands of feet plus 500 feet (e.g., 4,500 feet, 6,500 feet, 8,500 feet). Flights above 18,000 feet maintain whatever altitude or flight level air traffic control assigns them.

Pilots in controlled airspace flying by instrument flight rules also stick to the altitude assigned by air traffic control. But in uncontrolled space the rules vary based on the range of feet flown below or above a certain flight level and magnetic course. For example, below 18,000 feet on a magnetic course of 0 degrees through 179 degrees, maintained altitudes should be any odd thousand feet such as 3,000 feet, 5,000 feet, and so forth. And on a magnetic course of 180 degrees through 359 degrees, the altitudes flown should be any even thousand feet such as 2,000 feet, 4,000 feet, and so forth.

Wind isn't a factor on this sample trip; it's only blowing at 10 knots. Because this particular aircraft is capable of flying above the noted airspaces, its cruising altitude will be 5,500 feet. Next the pilot measures the flight's total distance as well as the distances between checkpoints. The total distance is 53 nautical miles. Using an instrument called a *plotter*, which you'll read about in Chapter 4, Lesson 5, the pilot can calculate direction. In this case the true course is 031 degrees, and taking into account the slight wind and isogonic lines, the compass heading will be 23 degrees.

Finally the pilot figures out groundspeed, which for this flight will be 106 knots. The total time is 35 minutes (30 minutes plus five minutes for climb) with a fuel burn of 4.7 gallons. When the plane eventually gets into the air, the pilot can adjust the heading, groundspeed, and time based on the real progress he or she is making during flight. Flight planning is very important for flight safety.

## The Procedures to Perform When Lost

Even when he or she draws up a plan, a pilot may sometimes get lost. Wind, weather, and faulty instruments can throw the plane off course. Being lost can be dangerous, especially if the aircraft runs low on fuel. But the pilot can follow some commonsense procedures to get back on track.

If unable to see a town or city, the pilot's first step is to climb. In doing this, he or she must be aware of weather conditions and any other air traffic. Greater altitudes increase radio and navigation reception ranges as well as radar coverage. If flying over a town, the pilot may be able to read the town's name on a water tower. He or she can use a navigational radio to find the aircraft's position. Or if there's a GPS on board, the pilot can use it to find the aircraft's position and the location of the nearest airport.

A lost pilot can also get in touch with personnel at any available facility using radio frequencies shown on the sectional chart. If the pilot makes contact with an air traffic controller, the controller may help the pilot determine his or her position. Other facilities may offer direction-finding assistance. The controller may ask the pilot to hold down the transmit button on the radio for a few seconds and then release it. The controller may also ask the pilot to change directions a few times and transmit again each time. This may give the controller enough information to plot the aircraft's position and give vectors to a landing site. If the situation grows too risky, the pilot should alert others to it by transmitting on the emergency frequency 121.5 megahertz (MHz) VHF and setting the transponder to Squawk 7700. Most facilities and airliners monitor this frequency all the time. A megahertz is *a frequency of one million cycles per second.*

Pilots have many options for orienting themselves when they get lost.
© iStockphoto/Thinkstock

This lesson has reviewed some of the difficulties early navigators faced because they lacked ways to determine accurate time and direction. It also looked at map elements, how to use aeronautical charts, and the basics of flight planning. It also covered what a pilot can do if he or she ever gets lost. The next lesson will consider a navigation technique called dead reckoning and the effects of wind on navigation.

# ✓ CHECK POINTS

## Lesson 2 Review

Using complete sentences, answer the following questions on a sheet of paper.

1. How can a pilot work out his or her longitude?

2. What does a compass needle line up with?

3. Where does a pilot begin his or her research on a map?

4. How do maps specify elevations?

5. Why do charts go through frequent revisions?

6. What kind of chart do jet pilots use?

7. Where can a pilot find weight and balance information?

8. A flight plan gathers enough information about an aircraft and its route to enable the FAA to do what?

9. Why should a pilot check the areas on either side of a planned route?

10. When can a pilot pick a flight altitude?

11. What is a pilot's first step—when lost—if he or she can't see a town or city?

12. With whom should a pilot get in touch when lost?

## APPLYING YOUR LEARNING

13. A pilot is about to fly 100 nautical miles in a small aircraft under VFR. The sky has scattered clouds. The route will take the aircraft through Class C airspace on a magnetic course of 145 degrees. What does the pilot need to do when nearing the Class C airspace? How do the clouds affect the flight path? What is the correct cruising altitude?

# ✈ Dead Reckoning and Wind

## ✎ Quick Write

_____

_____

How might a pilot deal with a strong wind blowing his or her aircraft off course?

## ✦ Learn About

- the basic principles of dead reckoning
- the wind triangle and its application in air navigation
- how the principles of dead reckoning relate to inertial navigation systems

**O**n 10 July 1943, during World War II, some 225 aircraft and 3,400 paratroopers under Col James Gavin headed from Tunis, Tunisia, in North Africa for Sicily in the Mediterranean Sea. They were about to invade the Axis-held Italian island to prepare the beaches for the American and British Soldiers that would follow.

But northwesterly winds introduced an unwelcome wrinkle to the already difficult operation. According to historian Ed Ruggero's book _Combat Jump_, before the paratroopers took off from Tunisia, someone warned the colonel that the wind was blowing at 35 miles per hour. If such winds continued all along the flight, they would affect navigation and endanger the paratroopers when they jumped.

This Allied force was rather inexperienced: the invasion was Colonel Gavin's first taste of combat, and he was leading a large force. Many of the pilots had few hours under their belts, particularly in wartime conditions. They were facing a threefold challenge besides—strong winds on a night mission over water.

This was a historic assignment for paratroopers because it was the first ever for an airborne division. The US Army's 505 Regimental Combat Team had trained hard but had only practiced night jumps twice in the previous two to three months while stationed in North Africa. They had never jumped in winds of more than 15 miles per hour.

The colonel's aircraft—a C-47 Skytrain—was in the lead. His pilots were to find their way east and then north across the Mediterranean using a compass, clock, and airspeed over a known distance. They would also use two landmarks—the islands of Linosa and Malta—to check their progress. Big bodies of water can be confusing stretches of flat sameness, so it was fortunate to have such landmarks.

But with the winds blowing the aircraft off course and the pilots' inexperience, the pilots failed to find either critical checkpoint.

By the time the aircraft reached Sicily, the colonel didn't know for certain that they had reached their target. But with antiaircraft fire and land-based searchlights scouring the skies, he could make a pretty good guess that they had at least found hostile territory. He gave the signal for everyone to jump. Around 80 percent of the paratroopers missed their drop zones by anywhere from one to 60 miles.

Although tens of thousands died in the invasion on both sides, the operation was a success. On 16 August 1943 the Allies seized final control of Sicily from the Axis Powers.

## Vocabulary

- dead reckoning
- pilotage
- fix
- drift angle
- wind correction angle
- wind triangle
- protractor
- inertial navigation system
- waypoints
- accelerometer
- gyroscope

The Douglas C-47 of the Troop Carrier Command was the workhorse of the US airborne forces. In addition to its main task of hauling cargo, the Army used it to transport paratroopers. This photo shows a C-47 with a glider in tow training in North Africa.
*Courtesy of USAMHI*

## The Basic Principles of Dead Reckoning

Pilots have many ways to double-check their position during a flight. The modern era has introduced numerous electronic instruments for this purpose. But equipment can fail. This is one reason older traditions in navigation have survived into the twenty-first century.

Dead reckoning is *navigation based on computations using time, airspeed, distance, and direction.* It can't pinpoint an aircraft's position as accurately as electronic instruments, but it's a very useful tool pilots can use to estimate their location fairly precisely. Other variables a pilot will take into account in his or her dead-reckoning calculations are wind speed and velocity. By working with all of these elements, a pilot can estimate heading and groundspeed and, ultimately, the aircraft's position. The heading takes the aircraft along the intended path, and the groundspeed determines what time it will arrive at the checkpoints and destination. An accurate clock plays a critical role in dead-reckoning calculations.

### Wing TIPS

The term *dead reckoning* comes from *deduced reckoning,* in other words, an estimate of one's position.

Dead reckoning by itself works best over short distances but grows ever more inaccurate over longer distances and time. So when flying over land, a pilot can combine dead reckoning with pilotage—*navigation by visual reference to landmarks or checkpoints.* By referring to these, pilotage lets a pilot constantly monitor and correct heading and groundspeed.

Navigating from point A to point B using dead reckoning involves five steps:

1. The pilot charts a course of straight lines between checkpoints on an air navigation chart, starting at point A and ending at point B. He or she then adds up the distances between the checkpoints to get the total distance.

2. The pilot then figures out the true course by measuring the journey's angle with reference to a meridian, which is true north.

3. At this point, the pilot uses the true course to work out which compass heading to follow with the magnetic compass. He or she factors in wind, variation, and deviation.

4. To find the *estimated time en route*—or the time it will take to complete the entire flight, as well as the distances from checkpoint to checkpoint—the pilot estimates airspeed and divides that into the distance.

5. The pilot then calculates fuel use in gallons per hour to figure out how much fuel will be needed. It's important to ensure there's enough fuel in case he or she gets lost.

## Pilotage

A pilot can use pilotage on any route that has reliable checkpoints, although it is more often used along with dead reckoning and VFR radio navigation.

Pilots who use pilotage are generally flying small aircraft in good weather when they can clearly see landmarks such as roads and bridges (Figure 3.1). Sometimes a pilot can simply follow a road from the starting point to the destination. Bad weather and landscapes with few features, however, can throw a pilot relying on pilotage off course.

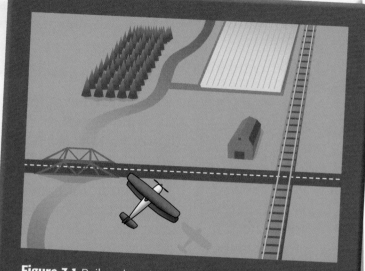

**Figure 3.1** Railroad tracks, tree lines, bridges, roads, even familiar landmarks such as a bright red barn can all play a role in pilotage.
*Reproduced from NASA*

Dead reckoning is particularly useful for pilots who fly the same routes time and again. They can become familiar with certain checkpoints and how long it takes to get from one checkpoint, or fix, to another. A fix is *an aircraft's position in the sky over a checkpoint*. By monitoring the time it takes to fly from fix to fix (Figure 3.2), a pilot knows whether he or she is on schedule and on the right path.

### Wing TIPS

Dead reckoning does not use celestial navigation to determine position.

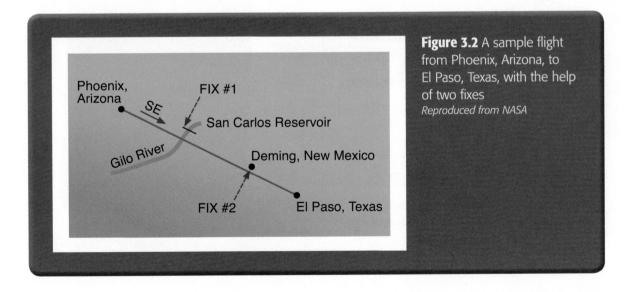

**Figure 3.2** A sample flight from Phoenix, Arizona, to El Paso, Texas, with the help of two fixes
*Reproduced from NASA*

A pilot monitors distance using fixes, direction with a magnetic compass, time with an accurate clock, and speed based on calculations of distance divided by time or by referring to instruments in the cockpit.

## The Wind Effect

As mentioned earlier, wind is one of the variables when a pilot is using dead reckoning. When the wind blows from the north at 25 knots, it means that the air is moving south over Earth's surface at a rate of 25 nautical miles per hour.

Under these conditions, the wind carries any motionless object in the air 25 nautical miles in one hour. Even an aircraft driven by powerful engines feels wind's effect. The aircraft moves through the air at the same time the air moves over the ground. So, at the end of an hour-long flight, the plane is in a position that's the result of two motions—the movement of the air mass over the ground and the aircraft's forward movement through the air mass.

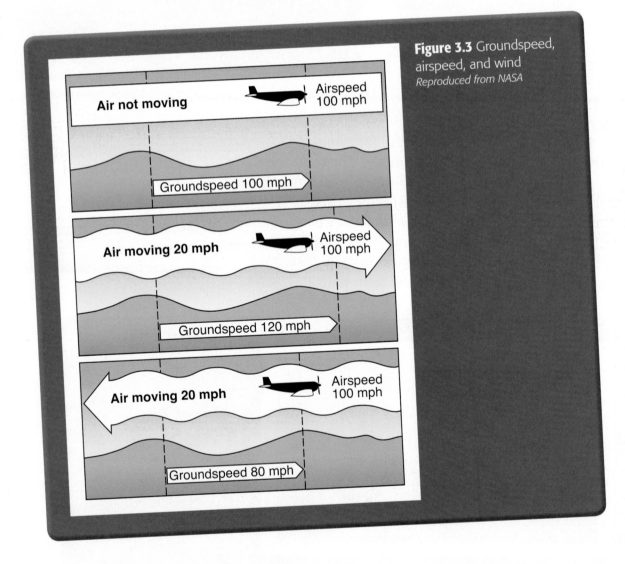

**Figure 3.3** Groundspeed, airspeed, and wind
*Reproduced from NASA*

**CHAPTER 4** Flying From Here to There

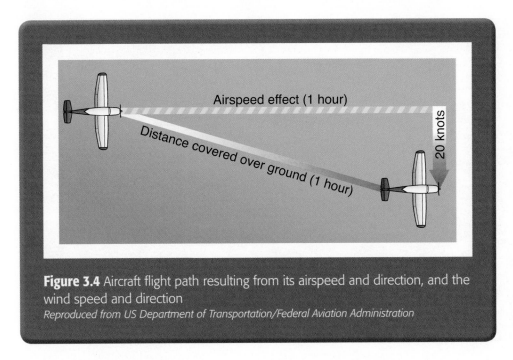

**Figure 3.4** Aircraft flight path resulting from its airspeed and direction, and the wind speed and direction
*Reproduced from US Department of Transportation/Federal Aviation Administration*

A pilot generally won't notice whether he or she is flying through wind, unless the aircraft experiences turbulence or the pilot can study his or her progress over the ground. If the pilot can see the ground, he or she might notice that the aircraft appears to fly faster with a tailwind, slower with a headwind, or drift right or left with a crosswind.

Back in Chapter 4, Lesson 1, you read about an aircraft flying east at an airspeed of 100 knots in still air (Figure 3.3). Its groundspeed was exactly the same—100 knots. But if an air mass were moving east at 20 knots, the aircraft's airspeed would remain unaffected at 100 knots while its progress over the ground increased to 120 knots (100 knots plus 20 knots). And if the mass of air were moving west at 20 knots, the airspeed would again remain the same while its groundspeed would decrease to 80 knots (100 knots minus 20 knots).

Assuming the pilot makes no corrections for wind effect, if the aircraft heads east at 100 knots and the air mass moves south at 20 knots, the aircraft will be almost 100 miles east of its point of departure at the end of one hour because of its progress through the air. It will also be 20 miles further south because of the air's motion (Figure 3.4). Both motions affect the aircraft's position.

In these wind conditions, the airspeed remains 100 knots but the groundspeed is a combination of the aircraft's movement along with the mass of air's movement. A pilot can measure groundspeed as the distance from the point of departure to the aircraft's position after one hour. This speed is calculated by dividing the known distance between two points by the time it takes to fly between them (e.g., 120 miles/2 hours = 60 miles per hour). You will learn another way to determine groundspeed in the next lesson.

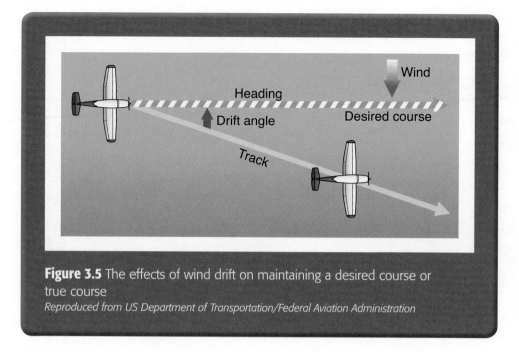

**Figure 3.5** The effects of wind drift on maintaining a desired course or true course

*Reproduced from US Department of Transportation/Federal Aviation Administration*

### Wing TIPS

Charles Lindbergh planned to navigate by dead reckoning on his famed trip across the Atlantic Ocean to Europe. A naval officer attending a lecture given by Lindbergh before the flight asked him, "Suppose you strike a wind change in the night, and it drifts you far off course?" Lindbergh replied, "A navigating error wouldn't be too serious; this flight isn't like shooting for an island. I can't very well miss the entire European coast." His audience laughed.

The direction in which an aircraft points as it flies is known as *heading*. An aircraft's path over the ground, a combination of the aircraft's motion and the air's motion, is its track. *The angle between the heading and the track* is the drift angle. If the heading and true course agree but the wind is blowing from the left, then the track won't follow the true course. Instead the wind causes the aircraft to drift to the right so the track moves to the right of the true course (Figure 3.5).

### How to Find the Compass Heading

A pilot follows a series of steps to find his or her compass heading. The first step is to find the true course (TC) from the chart. The pilot then applies wind correction for a true heading (TH). He or she then follows the formula (TH ± magnetic variation [V] = magnetic heading MH ± compass deviation [D] = compass heading [CH]) (Figure 3.6).

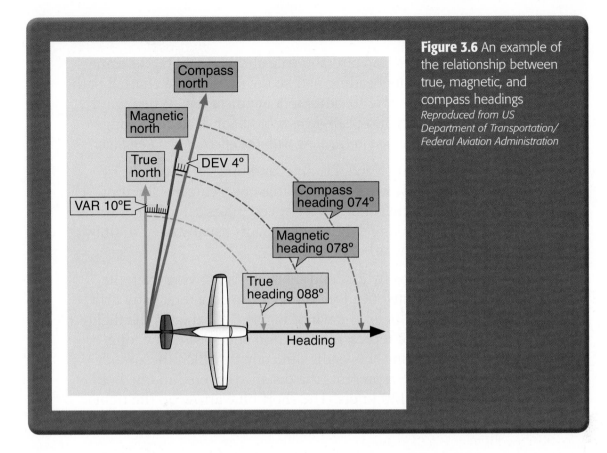

**Figure 3.6** An example of the relationship between true, magnetic, and compass headings
*Reproduced from US Department of Transportation/ Federal Aviation Administration*

By figuring out the amount of drift, a pilot can counteract the wind's effect and make the aircraft's track agree with the desired course. If the air mass moves across the course from the left, the aircraft drifts to the right. The pilot then corrects for this wind drift by pointing the aircraft to the left a certain number of degrees. This is the wind correction angle, which is *the angle between an aircraft's desired track and the heading needed to keep the aircraft flying over its desired track.* It is expressed in terms of degrees right or left of true course (Figure 3.7).

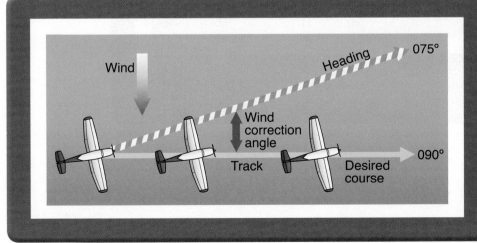

**Figure 3.7** Establishing a wind correction angle that will counteract wind drift and maintain the desired course
*Reproduced from US Department of Transportation/ Federal Aviation Administration*

## The Wind Triangle and Its Application in Air Navigation

When flying with no wind, an aircraft's true airspeed will be the same as the groundspeed and its heading will be the same as the charted path along the ground. But it's rare to run into a windless day. When wind is present, pilots use the wind triangle as a basis for dead reckoning. The wind triangle is *a method for dealing with the effect of wind on flight*.

A pilot can find the groundspeed, heading, and time for any flight by using the wind triangle. It works for a simple cross-country flight or a very complex instrument flight. Experienced pilots can often estimate calculations for a wind triangle, but rookie pilots usually need to draw diagrams to understand wind effects.

Imagine a pilot is going to fly east and the wind is blowing from the northeast. The aircraft must head north of east to counteract any drift from the intended path due to the wind. Figure 3.8 explains the flight with a diagram. Each line represents direction and speed. The long blue-and-white hashed line shows the direction in which the aircraft is heading, and its length represents the distance covered in one hour based on an airspeed of 120 knots. The short blue arrow on the right shows wind direction, and its length represents wind velocity for

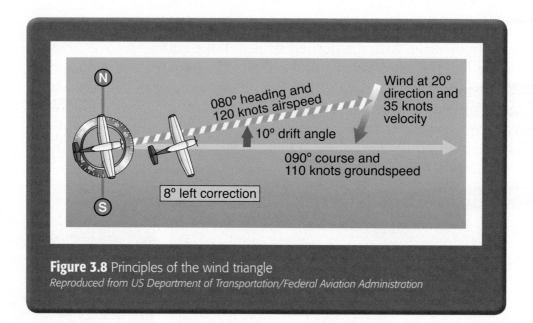

**Figure 3.8** Principles of the wind triangle
*Reproduced from US Department of Transportation/Federal Aviation Administration*

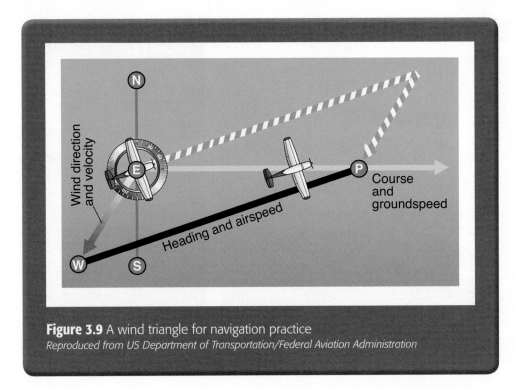

**Figure 3.9** A wind triangle for navigation practice
*Reproduced from US Department of Transportation/Federal Aviation Administration*

one hour. The solid yellow line shows the direction of the aircraft's track, and its length represents the distance traveled in one hour, that is, the groundspeed.

While Figure 3.8 explains the wind triangle's principles, pilots actually diagram their wind triangle in a slightly different way. The next hypothetical flight will travel east, from point E to point P (Figure 3.9). First the pilot draws a line on an air navigation chart to connect the two points, then measures its direction with a protractor in reference to a meridian. A protractor is *a semicircular tool for measuring angles; it is also known as a plotter.* This is the true course, which in this example is 90 degrees, which is due east. The wind will blow at 40 knots from the northeast (45 degrees) at the flight's intended altitude.

Now the pilot takes a clean sheet of paper and draws a vertical line from north to south. To calculate heading and groundspeed, the pilot follows these steps using a protractor and a ruler (Figure 3.10):

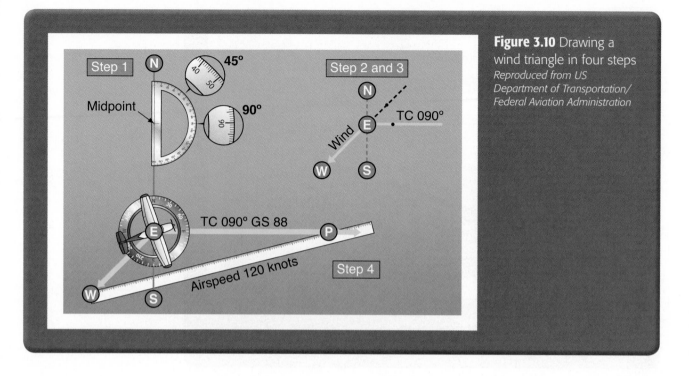

**Figure 3.10** Drawing a wind triangle in four steps
*Reproduced from US Department of Transportation/ Federal Aviation Administration*

1. The pilot places the protractor's straight edge on the vertical north-south line and its curved edge facing east. He or she places the straight edge's center point at the point of departure, marks a dot, and labels it *E*. At the curved edge, the pilot makes a dot at 90 degrees. This mark indicates the direction of the true course. He or she makes another dot at 45 degrees to indicate wind direction.

2. The pilot then uses the protractor's straight base to draw the true course line from E. He or she extends it beyond the dot at 90 degrees and labels it *TC 090.°*

3. Next the pilot lines up the ruler with E and the dot at 45 degrees. He or she draws a line downward at 45 degrees from E, not toward the dot at 45 degrees. This line represents the direction the wind is blowing (downwind). The pilot makes it 40 units long to stand for the wind velocity of 40 knots. It's best to use simple units, such as ¹/₄ inch equals 10 knots. Once the pilot has chosen a scale (like the map scale you read about in the previous lesson), he or she must continue to use this same scale for each linear movement in this particular diagram. He or she identifies the wind line by placing the letter W at the end to show wind direction.

4. Now the pilot measures 120 units on the ruler for airspeed and marks it with a dot on the ruler. He or she places the ruler so its end is on the W and the 120-knot dot crosses the true course line. The pilot draws a line between these two points and labels it *AS 120* for an airspeed of 120 knots. The point P at the intersection represents the aircraft's position at the end of one hour. The diagram is complete.

**CHAPTER 4** Flying From Here to There

With this diagram a pilot can determine the distance flown in one hour (the groundspeed) by looking at the number of units on the true course line (88 nautical miles per hour, or 88 knots). The true heading needed to offset drift is indicated by the direction of the airspeed line, which can be determined in one of two ways:

- The pilot places the protractor's straight side along the north-south line. Its center point should be at the intersection of the airspeed line and the north-south line. He or she then looks at the protractor to find the true heading directly in degrees (76 degrees) (Figure 3.11).

- The pilot places the protractor's straight side along the true course line with its center at P (Figure 3.12). He or she reads the angle between the true course and the airspeed line. This is the wind correction angle. The pilot then applies the wind correction angle to the true course to find the true heading. If the wind is blowing from the right of the true course, he or she adds the angle to the true course. If the wind is coming in from left of the true course, the pilot subtracts the angle from the true course. In this example, the wind correction angle is 14 degrees and the wind is from the left. So the pilot subtracts 14 degrees from the true course of 90 degrees to get the true heading of 76 degrees.

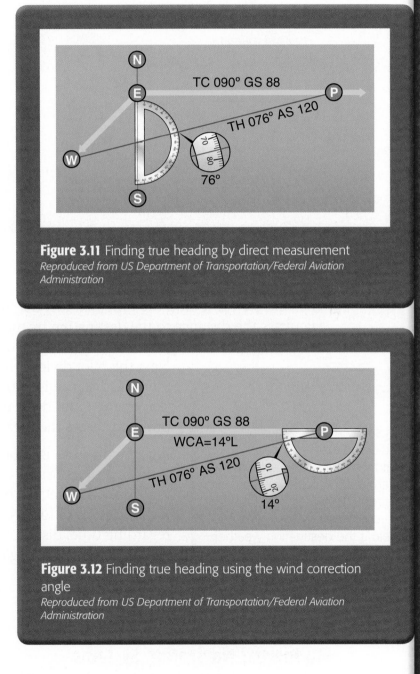

**Figure 3.11** Finding true heading by direct measurement
*Reproduced from US Department of Transportation/Federal Aviation Administration*

**Figure 3.12** Finding true heading using the wind correction angle
*Reproduced from US Department of Transportation/Federal Aviation Administration*

Once the pilot has the true heading, he or she applies corrections for variation to find the magnetic heading, and applies deviation to find the compass heading. The pilot then uses the compass heading to fly to the destination by dead reckoning.

## Time and Fuel

The next step is to find the time and fuel the flight will require. First the pilot calculates the distance to the destination by measuring the length of the course line drawn on the air navigation chart. He or she uses the scale located on the chart.

Next the pilot divides the distance by the groundspeed. So if the distance is 220 nautical miles, the pilot divides that by the groundspeed of 88 knots. This simple division problem shows the flight will take 2.5 hours.

Figuring fuel use is another easy math problem to solve. If the aircraft consumes eight gallons an hour, the pilot multiplies the fuel use by the flight time, or 8 × 2.5 for 20 gallons of fuel. A pilot can make many calculations regarding time, distance, speed, and fuel consumption using a tool called a *flight computer*, which you'll read about in Chapter 4, Lesson 5.

### Wing TIPS

As a safety measure, a pilot should add extra fuel beyond what the flight seems to demand. A standard fuel reserve would be about 10 percent, plus the fuel required to fly to a suitable airfield if the destination weather is questionable.

## How the Principles of Dead Reckoning Relate to Inertial Navigation Systems

A more modern way to compute dead reckoning without the protractor, ruler, and paper diagram is the inertial navigation system (INS). The INS is *a computer-based navigation system that tracks an aircraft's movements with signals produced by onboard instruments.* The system is fully self-contained so it doesn't call for any input—such as radio signals—from outside the aircraft.

A pilot starts the INS before taking off. The INS can take anywhere from 2.5 to 45 minutes to warm up. The pilot enters the aircraft's exact location (latitude and longitude) and the aircraft's destination on the ground into the system. The system uses these starting coordinates to then track the aircraft's progress and update position. A pilot can also program the INS with waypoints—*geographical locations in latitude and longitude along a route to mark a route and follow an aircraft's advance along that desired route.*

**Wing TIPS**

The inertial navigation system tells a pilot the aircraft's track, drift angle, distance, and flight time.

## Parts of the INS

One of the main parts of an INS is the accelerometer. An accelerometer is *an instrument that measures acceleration in one direction.* It calculates velocity using acceleration readings and time en route. A typical INS contains three accelerometers to record three different movements: north-south, east-west, and up-down.

An accelerometer has two components: one part moves and the other part doesn't. Between them they generate a magnetic field. The moving part mirrors the aircraft's movements, and whenever it changes position along with the aircraft this sends a signal via the altered magnetic field to the INS. The onboard system then calculates exactly how much the aircraft's position has changed.

The gyroscope is another important part of inertial navigation systems. It is *an instrument with a wheel spinning about a central axis that helps with stability.* The gyroscope measures direction in the INS.

The gyroscope's spinning wheel and axis sit inside a number of gimbals—or rings— on top of a base. This setup gives the spinning wheel freedom of movement. As an aircraft maneuvers through the air, the wheel can pitch, roll, and yaw without restraint within the gimbals. For example, an aircraft may point its nose up and the gyro's fixed base will tilt up along with it, but the freely moving spinning wheel will turn about within the rings so that it always maintains its same attitude. In other words, the gyroscope remains in the plane in which it was originally spinning (Figure 3.13).

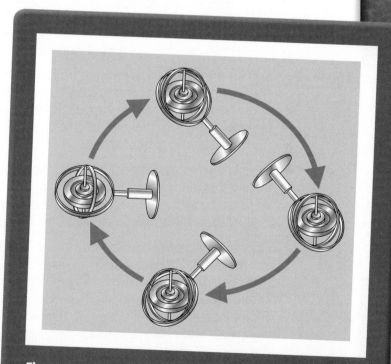

**Figure 3.13** Regardless of the position of its base, a gyroscope tends to remain rigid in space, with its axis of rotation pointed in a constant direction.
*Reproduced from US Department of Transportation/Federal Aviation Administration*

More recent versions of the INS use *ring laser gyroscopes*—gyroscopes that operate with lasers. This change means that gyroscopes and accelerometers no longer have to stay fixed with respect to the horizon and true north. More-powerful computers in these new systems are able to calculate corrections for the horizon and true north. The INS measures change in aircraft position in nautical miles per hour.

## System Errors

Time is an inertial navigation system's greatest enemy. The system figures out position by starting with the accurate position input by the pilot while still on the ground. The position updates continuously as the accelerometers and gyroscopes provide speed and direction data. But accelerometers and gyroscopes make small errors that add up to bigger errors over the course of a long flight.

### The Shootdown of KAL 007

The human element in navigation remains critical, even in these days of electronic systems. Possibly the most tragic navigational error in aviation history occurred on 1 September 1983, when a Korean Air Boeing 747 strayed into Soviet airspace. A Soviet fighter jet shot it down, killing all 269 passengers and crew, including an American member of Congress. The incident led to an international showdown between the Soviet Union and the United States at a tense moment in the cold war.

Korean Air Flight 007 was flying from New York City to Seoul, South Korea, via Anchorage, Alaska. Before taking off, the crew incorrectly set the autopilot system, which gave incorrect information to the inertial navigation system. This caused the aircraft to deviate several degrees west of its planned course over international airspace. It entered Soviet airspace over the islands north of Japan.

In violation of international air rules, the Soviet air force shot the plane down instead of forcing it to land. International public opinion was outraged, and the commercial airlines of several countries refused to fly to the Soviet Union for two weeks. The International Civil Aviation Organization launched an investigation.

Over the following years, several changes were made to prevent a recurrence of the incident. The autopilot system for large airliners was redesigned to prevent a similar misprogramming. The United States decided to use military radars to help with air traffic control near Alaska. In addition, the United States, the Soviet Union, and Japan established a joint air traffic control system in the North Pacific, ensuring that the three countries' controllers were always in touch with each other. Finally, President Ronald Reagan ordered the military's global positioning system be made available for civilian air traffic use.

The best inertial navigation systems may be off by 0.1 to 0.4 nautical miles over the course of a four- to six-hour flight over the Atlantic Ocean, for example. Less-expensive, smaller models might be off by one to two nautical miles in just one hour. However, the less-accurate systems are fine if they are used with a global positioning system (GPS). You'll read more about the GPS in Chapter 4, Lesson 5.

In this lesson you've read about a method of navigation called dead reckoning. To calculate position, heading, and groundspeed, it requires time, airspeed, distance, and direction. Wind also greatly affects navigation. Therefore pilots usually have to include wind effect in their navigation computations. The next lesson will cover four very important flight instruments that aid with navigation. These particular tools help a pilot figure out airspeed, altitude, direction, and attitude.

# ✔ CHECK POINTS

## Lesson 3 Review

Using complete sentences, answer the following questions on a sheet of paper.

1. What plays a critical role in dead reckoning calculations?

2. When does dead reckoning work best and when does it grow ever more inaccurate?

3. When do pilots use the wind triangle?

4. What can a pilot find for any flight by using the wind triangle?

5. When does a pilot start the inertial navigation system?

6. How does an accelerometer calculate velocity?

## APPLYING YOUR LEARNING

7. An aircraft is at Lambert-St. Louis International Airport. The pilot wants to fly south to Louis Armstrong New Orleans International Airport. The distance is about 525 nautical miles or 605 statute miles. The aircraft will fly at 180 knots. A mass of air is blowing south at 30 knots. The pilot plans to take off from Lambert at 9:30 a.m. The aircraft gets 10 gallons per hour. What is the groundspeed, estimated time en route, and estimated time of arrival, and how much fuel will the pilot need?

# LESSON 4

## Flight Instrumentation

### Quick Write

_____

_____

Since the Miracle on the Hudson, Captain Sullenberger has become a hero to many. Why do you think that is?

### Learn About

- the functions of airspeed indicators
- the functions of the altimeter
- the functions of a horizontal situation indicator
- the functions of attitude indicators

About two minutes after US Airways Flight 1549 took off from New York's LaGuardia Airport on 15 January 2009, a flock of geese struck the engines. The Airbus A320 lost thrust. The pilot, Captain Chesley "Sully" Sullenberger, had to make some quick decisions. As he later said, "Not only did we not have time to go through a ditching checklist, we didn't have time to even finish the checklist for loss of thrust in both engines."

Captain Sullenberger had 155 passengers and crew onboard. He ran through a few options in his head—try to return to LaGuardia Airport, try to reach Teterboro Airport in New Jersey, or ditch the aircraft in the Hudson River. To ditch is *to land an aircraft on water after a loss of power or some other emergency*. The decision depended on the aircraft's "energy state," Sullenberger said, that is, its ability to reach one of these targets.

In an interview with *Air & Space* magazine, the captain said, "I knew the altitude and speed were relatively low, so our total energy available was not great. I also knew we were headed away from LaGuardia, and I knew that to return to LaGuardia I would have to take into account the distance and the altitude necessary to make the turn back." He further concluded Teterboro was too far away even though it was in the right direction. That meant only one option remained: the Hudson River.

The navigation elements Sullenberger cites—altitude, airspeed, direction, and distance—all played a role in his decision making. The information would have been available from the flight instruments in the cockpit and through the view out his window.

One of his first actions was to take over control of the aircraft from First Officer Jeff Skiles. Sullenberger was more familiar with the A320 while Skiles had just been through training to handle emergencies. So the captain flew while Skiles attempted to restart the engines.

Fortunately Sullenberger had lots of flying experience to draw from. He had graduated from the US Air Force Academy in Colorado in 1973, where as a cadet he put in more than 1,000 flight hours as an Academy glider instructor pilot. That number of flying hours is generally unheard of for a cadet. He's a former Air Force F-4 Phantom fighter pilot. He was a pilot for nearly 30 years with US Airways.

Sullenberger had also involved himself during his career in improving flight safety. He was safety chairman of the Air Line Pilots Association (ALPA), the airline pilots' union. He spent many hours volunteering through ALPA as an accident investigator for the National Transportation Safety Board as it looked into two California airplane crashes. He also took part in US Air Force accident investigations. He co-wrote a paper on flight with NASA scientists. In 2007 he founded a consulting firm that helps business, government, the aviation industry, and others improve their safety records, among other areas. In short, Sullenberger

## Vocabulary

- ditch
- airspeed indicator
- dynamic pressure
- static pressure
- indicated airspeed
- calibrated airspeed
- true airspeed
- aneroid
- indicated altitude
- altimeter setting
- true altitude
- absolute altitude
- pressure altitude
- density altitude
- horizontal situation indicator
- glide slope indicator
- lubber line

Captain Chesley "Sully" Sullenberger is known as the hero of the Hudson for his excellent piloting skills in the 2009 crash landing on the Hudson River. In April that same year, the Air Force Academy presented Sullenberger with the Col James Jabara Award for Airmanship.
*Courtesy of USAF/Mike Kaplan*

had spent many years thinking about what causes accidents and how to prevent them. "The way I describe this whole experience … is that everything I had done in my career had in some way been a preparation for that moment," the captain told *Air & Space*.

At the time Flight 1549 lost thrust, the aircraft was doing under 250 knots per hour. Its altitude was about 2,800 feet. The captain said, "our speed began to decay very rapidly because the nose was still up in a climb attitude, but without climb thrust on the airplane." He then pushed the nose down to "maintain our best lift-over-drag airspeed." As soon as the captain and Skiles decided to ditch, they had to slow the plane enough to deploy the flaps. So they raised the nose. These navigation decisions, including changes in attitude, took place in the space of about three minutes.

As a result of the captain's and copilot's efforts, the ditch was successful and everyone aboard survived and was able to evacuate. Ferries as well as boats from the fire department and Coast Guard rushed to the rescue. The news media hailed the happy ending the "Miracle on the Hudson."

Passengers wait for rescue on the wings of Flight 1549's Airbus 320. Because everyone survived the crash landing on the Hudson River, the 2009 incident is often hailed as the "Miracle on the Hudson."
© Brendan McDermid/Reuters

## The Functions of Airspeed Indicators

To safely fly any aircraft, a pilot needs to understand how to interpret and operate the flight instruments, as Captain Sullenberger's experience shows. A pilot must also know how to recognize when a device is displaying flawed data. For example, in Chapter 4, Lesson 2, you read about correcting magnetic compass readings for variation and deviation. Once a pilot is comfortable working with flight instruments, he or she will be able to fly confidently and put the equipment to the best possible use.

Flight instruments feed a pilot lots of important information, from how fast the plane is moving to its attitude. When used properly they keep the aircraft on the desired route.

An airspeed indicator is one such tool. It is *an instrument in the cockpit that displays an aircraft's airspeed, typically in knots.* An airspeed indicator determines airspeed by measuring the difference between two types of pressure. One of these is dynamic pressure, which is *a type of pressure that is present when an aircraft is in motion.* Therefore, it is a pressure due to motion. Wind also generates dynamic pressure. It doesn't matter if an aircraft is moving through still air at 70 knots or if the aircraft is facing a wind that is blowing at 70 knots. Each generates the same dynamic pressure.

The other type of pressure used to calculate airspeed is static pressure, which is also known as *ambient pressure, a type of pressure that is always present whether an aircraft is moving or at rest.* Static pressure is, in other words, the barometric pressure in the air surrounding the aircraft.

An airspeed indicator gets its data from a system called the *pitot-static system* that uses static air pressure and dynamic pressure to calculate airspeed (Figure 4.1). The system includes a pitot tube— which you read about in Chapter 2, Lesson 5—and a static port. The pitot tube captures dynamic as well as static pressure. The static port captures only static pressure.

The pitot tube has an opening in front where the dynamic and static pressures— referred to together as *total pressure*— enter a pressure chamber. The pressures are a result of *ram air*—the air from the airstream generated as the aircraft moves or wind blows—striking the opening.

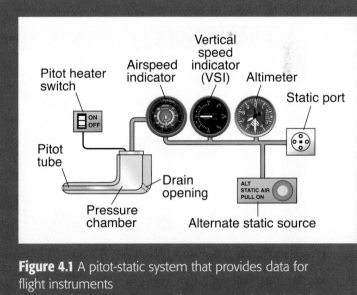

**Figure 4.1** A pitot-static system that provides data for flight instruments
*Reproduced from US Department of Transportation/Federal Aviation Administration*

A small hole at the back of the chamber called a *drain hole* lets moisture from precipitation drain from the system. Pilots must keep both holes clear of anything that might block them, such as insects and ice. Otherwise, the system could malfunction. A pitot heater is vital in icy conditions.

## Wing TIPS

If a static port gets blocked, some aircraft include an *alternate static source* to provide static pressure. However, the alternate may cause an airspeed indicator to show a greater airspeed than the actual airspeed.

The pitot tube's pressure chamber sends the total pressure to the airspeed indicator through a small tube. The static port sends static pressure to the airspeed indicator as well. The two static-pressure readings cancel each other out, leaving the airspeed indicator with only a dynamic pressure reading. When the dynamic pressure changes, the airspeed indicator shows either an increase or decrease.

The total and static pressures are equal when an aircraft is parked on the ground in calm air. But when an aircraft moves through the air, the pressure on the pitot line becomes greater than the pressure on the static lines. A pointer on the airspeed indicator's face registers this pressure difference in knots per hour, miles per hour, or both.

An airspeed indicator is actually the only flight tool that uses both the pitot and static systems (Figure 4.2). Static pressure enters the airspeed instrument through a static air line. Pitot pressure enters into a part called the *diaphragm*. The dynamic pressure expands or contracts one side of the diaphragm. The movement sets in motion other parts of the airspeed indicator—the long lever, sector, and handstaff pinion—which then move the airspeed needle accordingly. By reading the needle, a pilot knows how fast the aircraft is moving.

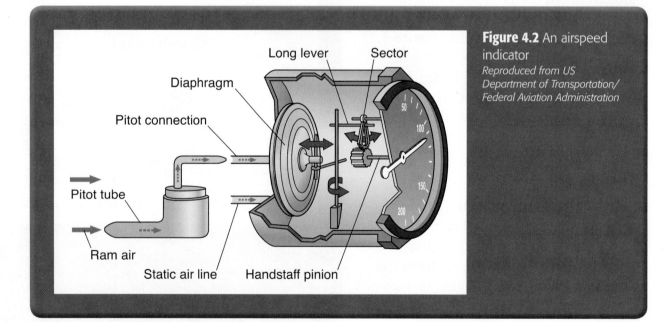

**Figure 4.2** An airspeed indicator
*Reproduced from US Department of Transportation/ Federal Aviation Administration*

## Types of Airspeed

Pilots work with several types of airspeed. They should be familiar with them all. The airspeeds are:

- *Indicated airspeed*—*The airspeed shown on the airspeed indicator.* This direct reading is uncorrected for changes in atmospheric density, installation error, and instrument error. Two important manuals that pilots use—the airplane flight manual and the pilot's operating handbook, which are published by the manufacturer and different for each type of aircraft—list takeoff, landing, and stall speeds in terms of indicated airspeed.

- *Calibrated airspeed*—*The indicated airspeed corrected for installation error and instrument error.* Readings may be off by several knots at certain airspeeds and flap settings. Errors are greatest at low airspeeds. But indicated airspeed and calibrated airspeed generally agree at cruising speed and higher airspeeds. A pilot makes corrections to the indicated airspeed with the help of an airspeed calibration chart.

- *Equivalent airspeed*—As a plane flies faster than 200 knots, the air being squeezed into the pitot tube becomes more compressed. This leads to an error that indicates the plane is flying faster than it really is. This error grows particularly large near the speed of sound (Mach 1). Pilots can correct for the error using a chart to find the *equivalent airspeed* (EAS)—the speed at sea level that would produce the same dynamic pressure as the true speed at the altitude the plane is flying.

- *True airspeed*—*The calibrated airspeed corrected for altitude and temperature.* Because air density decreases with an increase in altitude, a pilot has to fly an aircraft faster at higher altitudes to create the same difference in pressure between the pitot pressure and the static pressure. So for a given calibrated airspeed, true airspeed increases as altitude increases. A pilot can figure out true airspeed in two ways. The most accurate method is to use a flight computer. The computer calculates the true airspeed when the pilot enters the calibrated airspeed, pressure altitude, and temperature. The second method gives the pilot a ballpark true airspeed. The pilot just adds 2 percent to the calibrated airspeed for every 1,000 feet gained in altitude. The true airspeed is important because the pilot uses it for flight planning and when filing a flight plan.

- *Groundspeed*—An airplane's actual speed over the ground. It is true airspeed adjusted for wind. The groundspeed decreases with a headwind and increases with a tailwind.

As an example of how all this works together, say a plane is flying at 10,000 feet above sea level with an indicated airspeed of 250 knots and a 50-knot tailwind. What is the aircraft's groundspeed? (This example assumes that indicated airspeed = calibrated airspeed.) To find the true airspeed, the pilot adds 5 knots per each

1,000 feet of altitude (2 percent of 250 knots indicated airspeed per 1,000 feet = 5 knots per 1,000 feet) or 50 knots to the indicated airspeed. This gives a true airspeed of 300 knots. The pilot then applies the tailwind speed (50 knots) to arrive at a groundspeed of 350 knots. Thus, while the indicated airspeed in the cockpit is 250 knots, the actual speed across the ground is 350 knots—fully 100 knots higher.

## Airspeed Indicator Markings

The airspeed indicator has color-coded markings. The color system lets a pilot see at a glance whether the aircraft is exceeding a safe airspeed. Some colors indicate danger or that the aircraft is approaching a dangerous speed. For example, if while the aircraft is performing a certain maneuver the airspeed needle is in the yellow arc but quickly nearing the red line, the pilot should immediately reduce airspeed.

The following are some standard color-coded markings (Figure 4.3):

- **White arc**—An aircraft's operating range for approaches and landings. Pilots generally operate their flaps within this arc.

- **Green arc**—An aircraft's normal operating range. Most flying takes place within this range.

- **Yellow arc**—A range in which a pilot should be cautious. Pilots should fly within this range only in smooth air, and even then, only with caution.

- **Red line**—A pilot should never exceed this speed. Doing so may damage an aircraft or even cause structural failure.

### Wing TIPS

Before takeoff the airspeed indicator should read zero. However, the device may read higher than zero if strong winds are blowing into the pitot tube. If this is the case the pilot must make sure his or her airspeed increases at an appropriate rate as he or she begins the takeoff.

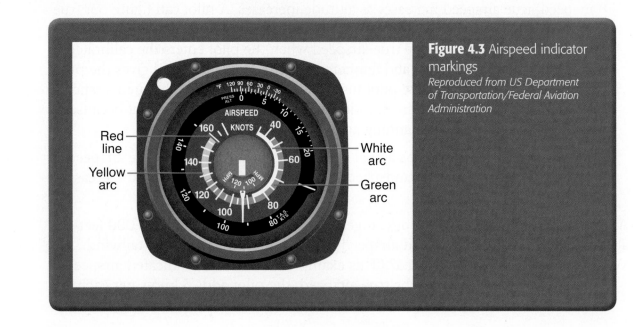

**Figure 4.3** Airspeed indicator markings
*Reproduced from US Department of Transportation/Federal Aviation Administration*

## The Functions of the Altimeter

An aircraft's altitude is another important piece of navigation data. The aircraft's altimeter is able to gauge altitude by sensing pressure changes. Recall that pressure decreases with altitude. Because the altimeter is the only navigation instrument that can indicate altitude, it is one of an aircraft's most important devices.

A common type of altimeter is the pressure altimeter—so named because it reacts to changes in atmospheric pressure. One of the main parts of this type of altimeter is a stack of *sealed aneroid wafers*; an aneroid is *a sealed flat capsule made out of thin metal disks and emptied of all air* (Figure 4.4). In fact this type of altimeter is also known as an *aneroid barometer*. The pressure inside the sealed wafers is 29.92 inches of mercury. A static port captures static pressure and introduces it into the sealed altimeter case. The wafers expand or contract because of changes in the static pressure in the surrounding atmosphere.

**Wing TIPS**

A *vertical speed indicator* (pictured in Figure 4.1) is a flight instrument that indicates whether an aircraft is climbing, descending, or in level flight. Like the altimeter, it uses a static port to capture static pressure.

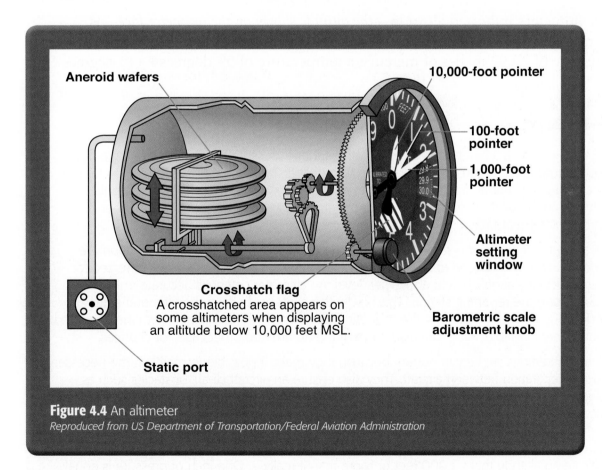

**Figure 4.4** An altimeter
*Reproduced from US Department of Transportation/Federal Aviation Administration*

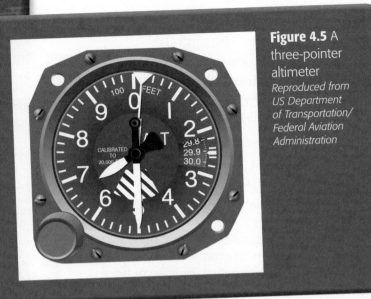

**Figure 4.5** A three-pointer altimeter
*Reproduced from US Department of Transportation/ Federal Aviation Administration*

A static pressure higher than 29.92 inches of mercury presses down on the wafers and they contract. A lower static pressure lets the wafers expand. Mechanical parts connect the wafers' movements to needles on the altimeter's face. When the wafers contract, the needles show a decrease in altitude. When the wafers expand, the needles indicate an increase in altitude.

Altimeters show altitude in terms of feet or meters above a selected pressure level. Some altimeters have only one needle to show altitude. Others have two or more pointers (Figure 4.5). The typical altimeter face features numbers arranged clockwise from zero to nine.

## Standard and Nonstandard Pressure and Temperature

An altimeter's indicated altitude—*the altitude shown on the altimeter face*—is correct only when certain standards are met. These are a sea level barometric pressure of 29.92 inches of mercury, a temperature of 59 degrees F (15 degrees C),

### Setting the Altimeter

Pilots adjust the altimeter for changes in atmospheric pressure many times during a flight, from before takeoff to just before landing. The pilot first obtains a pressure reading for the location from either air traffic control or an automated system. Then he or she inputs that setting into the altimeter's pressure scale by turning a knob or adjusting some other device on the instrument face.

An altimeter setting is *the station pressure—the barometric pressure at the location the reading is taken—reduced to sea level*. The station pressure will have been corrected for that particular station's height above sea level. An altimeter setting is accurate only in the area around the reporting station. This is why the pilot must adjust the altimeter setting while flying from one station to the next. Air traffic control will let a pilot know when updated settings are available. Pilots can also monitor the automated broadcasts for changes.

Altimeter settings are very important because they make it possible to maintain the necessary vertical separation between aircraft. They also ensure an aircraft clears obstacles such as towers and terrain such as mountains. However these settings won't always be entirely accurate at all altitudes. This is because of irregularities at higher altitudes. It is also due to the effect of nonstandard temperatures—anything above or below 59 degrees F—at altitude. For example, conditions when flying over a mountain range may cause the altimeter to indicate an altitude that's 1,000 feet or more than it really is. One inch of pressure is equal to about 1,000 feet of altitude. So even a small error in pressure, for instance, 0.25 inches of mercury, could cause an aircraft to be 250 feet from the desired altitude.

## Types of Altitude

Normally altitude refers to a height above sea level. This is how it is used to describe airspace and obstacles as well as separate air traffic. However, altitude also refers to the height above any reference level. Pilots use five types of altitude:

- *Indicated altitude*—The altitude shown on the altimeter face.
- *True altitude*—*An aircraft's vertical distance above sea level.* True altitude is often expressed in feet above *mean sea level (MSL).* The elevations of airports, terrain, and obstacles on air navigation charts are true altitudes.
- *Absolute altitude*—*An aircraft's vertical distance above the terrain, that is, above ground level (AGL).*
- *Pressure altitude*—*The altitude indicated on the altimeter when the altimeter-setting window is adjusted to 29.92 inches of mercury.* A pilot needs to know pressure altitude to figure out such things as true altitude, true airspeed, and density altitude.
- *Density altitude*—*Pressure altitude corrected for changes from standard temperature.* When conditions are standard, pressure altitude and density altitude are the same. But if the temperature is above standard, the density altitude is higher than pressure altitude. If the temperature is below standard, the density altitude is lower than pressure altitude.

Density altitude is important because it directly relates to an aircraft's performance. Air density affects engine power and airfoil efficiency. When flying through lower pressures where there are fewer air molecules, performance decreases and it takes longer to get airborne during takeoffs. When flying through higher pressures where there are more air molecules, performance increases and aircraft perform better.

and a standard rate of decrease in pressure and temperature with an increase in altitude.

### Pressure

But barometric pressure and temperature generally don't remain constant throughout an entire flight. So pilots must adjust their altimeters even during the course of short flights. For example, if a pilot flies from a high-pressure area to a low-pressure area without adjusting the altimeter, the aircraft's actual height above the ground will be lower than the indicated altitude. Aviators have an old saying: "Going from a high to a low, look out below." The reverse is also true. If a pilot travels from a low-pressure area to a high-pressure area without adjusting the altimeter, the aircraft's actual altitude will be higher than the indicated altitude. If a pilot doesn't pay attention to these types of changes, flight becomes dangerous.

### Wing TIPS

The radio altimeter, also known as a radar altimeter, is another type of altimeter. It calculates an aircraft's absolute altitude (distance above the ground) by sending a signal to a receiver on the ground and measuring how long it takes that signal to return. Its main role is to provide accurate absolute altitude during approach and landing.

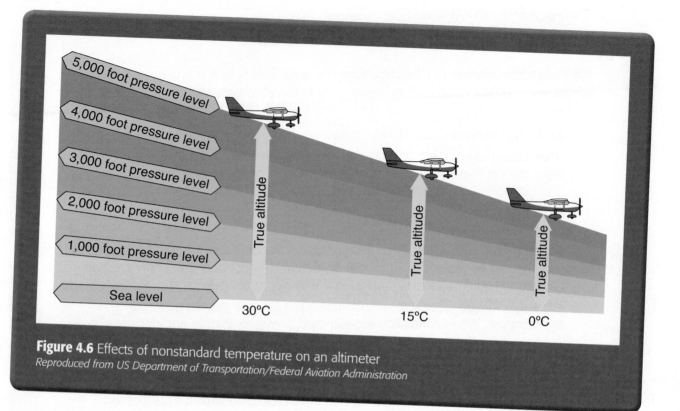

**Figure 4.6** Effects of nonstandard temperature on an altimeter
*Reproduced from US Department of Transportation/Federal Aviation Administration*

## Temperature

A pilot must also be aware of the effects of temperature changes on the altimeter. This is because cold air is denser than warm air. When flying in temperatures that are colder than standard, the actual altitude will be lower than the altimeter indicates (Figure 4.6). The size of an unadjusted altimeter's error will depend on how great the difference in temperature is from the standard. High terrain and obstacles can be a particular concern when flying through a cooler air mass. In such a case, a pilot should fly at a higher indicated altitude to be sure to clear the mountain, radio tower, or other tall obstacle.

**Wing TIPS**

Before flying, a pilot should always check how accurately the altimeter is performing. The pilot should set its barometric scale to the current reported altimeter setting sent by the local flight station. The altimeter pointers should then indicate the airport's known elevation. If the reading is off by more than 75 feet, the pilot should take the altimeter to an instrument repair shop. Flying with a faulty altimeter is too dangerous to attempt.

Extremely cold temperatures can also interfere with an altimeter's performance. As one example if the temperature is −58 degrees F (−50 degrees C), the altimeter reading will by off by as much as 300 feet for someone flying at 1,000 feet. On the other hand, when the air is warmer than standard, the aircraft will be higher than the altimeter says. The pilot can correct for temperature with the help of a computer.

## The Functions of a Horizontal Situation Indicator

Direction is another key piece of data to have when flying. A horizontal situation indicator (HSI) is *a flight instrument that indicates an aircraft's position and direction in relation to the desired route.*

This one device is able to relay a lot of different data, all of it involving an aircraft's location and direction. The information includes magnetic heading, the course the pilot has selected, the aircraft's deviation from the intended course, the runway's magnetic course, the aircraft's deviation from the intended descent to the runway, and red warning flags to indicate when the HSI isn't working properly.

The HSI combines a magnetic compass with navigation signals and a glide slope indicator to find out all of these things and more. A glide slope indicator is *an instrument that vertically guides an aircraft on its final approach to a runway.* One of the main features on an HSI's face is the *compass card.* The compass card shows direction from zero to 360 degrees. It also includes markings to show north, east, south, and west. At the face's center is the figure of an airplane. This symbolic aircraft doesn't move. It stands for the aircraft's position in relation to the desired route. The compass card rotates freely about the fixed airplane.

Figure 4.7 shows an HSI with some sample settings. The lubber line—*a fixed line on a compass or other flight instrument aligned with an aircraft's longitudinal axis*—shows that the aircraft's magnetic heading is 184 degrees. A hypothetical pilot working this HSI has set the *course select pointer*, the desired course, to 295 degrees. He or she entered the course by turning the *course select knob.* The knob rotated the course select pointer until it pointed to the desired course on the compass card.

**Figure 4.7** A horizontal situation indicator
*Reproduced from US Department of Transportation/Federal Aviation Administration*

The HSI also shows how much an aircraft has strayed from its intended course. Navigation aids send signals to the HSI that the flight instrument uses when figuring out an aircraft's position. A *course deviation bar* illustrates with a yellow line through the fixed aircraft on the face how much the aircraft has wandered to the left or right of the selected course. The *to/from indicator* tells the pilot whether the aircraft's current heading will lead to or away from the intended fix or destination.

The *glide slope pointers* show the aircraft's distance above or below the desired glide slope as the plane descends to a runway. When a plane is above its desired glide slope, the pointers deflect down. When an aircraft is below its desired glide slope, the pointers deflect up.

An HSI also includes two warning flags when things aren't working as they should. The red *NAV warning flag* alerts the pilot that the HSI is no longer receiving the navigation signals it needs to supply direction and position. The red *compass warning flag* informs him or her that the compass card is malfunctioning. If all else fails a pilot can fall back on a magnetic compass.

## The Functions of Attitude Indicators

Speed, height, and direction are all very important for safe flight. But knowing the aircraft's attitude is of utmost importance. The attitude indicator—a flight instrument that displays an aircraft's attitude—is a primary flight instrument.

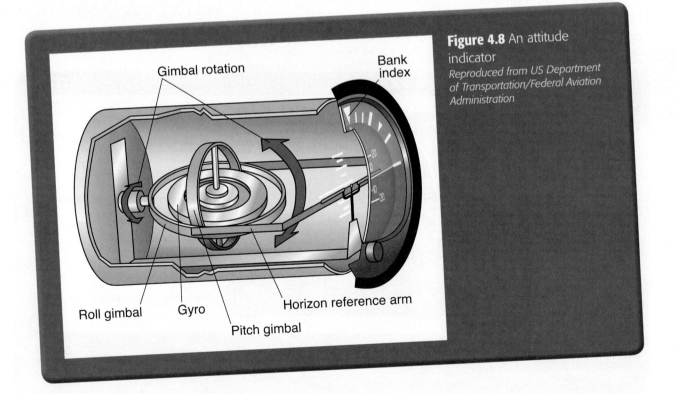

**Figure 4.8** An attitude indicator
*Reproduced from US Department of Transportation/Federal Aviation Administration*

While airspeed indicators, altimeters, and horizontal situation indicators are also primary flight instruments—that is, *essential* to flight—the attitude indicator is the most realistic primary flight instrument in the cockpit. It shows a very close approximation of the aircraft's actual attitude. It can alert the pilot to the slightest change in attitude.

In addition to relaying the aircraft's relationship with the horizon, an attitude indicator can also alert the pilot to possible changes in airspeed, altitude, and direction. Nose down can mean the aircraft is gaining airspeed. Nose up and it may be gaining altitude. If the aircraft banks, it may be changing direction or altitude or both. The reverse can also be true—a change in speed, altitude, or direction can also serve to alert the pilot to a change in orientation.

An attitude indicator (Figure 4.8) works according to gyroscopic rules. An attitude indicator features a fixed aircraft on its face and a freely moving horizon bar. The horizon bar represents the true horizon. The miniature aircraft's relationship to the horizon bar is the same as the real aircraft's relationship to the real horizon. If an aircraft pitches or banks, the horizon bar remains in its original horizontal plane. This is because the bar is fixed to the gyroscope so it can maintain its horizontal plane even as the aircraft moves about its lateral or longitudinal axis.

A pilot can move the fixed aircraft up or down with a knob to align it with the horizon bar to match what the pilot sees out his or her window. But normally pilots adjust the fixed aircraft so that its wings overlap the horizon bar when the real aircraft is in straight-and-level cruising flight.

While not all attitude indicators will behave the same, Figure 4.9 shows how to read a typical version of this flight instrument. When the aircraft is banking left, the horizon bar will tilt right. This is the same view that the pilot should see out the windshield—as the plane banks left the real horizon will appear to tilt to the right. And when the aircraft is banking right, the horizon bar will tilt left. When it's climbing, the horizon bar will fall below the fixed aircraft's wings. And when it's descending, the horizon bar will rise above the wings.

An attitude indicator generally has limits as to how much pitch and roll it can indicate. Beyond those limits, the instrument will tumble inside the case and no longer give a correct reading until the pilot realigns the fixed aircraft with the artificial horizon. Banking limits are usually from 100 to 110 degrees. Pitch limits are from about 60 to 70 degrees.

This lesson described four primary flight instruments and their uses. They are the airspeed indicator, altimeter, HSI, and attitude indicator. They tell a pilot the aircraft's airspeed, altitude, direction, and attitude. These are four key navigation elements. The next lesson will look at advances in navigation tools. These include radio aids and computers. It will also take you through how the Air Force uses GPS and INS to fulfill its duties in the air.

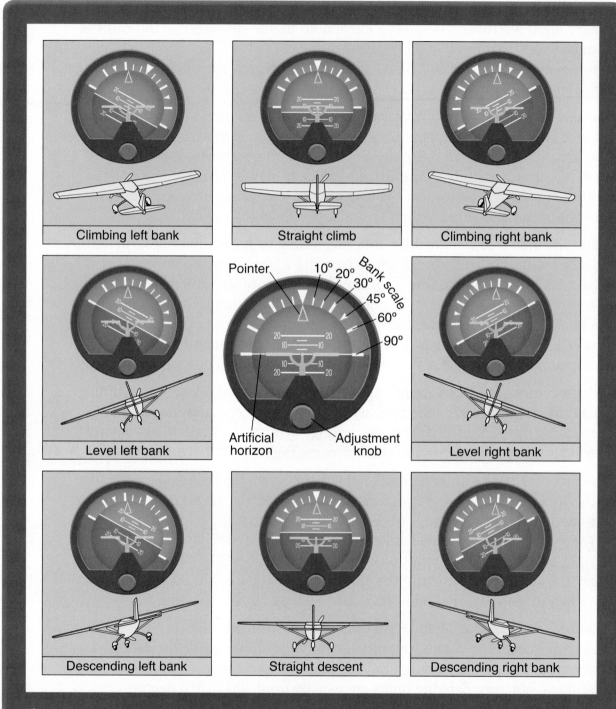

**Figure 4.9** An attitude indicator shows attitude in several situations. This attitude matches the aircraft's relation to the real horizon.
*Reproduced from US Department of Transportation/Federal Aviation Administration*

# ✔CHECK POINTS

## Lesson 4 Review

Using complete sentences, answer the following questions on a sheet of paper.

1. Where does an airspeed indicator get its data from?

2. If the airspeed needle is in the yellow arc but quickly nearing the red line, what should a pilot do?

3. How is an altimeter able to gauge altitude?

4. What two things generally don't remain constant throughout an entire flight?

5. What is a horizontal situation indicator?

6. What does a course deviation bar illustrate?

7. What does an attitude indicator's horizon bar represent?

8. When the aircraft is banking left, which way will the horizon bar tilt?

# APPLYING YOUR LEARNING

9. A pilot is in a descending left bank, turning from north to west. What effect will this have on the four primary flight instruments?

# LESSON 5

## ✈ Navigation Technology

### Quick Write

_____

_____

Can you think of any ways that GPS jammers could be useful?

### Learn About

- the uses of the plotter
- the uses of the dead reckoning computer
- how and when to use radio aids to navigation
- current developments in navigation technology
- how the Air Force uses GPS and inertial navigation in air operations
- the purpose of computer flight-planning tools

The FAA had a mystery to solve at Newark International Airport in 2009. The federal agency was just setting up a GPS system at the airport to help aircraft during landings. It was part of a new air traffic control technology called NextGen, which you'll read more about later in this lesson. GPS ground stations would receive satellite signals they could then feed to aircraft about to land. But something was jamming the signals.

Deepening the mystery was that whatever was interfering with the GPS radio signals did so randomly and usually only on weekdays. If aircraft were to rely on this navigation technology, however, it had to work reliably and at all times.

The FAA launched an investigation. It took several months, but by spring 2010 the agency had figured out that the source of interference was a small GPS jammer. Some companies use GPS to track their vehicles and goods—anything from rental cars to truck cargo. Anyone wanting to block this tracking ability needs a GPS jammer. GPS jammers emit *radio frequency interference*, which prevents signals from reaching GPS devices within a certain range. So they interfere not only with the intended device but with any other nearby GPS receivers, too.

The FAA found that a trucker driving along the New Jersey Turnpike used a jammer. Although these devices are generally very small—some of them plug into a car's cigarette lighter just as do GPS devices—it had enough power to obstruct air traffic GPS signals. Authorities seized the device, which is illegal. Some advisers to the federal government now warn that if airports and others are going to rely on GPS, then they must find ways to protect the technology from outside interference.

# The Uses of the Plotter

Pilots use navigation technology to draw up flight plans and follow the planned route once in the air. The equipment they use ranges from the elementary to the very modern. Some of these devices have been around for decades. Others are quite new and are still undergoing development. Each plays an important role in safely and accurately getting an aircraft from here to there.

Like the navigational aids and flight instruments you read about in the previous lessons, navigation technology is generally used along with other tools. A pilot applies navigation technology to plot out a route on a navigational aid such as a chart. Or a pilot will use navigation technology such as a radio aid to stay true to a course, even while referring to flight instruments in the cockpit to monitor an aircraft's performance.

One of the more basic navigation technologies a pilot uses is the plotter (Figure 5.1). A plotter is a combination protractor and ruler that helps determine true course and measure distance. Pilots use plotters for flight planning. Most plotter rulers measure in both nautical miles and statute miles. They also include a sectional-chart scale on one side and a world-aeronautical-chart scale on the other.

## Vocabulary

- dead reckoning computer
- slide rule
- crosswinds
- radial
- azimuth
- hertz
- distance-measuring equipment
- tactical air navigation
- kilohertz
- relative bearing
- instrument landing system
- general aviation
- long-range navigation
- automatic dependent surveillance-broadcast
- multi-function display
- primary flight display
- situational awareness
- National Airspace System
- smart bomb
- flight management computer

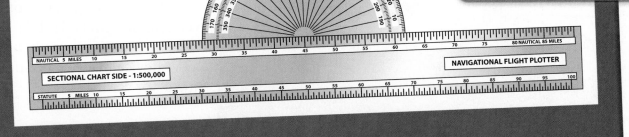

**Figure 5.1** A plotter
*Reproduced from US Department of Transportation/Federal Aviation Administration*

You first read about the plotter in Chapter 4, Lesson 2. The scene set up in that lesson calculated true course—a route's angle relative to true north—for a short hop between two airports in Oklahoma. With the true course gathered from the plotter and chart, a pilot could then find the compass heading, which takes wind and magnetic variation into account. Chapter 4, Lesson 3 took you step-by-step through an exercise in using a protractor and ruler (the plotter) along with paper and pen to determine true course and distance.

The protractor part of a plotter includes an outer half circle and an inner half circle to indicate degrees for direction. The outer half circle marks off degrees in a counterclockwise direction from zero to 180. The inner half circle also marks off degrees from right to left from 180 to 360 degrees. These are for east-west routes.

For north-south routes, pilots use the two smaller scales on the protractor. These mark off degrees from right to left as well—from 150 to 210 degrees, and from 330 to 30 degrees. The plotter's ruler aids in determining distance. The nautical and statute mile marks on the ruler can be used for airspeed-to-distance calculations.

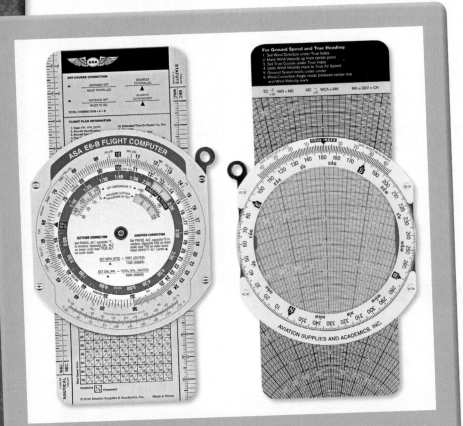

An E6B flight computer allows a pilot to figure out time, distance, speed, fuel use, and more on the slide rule shown at left, and wind correction angle on the slide rule shown at right.
*Courtesy of Aviation Supplies & Academics, Inc.*

## The Uses of the Dead Reckoning Computer

Pilots use another basic navigation tool called a dead reckoning computer— also known as a *flight computer*. This is *a device pilots use to solve equations associated with flight planning and navigation*. The types of equations a pilot can solve with a flight computer are calculations of time, distance, speed, fuel consumption, and wind drift.

The typical dead reckoning computer is not an electronic computer with a colorful monitor—although electronic versions are available. Like the plotter, it is more often a flat device made of plastic, metal, or cardboard.

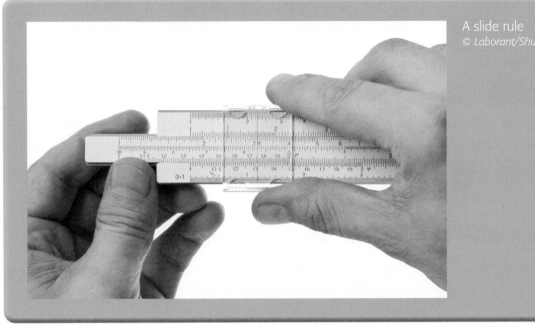

A slide rule
© Laborant/ShutterStock, Inc.

Dead reckoning computers and plotters are manual instruments. That is, pilots operate them by hand. A flight computer has two moving parts that a pilot slides and rotates about each other to make calculations. These parts include a wide ruler attached to a round component with different sets of scales on it. The scales represent such flight factors as time, speed, and temperature. The two parts combined make up a math instrument known as a slide rule—*a device with two moving parts used to make calculations*. It is a forerunner of today's electronic calculators and computers. A pilot can run the round slide up and down the ruler or spin it to make calculations.

**Wing TIPS**

Lt JG Philip Dalton (1903–1941) was an inventor. Over the course of many years, the US naval officer perfected the flight computer. His final model was the E6B Dalton dead reckoning computer, completed in 1940. British Royal Air Force pilots used one version of his invention during the Battle of Britain, also in 1940. The US Navy and Army adopted the E6B about the same time. Military and civilian pilots still use it today.

Pilots use both sides of a flight computer. One side is sometimes called the *slide rule side* and the other the *wind side*—the name depends on the flight computer's manufacturer. On the slide rule side, pilots can figure out a trip's time, distance, speed, and fuel use, among other information. This same side also helps work out problems involving changes in temperature and air pressure to find true airspeed and true altitude. If a pilot has problems with wind drift, some quick work on the flight computer will indicate which changes to make to return to the desired course. It can calculate as well the speed of headwinds, tailwinds, and crosswinds—*winds that blow at an angle to the direction of travel*.

Pilots use a flight computer's flip side to solve wind problems. Its wind side is easy to recognize because manufacturers generally print directions on how to use it at the top of the ruler. The steps require the pilot to input the true course, true airspeed, and wind data (direction and velocity). With this information the pilot can determine the aircraft's groundspeed and the wind correction angle. From the wind correction angle, he or she can figure out the true heading, the direction in which the aircraft's nose should point to reach the destination. The pilot can also reverse the process to find the wind velocity and direction. The pilot just needs to know the groundspeed, true heading, true course, and true airspeed.

## How and When to Use Radio Aids to Navigation

Radio aids are more sophisticated pieces of technology than the plotter and flight computer. They help pilots navigate with precision to almost any point desired. As you read in Chapter 4, Lesson 1, the US government developed the earliest radio navigation system during the 1920s—a four-course radio range that emitted Morse code. By the mid-1940s engineers had introduced the very high frequency omnidirectional radio range (VOR), which fed information to an aircraft's instrument panel in the cockpit.

Pilots use radio aids for many reasons, such as when the weather isn't clear enough to fly by visual flight rules alone. They also turn to them when flying at altitudes too great to follow landmarks on the ground. Certain airspaces require the use of radio aids to maintain safe distances between aircraft. Radio aids are also an important supplement to pilotage.

A radio aid is a very specific type of navigation technology. To use it, a pilot tunes in to a broadcast frequency to gather information sent by a transmitter located on the ground. Operating a radio aid requires:

- Installing a radio receiver in the aircraft

- Referring to air navigation charts that show the exact location of ground transmitting stations and their frequencies

- The ground stations themselves

- A flight deck with all of its flight instruments for the pilot to check aircraft performance.

### Types of Radio Aids

The VOR receiver finds the line or radial transmitting from a VOR station located on the ground. A radial is *an aircraft's position from a station*. The VOR receiver measures the radial in degrees clockwise from magnetic north (Figure 5.2).

A VOR does not indicate the aircraft's heading. It only tells the pilot the aircraft's direction in degrees from a chosen station. It will relay the same direction indications no matter which way the aircraft nose is pointing. The system provides the pilot with 360 courses to and from a VOR station—because there are 360 degrees in a circle, and the VOR transmitter sends signals on each and every degree from the station.

## How VOR Works

A VOR sends the pilot azimuth data—*a measurement of distance in degrees in the horizontal plane from a fixed point.* In this case, the fixed point is a VOR ground station. Picture a VOR station as the hub of a wheel and the radials as the spokes. How far the VOR station broadcasts the radials depends upon the transmitter's power but it can be as far as 200 nautical miles or more. Radials projecting from a station are assigned numbers starting with 001, which is 1 degree east of magnetic north. They progress in order through all the degrees of a circle until reaching 360 degrees. Charts generally include a compass rose at each station so pilots can orient themselves.

VOR ground stations broadcast at very high frequency, or VHF. VHF is a band of radio frequencies that falls between 30 and 300 megahertz (MHz), which is a frequency of one million cycles per second. A hertz is *the unit for measuring frequency.*

**Wing TIPS**

The parts of a word can often help you figure out what that word means. The word *omnidirectional* includes the prefix *omni*, which means "all" in Latin. So an omnidirectional range is a VHF-radio-transmitting ground station that sends radials—or straight-line courses—from the station in all directions.

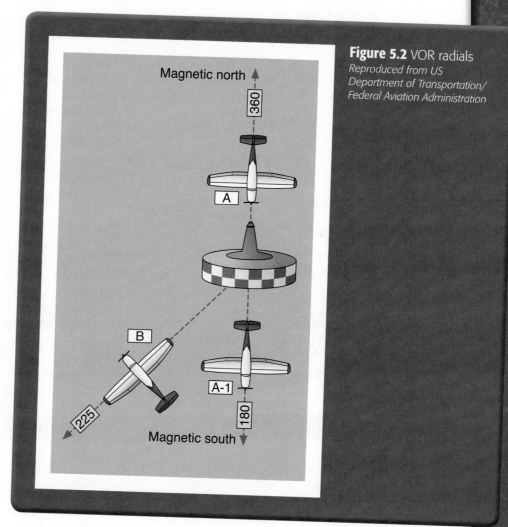

**Figure 5.2** VOR radials
*Reproduced from US Department of Transportation/ Federal Aviation Administration*

Pilots often use the acronym TIMS to help them remember the steps to take when using VOR and other radio navigation aids during an approach:

- Tune to the desired frequency
- Identify the station to be sure it's the correct one
- Monitor the station identification while navigating to continue toward the correct one
- Select the proper input into the navigation system.

VOR ground stations transmit within a VHF frequency band from 108.0 to 117.95 MHz. VHF can only broadcast signals in straight, unobstructed lines. So if a big obstacle such as a mountain or skyscraper sits in the way, the VHF signal cannot go through it. This is why VHF generally works better for aircraft at higher altitudes (Figure 5.3).

The pilot's first step when using VOR is to tune the VOR receiver to the selected VOR station's frequency. If he or she is flying directly to or from the station using a course direction indicator (CDI), the course deviation needle—or *CDI needle*—deflects either left or right of the *course index* (Figure 5.4). The needle shows the radial's position in relation to the aircraft. It moves left or right if the aircraft drifts from the radial.

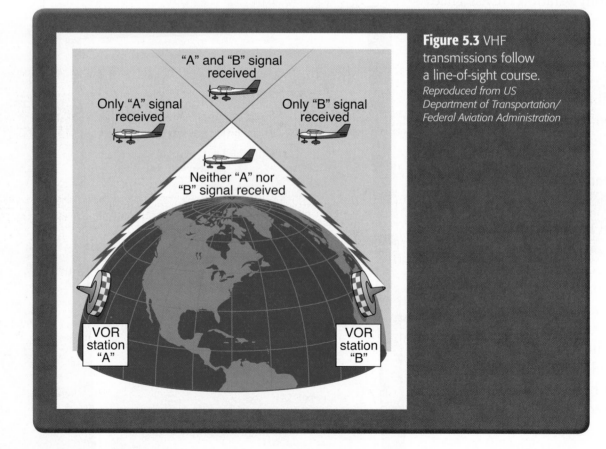

**Figure 5.3** VHF transmissions follow a line-of-sight course. *Reproduced from US Department of Transportation/ Federal Aviation Administration*

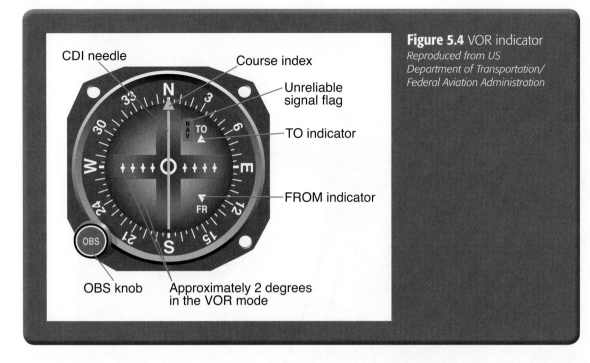

**Figure 5.4** VOR indicator
*Reproduced from US Department of Transportation/ Federal Aviation Administration*

Next the pilot rotates the *omnibearing selector* knob until the CDI needle centers on the course index. An omnibearing selector is also known as a *course selector*. Once the needle is centered, the *to/from indicator* should say "to." A to/from indicator tells the pilot whether the selected course will take the aircraft toward or away from the selected station. It does *not* indicate whether the aircraft is heading to or from the station, only whether the selected course will get the aircraft there. An *unreliable signal flag* (the red NAV flag) indicates a weak or unreliable signal.

### More on VORs

Some VORs include distance-measuring equipment (DME). DME is *an electronic navigation system that determines the number of nautical miles between an aircraft and a ground station or waypoint.* To use this system the aircraft requires a tuner for it. This tuner is more expensive than the one a VOR alone uses. It is an ultra high frequency (UHF) tuner that also receives data from a ground-based station. When a VOR and DME are installed together, they are known as *VOR/DME*.

## Wing TIPS

A VOR acting alone can only give the pilot the line, or radial, the aircraft is on in relation to a station. But with DME the pilot can find his or her exact location on that line. This is because DME finds the *slant range distance* to or from a station. Slant range distance is the direct distance between an aircraft and a station, and so includes aircraft altitude. If the plane is at 5,000 feet, for example, the DME in the cockpit will read approximately one mile.

SrA Nelson Santiago checks a portable tactical navigation aid during Exercise Salitre, a Chilean-led multinational exercise between Chile, the United States, France, Argentina, and Brazil in Chile in 2009. The TACAN allows US F-15 Eagle pilots to locate the two operating airfields involved in the exercise and orient themselves for landing.
*Courtesy of USAF/TSgt Eric Petosky*

Besides civilian pilots, the US military also works with VOR equipment. The military installs it along with a piece of equipment called tactical air navigation, or TACAN. This is *an electronic navigation system used by military aircraft that provides distance and direction.* Virtually all US military installations with an established airfield have a TACAN system. The military also uses TACAN when setting up airfields in remote locations. By dialing into the TACAN's frequency, pilots can then find the airfields, figure out how far away they are, and orient themselves for landing. When TACAN is installed with a VOR and DME, it is known as *VORTAC.* Whichever type of VOR system an aircraft has, the VOR indicator works the same.

## The Automatic Direction Finder

The automatic direction finder (ADF) is an even older technology than the ground-based VOR. It uses the AM radio frequency band.

An ADF determines the bearing from the aircraft to a selected ground station. The station is a *nondirectional radio beacon* (NDB) station that transmits radio energy in all directions. NDB stations normally operate in a low- or medium-frequency band of 200 to 415 kilohertz (kHz). A kilohertz is *a frequency of 1,000 cycles per second*. The frequencies are listed on charts and in the Airport/Facility Directory.

The ADF consists of a tuner, two antennas, and an indicator (Figure 5.5). To navigate, the pilot tunes the receiving equipment to an NDB ground station. The station transmits a three-letter code at all times. The antennas pick up the station signal. The ADF indicator needle points to the NDB ground station to determine the relative bearing to the transmitting station. Relative bearing is *the angular difference between the aircraft heading and the direction to the station.* Pilots measure it clockwise from the aircraft's nose. Therefore a pilot uses an ADF to "home in" on a station. *Homing* is flying an aircraft on any heading required to keep the needle pointing directly to the 0-degree relative bearing position.

### Wing TIPS

AM radio stations have a broadcast band frequency above the NDB band (550 to 1650 kHz). While most ADFs can tune into the AM band frequencies, the FAA does not approve these frequencies for navigation because AM stations don't continuously identify themselves the way NDB stations do.

**Figure 5.5** ADF indicator and receiver
*Reproduced from US Department of Transportation/Federal Aviation Administration*

AM signals have an advantage over the signals broadcast by VOR stations. AM signals can curve with the Earth. They aren't restricted to straight lines. So as long as an aircraft is within a station's range, it can receive signals no matter the airplane's altitude. But AM signals have a disadvantage as well.

Electrical disturbances, such as lightning, can interfere with low-frequency signals. These disturbances create static, cause needles to move erratically, and weaken signals. Distant stations can also interfere with AM transmissions, especially at night. That's because at night, AM signals bounce off the ionosphere more and travel farther than they usually do during the day.

### Instrument Landing System

The instrument landing system (ILS) is another radio aid. Engineers designed it to work with VOR but at different frequencies. ILS is *an electronic system that provides an approach path to a specific runway.* It guides the pilot both horizontally and vertically by lining the aircraft up with a runway and helping the pilot safely descend to it in any kind of weather.

The ILS provides three types of information: guidance, range, and visual. Different components serve different functions:

- Guidance comes from two transmitters placed on the ground. They are a *localizer* and a *glide slope indicator*. The localizer lets the pilot know whether the aircraft is right or left of the runway (Figure 5.6). The glide slope indicator tells the pilot whether the aircraft is above or below the proper glide path down to the runway touchdown point. The ideal descent is usually at a 3-degree slope.

- Range comes from marker beacons. These let an aircraft know how close it is in its approach to a runway threshold. Marker beacons send out VHF signals. An *outer marker* sits from four to seven miles from an airport. A *middle marker* sits about 3,500 feet from the start of the runway. An *inner marker* sits between the middle marker and the runway start, also known as the *landing threshold*.

- Visual aids come from all kinds of lights, including approach lights and runway lights. These lights come in different colors to send different signals, and some flash.

To avoid conflict between the ILS and VOR frequencies, VOR stations between 108.0 and 112.0 MHz operate in even-tenth increments (108.2, 111.6, etc.). ILS frequencies use odd-tenths in this range (108.3, 111.7, etc.).

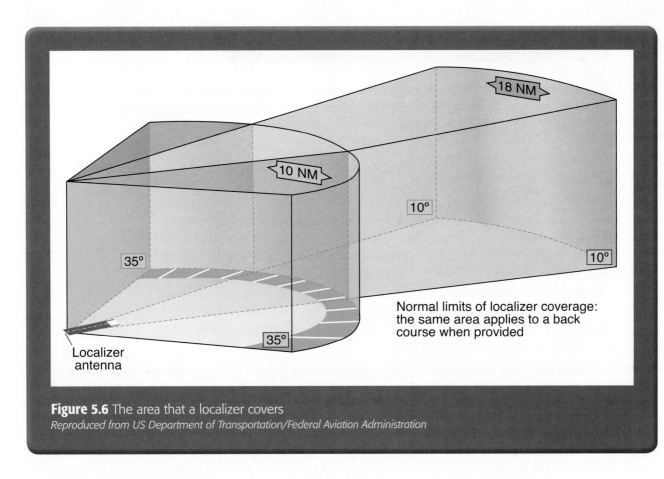

**Figure 5.6** The area that a localizer covers
*Reproduced from US Department of Transportation/Federal Aviation Administration*

## Global Positioning System

The global positioning system (GPS) was the next step forward in radio navigation. GPS is a satellite-based radio navigation system. It helps with en route navigation, departures, arrivals, and approaches.

A GPS receiver uses data from at least four satellites to pinpoint an aircraft's position. The receiver measures distance from a satellite by counting the length of time it takes a radio signal to travel from the satellite to the GPS set. Each satellite sends its own special code. The US Department of Defense operates the GPS satellite system because the armed forces rely on the technology for their missions and everyday functions. You will read more about this later in this lesson.

**Wing TIPS**

Like all electromagnetic waves, signals from satellites travel at 186,000 miles per second, the speed of light.

More and more aviators outside the military are turning to GPS. For instance, most commercial flights use it. In addition pilots in general aviation—*private flights that are not commercial or military*—are using GPS navigation sets in greater numbers. However these general-aviation pilots still rely heavily on ground-based radio aids. Some general-aviation organizations are pushing to maintain the old ground-based systems in the face of GPS's growing use.

## LORAN and Why It's Still Relevant

Some older systems eventually pass out of use. Long-range navigation (LORAN) is one of those systems. It is *a radio navigation system that determines position by measuring the difference in time that the system receives pulses from two transmitters*. The US Coast Guard operated it for more than 50 years but ended all US transmissions of its LORAN-C signals in 2010 as GPS use became more widespread. Aviators and mariners used the system to navigate. LORAN-C was a version of the system that operated in the 90 to 110 kHz frequency band.

Although out of use in the United States and Canada, LORAN still has value. Before terminating the system, the Coast Guard advised all mariners to switch to GPS, just as more and more pilots are using GPS. GPS relies on satellite transmission, however, which is a technology vulnerable to attack. According to the FAA, the federal government is so concerned about the possibility of losing GPS—which could affect the armed forces, transportation, and national security—that it "has renewed and refocused attention on LORAN." Radio signals are not as vulnerable as satellite transmissions.

## Current Developments in Navigation Technology

A projected 50 percent growth in air traffic in the United States by 2025 is spurring major investments and research into improved navigation technology. One of the main pushes is in a program developed by the FAA called NextGen, which stands for Next Generation air transportation system. You first read about this program in Chapter 1, Lesson 6.

### Wing TIPS

Currently 66 percent of commercial flight delays are due to weather. The delays cost the airlines $2 billion per year. They result in $9 billion in lost productivity each year due to goods not delivered on time, passengers who must be reimbursed, businesspeople who miss meetings, and more. NextGen will help decrease the number of flight delays.

NextGen will move air traffic from a radar-based system to a satellite-based technology. By turning to satellite transmissions, pilots and air traffic controllers can receive information in real time. Timely data means improved safety, fewer flight delays, and a greater ability to handle an increase in air traffic. It should also result in reduced effects on the environment because of more direct routes and other ways to be more efficient with time and fuel.

NextGen is also a boon to national security. According to the FAA, it gives the military, the Department of Homeland Security, and the FAA a "more effective means of monitoring our airspace." They can more easily track aircraft, including those that might not be flying according to flight plans.

## Wing TIPS

By 2018, if air traffic increases at the expected rate, the FAA predicts that NextGen will achieve the following:

- Reduce growth in flight delays by 21 percent per year
- Reduce CO$_2$ emissions by 14 millions tons per year
- Reduce fuel use by 1.5 billion gallons per year.

## Automatic Dependent Surveillance-Broadcast

One of the main components of NextGen is an invention called automatic dependent surveillance-broadcast (ADS-B). The FAA refers to ADS-B as the "backbone of the NextGen system." It is more accurate than radar. ADS-B is *a device used in aircraft that repeatedly broadcasts a message that includes the aircraft's position and velocity.* Other aircraft and air traffic control receive this information and can make decisions accordingly.

At the heart of ADS-B is an onboard GPS. It determines an aircraft's position. The ADS-B system then combines that data with flight details such as course, speed, and altitude. It also includes the aircraft's ID, which is a unique series of letters and numbers assigned to that plane. The ID tells air traffic control and other aircraft where a particular plane is. The ADS-B system broadcasts all of the data it has collected to nearby aircraft as well as to ADS-B ground stations on a regular basis (Figure 5.7).

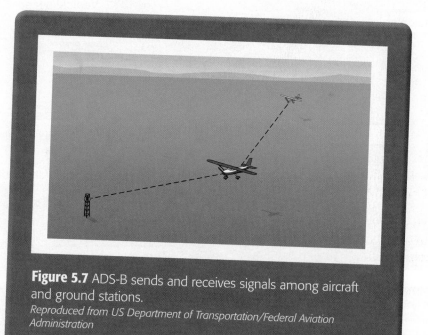

**Figure 5.7** ADS-B sends and receives signals among aircraft and ground stations.
*Reproduced from US Department of Transportation/Federal Aviation Administration*

Other ADS-B components to be installed in an aircraft include:

- A universal access transceiver (UAT), which operates at 978 MHz. The UAT sends and receives ADS-B broadcasts to and from one aircraft to another aircraft with ADS-B and to ADS-B ground stations. Aircraft flying above 18,000 feet need equipment that operates at 1090 MHz.

- An antenna to receive and send transmissions.

- A multi-function display (MFD) to receive traffic and weather broadcasts from the UAT. The MFD is *a small electronic screen at the center of the flight instrument panel that displays information such as traffic and weather.* Aircraft today also often have a primary flight display (PFD) on the instrument panel's left-hand side (Figure 5.8). The PFD is *a device that replaces traditional flight instruments with an electronic display of such data as airspeed, altitude, and horizon.*

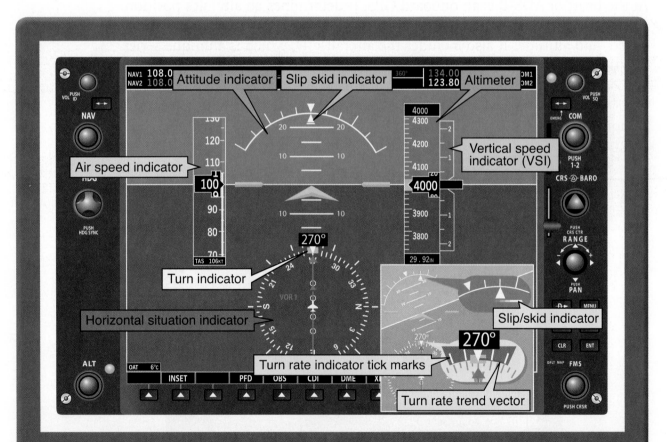

**Figure 5.8** A primary flight display. The location of indicators varies with the manufacturer.
*Reproduced from US Department of Transportation/Federal Aviation Administration*

All aircraft will need to be able to communicate with the ADS-B system by 2020 to fly in most airspace.

The FAA provides both traffic and weather reports free to pilots through its ADS-B broadcast services. The traffic data arrives through its traffic information service-broadcast (TIS-B). The service uses ADS-B ground stations to send the same radar-based information to the cockpit that air traffic controllers work with to track aircraft. Pilots can track other aircraft flying inside a 15-mile radius and within 3,500 feet above and below their own aircraft (Figure 5.9). With this data they can keep enough separation between aircraft to avoid midair collisions. So because of ADS-B, pilots have an improved situational awareness, *a pilot's understanding of where his or her aircraft is in relation to other aircraft, weather, location, and airspace regulations.*

The weather data comes via the flight information service-broadcast (FIS-B). It delivers timely weather data on the electronic multi-function display in the cockpit. It can help pilots avoid bad weather. The flight broadcast also includes some of the types of reports you read about in Chapter 2, Lesson 4—including Notices to Airmen (NOTAMs), aviation routine weather reports (METARs), and pilot weather reports (PIREPs). Ground stations will be able to deliver TIS-B and FIS-B across the entire country, including Alaska and Hawaii, by 2013.

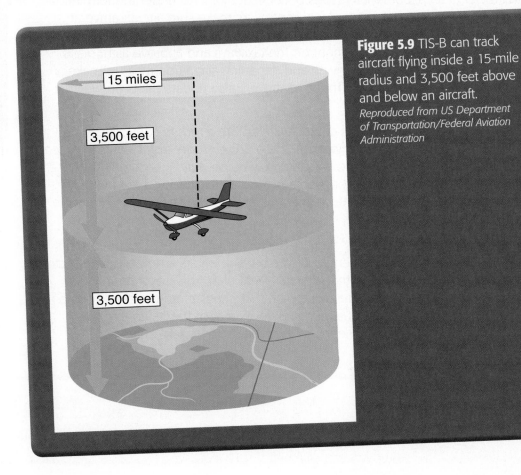

**Figure 5.9** TIS-B can track aircraft flying inside a 15-mile radius and 3,500 feet above and below an aircraft.
*Reproduced from US Department of Transportation/Federal Aviation Administration*

15 miles

3,500 feet

3,500 feet

Air traffic control's two main purposes are to prevent collisions between aircraft and to direct the flow of traffic. It relies on many layers of controllers in contact with aircraft, radar and NextGen technology, and aircraft complying with flight regulations within the National Airspace System (NAS). The NAS is *the network of people and equipment that make up the US airspace and ensure aircraft safety.* The FAA oversees the air traffic system.

Air traffic controllers work within the NAS to coordinate the movement of air traffic to ensure that planes stay a safe distance apart. Besides keeping in touch with pilots in the air and on the ground, controllers also keep in contact with one another to coordinate aircraft movements.

While safety is their greatest concern, controllers try to direct planes as efficiently as possible to reduce flight delays. Some controllers handle the traffic moving through certain airspace; others deal with airport arrivals and departures. When an aircraft moves out of one controller's area of responsibility, he or she transfers control of that aircraft to another air traffic controller. This ensures an aircraft is always supervised during key moments of flight.

*Terminal controllers* watch over all planes traveling in an airport's airspace. Their main job is to organize the flow of aircraft into and out of the airport. They often work in a control tower at an airport. Many types of controller jobs fall under the terminal controller category. For instance, *tower local controllers* make sure aircraft arrive at an airport in a certain order, and they clear aircraft for takeoff. Other controllers in the control tower direct aircraft movement on the taxiways and approve flight plans. *Terminal radar controllers* manage arriving and departing aircraft on radar to make sure they keep a safe distance apart. These controllers also update pilots on weather and runway conditions.

After each plane departs, the terminal controllers alert the *en route controllers* who take charge next. En route controllers work at one of 20 *air route traffic control centers* around the country. As many as 150 controllers may be on duty at each one of these centers during peak hours. Each center oversees a certain airspace containing many different routes. Controllers work alone or in teams of two. They generally will handle several planes at a time.

Aircraft may pass through quite a few airspace sectors during a flight. Control centers will hand off aircraft from center to center. Before a plane enters a new airspace sector, a *radar associate controller* reviews that aircraft's flight plan. This way if two planes are about to enter that team's sector along a path that brings them too close to one another, the controller can make adjustments. Before an aircraft exits an airspace district, the radar controller finds someone in the next sector to take charge of the plane.

# How the Air Force Uses GPS and Inertial Navigation in Air Operations

Advances in navigation technology are very important for national security, too. They allow fighter pilots, unmanned aircraft systems (UASs), bombs, and missiles to reach their targets with pinpoint accuracy. Two key navigation tools are the GPS and inertial navigation system (INS).

The Air Force is the Defense Department branch that operates the country's GPS satellites. It has an inventory of more than 30 satellites but generally tries to have 24 satellites up and running at any one time. It also keeps a few older satellites in working order so they can return to service if a newer satellite fails.

The military version of GPS is called *precise position service*. It is more accurate than the civilian service because it relies on two radio signals rather than just one to pin down a location. It is also encrypted and better able to ward off attempts to jam the system than is the civilian GPS.

The US military has a number of ways it uses GPS as well as INS. As you read in Chapter 4, Lesson 3, INS systems are self-contained onboard navigation systems. The military's guided bomb units (GBUs) use GPS/INS guidance systems. A GBU is also called a smart bomb—*a precision weapon guided by GPS, INS, and/or laser technology to its target.*

One example of a smart bomb is the 500-pound Joint Direct Attack Munition (JDAM) GBU-38. It is a guided free-fall bomb that includes a GPS-aided INS. This air-to-surface weapon has control surfaces on the tail fins that respond to the GPS/INS system. It can get within at least 15 feet (about 5 meters) of its target. When GPS isn't available, the JDAM can get within 90 feet (30 meters) of a target during a flight no more than 100 seconds long after an aircraft or UAS has released it.

## A Brief History of GPS

The early makings of a GPS system started during the cold war. In 1957 researchers William Guier and George Wiefenbach, who worked at Johns Hopkins Applied Physics Laboratory in Baltimore, Maryland, used radio waves emitted by *Sputnik*, the Soviets' first satellite, to find its orbit. Further studies by others reversed this concept to find an object's position on Earth. The Navy was particularly interested in navigational tools for its nuclear submarines. In the beginning, the technology was only available to the military. But after the Soviets shot down a Korean passenger airliner that had accidentally strayed into Soviet airspace in 1983, President Ronald Reagan ordered that GPS be made available for civilian use.

Another example of a precision weapon is the GBU-54. It lets a pilot choose either GPS or a laser system to guide the bomb to its target. The Air Force used this bomb for the first time in 2008 in Iraq and then in Afghanistan in 2010.

Besides releasing GPS/INS-guided bombs and missiles on targets, aircraft and UASs also rely on these navigation systems to get them where they need to go. For example, the F-22 Raptor stealth fighter uses INS and GPS to keep track of its location. The F-35, a stealth fighter that's still under development, will feature similar technology. UASs such as the Predator house GPS/INS in the vehicle's nose. One reason the military has adopted INS is that it is an onboard unit that depends on nothing outside the aircraft. So if hostile or natural forces knock out ground stations and satellites, a pilot can still rely on the INS system.

A GBU-54 rests on the wing of an F-16 Fighting Falcon at Bagram Airfield, Afghanistan, in 2010. The GBU-54 is the Air Force's newest 500-pound precision weapon, equipped with a special targeting system that uses a combination of GPS and laser guidance to accurately engage and destroy moving targets.
*Courtesy of USAF/SSgt Christopher Boitz*

## The Purpose of Computer Flight-Planning Tools

Before the launch of the first operational GPS satellites in 1989—a series known as Block II—many flight crews included a navigator. In the military, navigators were officers who played a particularly large role in flight planning for overseas flights. They would draft compass headings for great circle routes, use LORAN stations and celestial navigation to get from point to point, and study weather reports for wind effect. Often a navigator would suggest corrections in heading by just a few degrees every 15 to 20 minutes. As one former Air Force pilot, Wally Wethe, put it, when the aircraft reached its destination based on the planned course, "the crew would whoop and give the NAV a hearty round of applause!"

Crews don't include navigators as regularly these days. Technology has largely taken their place, playing ever larger roles in flight planning. Earlier in this chapter you read about the nonelectronic plotter and dead reckoning computer. Pilots still use these devices for flight planning. But they also use computer flight-planning tools. A flight management computer is *a computer system or software that lets aircrews electronically draw up and file flight plans.*

These systems can be very efficient. For example, a commercial airline that offers service between the same airports every day will generally have its pilots use a flight management computer with waypoints and ground stations already installed in the system's database. The airline may even assign a particular route its own code name that the pilot simply punches into the computer. Then up pops the flight plan, including accurate estimated arrival times, alternate airports, and more. This saves the airline time and money. Airlines and the military often employ flight-planning teams stationed on the ground who program the systems.

NASA flight navigator Manny Antimisiaris is responsible for directing the exact flight path of NASA's DC-8 airborne science laboratory during NASA's 2010 IceBridge campaign in the Arctic. Now in its second year, the five-week Operation IceBridge mission is the largest airborne survey ever flown of Earth's polar ice. While most commercial and military flights no longer employ a flight navigator, some flights such as this NASA research flight still require one.
*Courtesy of NASA/Tom Tschida*

These preprogrammed flight plans can take weather reports into account. They may be linked to ADS-B. Therefore they include predicted winds for the altitudes where the plane will fly. As a flight progresses the computer can also alter the heading for any changes in wind that the aircraft is encountering.

Some flight management computers can even suggest routes that reduce fuel use. They could eventually incorporate new fuel-efficient concepts, such as continuous descent approach, which you read about in Chapter 1, Lesson 6. Further, these flight computers increase safety because they integrate information into flight plans from such sources as NOTAMs, the reports that list airport and runway closings, along with other up-to-date hazards.

Some types of flights will be less automated. For example, for a nonstandard route a pilot will plug in all the desired checkpoints, as well as other items required on a flight plan—such as cruising speed and altitude. Even so, this is still quite efficient. Combat flights are another exception. While even combat pilots file flight plans, they may have to quickly change course if they run into enemy fire or hostile aircraft. Most noncombat military flights, however, can follow their flight plans to the letter.

This course has introduced you to many aspects of the science of flight. You learned about the principles and physics of flight, the various parts of an aircraft, and recent developments in aviation. You also studied about how atmospheric conditions affect flight, flight and the human body, and navigation basics.

Perhaps it has inspired you to pursue a career in aviation or to learn how to fly. There are many great opportunities for you on the horizon. The science, technology, engineering, and math fields offer some exciting careers, as does the world of aviation and flying. As you continue your journey, use this course as a foundation for pursuing those avenues and as a gateway to new adventures.

## Lesson 5 Review

Using complete sentences, answer the following questions on a

1. What does a plotter help a pilot find?

2. How does a plotter's outer half circle mark off degrees?

3. What types of equations can a pilot solve with a flight computer?

4. Why is a flight computer's wind side easy to recognize?

5. Operating a radio aid calls for what four things?

6. How does the military use TACAN?

7. What program will move air traffic from a radar-based system to satellite technology?

8. What device is at the heart of ADS-B?

9. Why is the military version of GPS more accurate than the civilian service?

10. What is one reason the military has adopted INS?

11. Whom do airlines and the military often employ and what do they do?

12. As a flight progresses, what can a flight management computer do?

## APPLYING YOUR LEARNING

13. If an asteroid destroyed a GPS satellite or two—knocking out GPS service in a wide area—what options does air traffic control have to get planes in that region safely to their destinations?

# References

## CHAPTER 1, LESSON 1   Principles of Flight

*Aircraft Pitch Motion*. (2010). NASA/Glenn Research Center. Retrieved 1 February 2011 from http://www.grc.nasa.gov/WWW/K-12/airplane/pitch.html

*Angle of Attack*. (n.d.). U.S. Centennial of Flight Commission. Retrieved 15 February 2011 from http://www.centennialofflight.gov/essay/Dictionary/angle_of_attack/DI5.htm

*Astronomy Picture of the Day: Isaac Newton Explains the Solar System*. (1998). NASA. Retrieved 31 January 2011 from http://apod.nasa.gov/apod/ap980920.html

*Bernoulli and Newton*. (2008). NASA/Glenn Research Center. Retrieved 31 January 2011 from http://www.grc.nasa.gov/WWW/K-12/airplane/bernnew.html

*Forces on Aircraft*. (n.d.). NASA Quest. Retrieved 31 January 2011 from http://quest.nasa.gov/aero/teachers/foa.html

Needham, J., & Wang, L. (1965). *Science and civilisation in China: Physics and physical technology, mechanical engineering*, Volume 4, Part 2, pp. 587, 589. Cambridge, UK: Cambridge University Press.

*Re-Living the Wright Way: Overview of the Wright Brothers Invention Process*. (2010). NASA. Retrieved 28 January 2011 from http://wright.nasa.gov/overview.htm

Ronan, C. A. (1994). *The shorter science and civilisation in China: an abridgement of Joseph Needham's original text*, Volume 4, Part 2, p. 285. Cambridge, UK: Cambridge University Press.

## CHAPTER 1, LESSON 2   The Physics of Flight

*Forces on Aircraft*. (n.d.). NASA Quest. Retrieved 4 February 2011 from http://quest.nasa.gov/aero/teachers/foa.html

*Four Forces on an Airplane*. (2010). NASA. Retrieved 2 February 2011 from http://www.grc.nasa.gov/WWW/K-12/airplane/forces.html

*How the Harrier Hovers*. (2004). Harrier.org.uk. Retrieved 4 February 2011 from http://www.harrier.org.uk/technical/How_Hovers.htm

*Oblique Shock Waves*. (2010). NASA/Glenn Research Center. Retrieved 4 February 2011 from http://www.grc.nasa.gov/WWW/K-12/airplane/oblique.html

*Octave Chanute*. (2011). American Institute of Aeronautics and Astronautics. Retrieved 7 February 2011 from http://www.aiaa.org/content.cfm?pageid=433

*Octave Chanute*. (n.d.). USAF. Retrieved 7 February 2011 from http://www.af.mil/information/heritage/person.asp?dec=&pid=123006452

*Octave Chanute—A Champion of Aviation*. (n.d.). US Centennial of Flight Commission. Retrieved 7 February 2011 from http://www.centennialofflight.gov/essay/Prehistory/chanute/PH7.htm

*Pilot's Handbook of Aeronautical Knowledge*. (2008). U.S. Department of Transportation/Federal Aviation Administration. Retrieved 2 February 2011 from http://www.faa.gov/library/manuals/aviation/pilot_handbook/

*Scalars and Vectors*. (2010). NASA/Glenn Research Center. Retrieved 2 February 2011 from http://www.grc.nasa.gov/WWW/K-12/airplane/vectors.html

*Similarity Parameters*. (2009). NASA/Glenn Research Center. Retrieved 4 February 2011 from http://www.grc.nasa.gov/WWW/K-12/airplane/airsim.html

*The Weight Equation.* (2010). NASA/Glenn Research Center. Retrieved 4 February 2011 from http://www.grc.nasa.gov/WWW/K-12/airplane/wteq.html

*What Is Thrust?* (2010). NASA. Retrieved 4 February 2011 from http://www.grc.nasa.gov/WWW/K-12/airplane/thrust1.html

*What Is Weight?* (2010). NASA. Retrieved 3 February 2011 from http://www.grc.nasa.gov/WWW/K-12/airplane/weight1.html

*The Wright Brothers: The Invention of the Aerial Age.* (2006). Smithsonian Institution: National Air and Space Museum. Retrieved 3 February 2011 from http://www.nasm.si.edu/wrightbrothers/index.cfm

## CHAPTER 1, LESSON 3    The Purpose and Function of Airplane Parts

*Airplane Parts and Function.* (2010). NASA. Retrieved 14 February 2011 from http://www.grc.nasa.gov/WWW/K-12/airplane/airplane.html

*Apple monocoque or not: Commercial Jets, Semi Monocoque.* (n.d.). Oobject. Retrieved 14 February 2011 from http://www.oobject.com/apple-monocoque-or-not/commercial-jets-semi-monocoque/4046/

*Beginner's Guide to Propulsion.* (2010). NASA. Retrieved 16 February 2011 from http://www.grc.nasa.gov/WWW/K-12/airplane/bgp.html

*Flaps and Slats.* (2010). NASA/Glenn Research Center. Retrieved 15 February 2011 from http://www.grc.nasa.gov/WWW/K-12/airplane/flap.html

*Frank W. Caldwell (1889–1974).* (2011). American Institute of Aeronautics and Astronautics. Retrieved 18 February 2011 from http://www.aiaa.org/content.cfm?pageid=424

*Fuselage.* (2010). NASA/Glenn Research Center. Retrieved 14 February 2011 from http://www.grc.nasa.gov/WWW/K-12/airplane/fuselage.html

*Gas Turbine Propulsion.* (2010). NASA/Glenn Research Center. Retrieved 17 February 2011 from http://www.grc.nasa.gov/WWW/K-12/airplane/turbine.html

*General Thrust Equation.* (2010). NASA/Glenn Research Center. Retrieved 16 February 2011 from http://www.grc.nasa.gov/WWW/K-12/airplane/thrsteq.html

*Glide Angle.* (2010). NASA/Glenn Research Center. Retrieved 14 February 2011 from http://www.grc.nasa.gov/WWW/K-12/airplane/glidang.html

*Jet Engines.* (n.d.). US Centennial of Flight Commission. Retrieved 18 February 2011 from http://www.centennialofflight.gov/essay/Evolution_of_Technology/jet_engines/Tech24.htm

*Pilot's Handbook of Aeronautical Knowledge.* (2008). U.S. Department of Transportation/Federal Aviation Administration. Retrieved 14 February 2011 from http://www.faa.gov/library/manuals/aviation/pilot_handbook/

*Propeller Propulsion.* (2008). NASA/Glenn Research Center. Retrieved 16 February 2011 from http://www.grc.nasa.gov/WWW/K-12/airplane/propeller.html

*Ramjet Propulsion.* (2008). NASA/Glenn Research Center. Retrieved 17 February 2011 from http://www.grc.nasa.gov/WWW/K-12/airplane/ramjet.html

*Rocket Propulsion.* (2008). NASA/Glenn Research Center. Retrieved 17 February 2011 from http://www.grc.nasa.gov/WWW/K-12/airplane/rocket.html

*Shape Effects on Lift.* (2010). NASA/Glenn Research Center. Retrieved 15 February 2011 from http://www.grc.nasa.gov/WWW/K-12/airplane/shape.html

*Spoilers.* (2010). NASA/Glenn Research Center. Retrieved 15 February 2011 from http://www.grc.nasa.gov/WWW/K-12/airplane/spoil.html

*Wing Geometry Definitions.* (2010). NASA/Glenn Research Center. Retrieved 14 February 2011 from http://www.grc.nasa.gov/WWW/K-12/airplane/geom.html

*Winglets.* (2010). NASA. Retrieved 14 February 2011 from http://www.grc.nasa.gov/WWW/K-12/airplane/winglets.html

*Wright 1903 Flyer.* (2010). NASA/Glenn Research Center. Retrieved 18 February 2011 from http://wright.nasa.gov/airplane/air1903.html

## CHAPTER 1, LESSON 4   Aircraft Motion and Control

*Aircraft Rotations: Body Axes.* (2008). NASA/Glenn Research Center. Retrieved 22 February 2011 from http://www.grc.nasa.gov/WWW/K-12/airplane/rotations.html

Christenson, S. (2010, December 10). 'Ordinary guy' honored for an extraordinary act. *The Houston Chronicle.* Retrieved 4 March 2011 from http://www.chron.com/disp/story.mpl/metropolitan/7334578.html

*Flaps and Slats.* (2010). NASA/Glenn Research Center. Retrieved 22 February 2011 from http://www.grc.nasa.gov/WWW/K-12/airplane/flap.html

Hansen, J. R. (2004). *The Bird Is on the Wing: Aerodynamics and the Progress of the American Airplane.* R. D. Launius (Ed.). College Station: Texas A&M University Press.

*Pilot's Handbook of Aeronautical Knowledge.* (2008). U.S. Department of Transportation/Federal Aviation Administration. Retrieved 14 February 2011 from http://www.faa.gov/library/manuals/aviation/pilot_handbook/

Siekert, A. (2010). Reservist awarded the Airman's Medal for off-duty heroism. Randolph Air Force Base: Joint Base San Antonio. Retrieved 4 March 2011 from http://www.randolph.af.mil/news/story.asp?id=123234560

*Slotted Wings, Flaps, and High Lift Devices.* (n.d.). US Centennial of Flight Commission. Retrieved 23 February 2011 from http://www.centennialofflight.gov/essay/Evolution_of_Technology/High_Lift_Devices/Tech6.htm

*SP-367 Introduction to the Aerodynamics of Flight.* (n.d.). NASA. Retrieved 23 February 2011 from http://history.nasa.gov/SP-367/chapt4.htm

*Spoilers.* (2010). NASA/Glenn Research Center. Retrieved 23 February 2011 from http://www.grc.nasa.gov/WWW/K-12/airplane/spoil.html

## CHAPTER 1, LESSON 5   Flight Power

*America by Air: A New Generation of Leaders.* (2007). Smithsonian: National Air and Space Museum. Retrieved 17 March 2011 from http://www.nasm.si.edu/americabyair/activities/newgeneration/index.cfm

*Boyle's Law.* (2011). NASA/Glenn Research Center. Retrieved 11 March 2011 from http://www.grc.nasa.gov/WWW/K-12/airplane/boyle.html

*Building a Better Plane.* (2010). NASA. Retrieved 17 March 2011 from http://www.nasa.gov/topics/aeronautics/features/openrotor.html

Demerjian, D. (2008, June 24). Greener Jet Engine Could Reduce Aviation's Carbon Footprint. *Wired.* Retrieved 17 March 2011 from http://www.wired.com/cars/futuretransport/news/2008/06/ecoaviation23

*Dictionary of Technical Terms for Aerospace Use.* (2001). NASA/Glenn Research Center. Retrieved 14 March 2011 from http://er.jsc.nasa.gov/seh/menu.html

*Engine Cooling System.* (2010). NASA/Glenn Research Center. Retrieved 14 March 2011 from http://wright.nasa.gov/airplane/cooling.html

*Equation of State (Ideal Gas).* (2011). NASA/Glenn Research Center. Retrieved 11 March 2011 from http://www.grc.nasa.gov/WWW/K-12/airplane/eqstat.html

*F-15 ACTIVE.* (2009). NASA/Dryden Flight Research Center. Retrieved 21 March 2011 from http://www.nasa.gov/centers/dryden/history/pastprojects/Active/index.html

*F-15 ACTIVE Flight Research Program.* (2008). NASA/Dryden Flight Research Center. Retrieved 21 March 2011 from http://www.nasa.gov/centers/dryden/history/pastprojects/Active/pub_online/setp_d6.html

*F119.* (n.d.). Pratt & Whitney: A United Technologies Company. Retrieved 21 March 2011 from http://www.pw.utc.com/vgn-ext-templating/v/index.jsp?vgnextoid=c94e34890cb06110VgnVCM1000004601000aRCRD

*First Pratt & Whitney PurePower® Engine Completes Initial Ground Tests, Exceeds Expectations.* (2011). Pratt & Whitney: A United Technologies Company. Retrieved 17 March 2011 from http://www.pw.utc.com/Media+Center/Press+Releases/First+Pratt+%26+Whitney+PurePower%28R%29+Engine+Completes+Initial+Ground+Tests%2C+Exceeds+Expectations

*GE and NASA Partner on Open Rotor Engine Testing.* (2009). NASA. Retrieved 17 March 2011 from http://www.nasa.gov/offices/oce/appel/ask-academy/issues/volume2/AA_2-3_F_ge_rotor.html

*GE Aviation Moving to Apply Ceramic Matrix Composites to the Heart of Future Engines.* (2009). General Electric Company. Retrieved 21 March 2011 from http://www.geae.com/aboutgeae/presscenter/other/other_20090309.html

*Hans van Ohain.* (2009). U.S. Centennial of Flight Commission. Retrieved 15 April 2011 from http://www.centennialofflight.gov/essay/Dictionary/vonOhain/DI126.htm

*How Does a Jet Engine Work?* (n.d.). Ultra-Efficient Engine Technology. NASA. Retrieved 15 April 2011 from http://www.ueet.nasa.gov/StudentSite/engines.html

*How Scramjets Work.* (2008). NASA. Retrieved 21 March 2011 from http://www.nasa.gov/centers/langley/news/factsheets/X43A_2006_5.html

*Internal Combustion Engine.* (2008). NASA/Glenn Research Center. Retrieved 14 March 2011 from http://www.grc.nasa.gov/WWW/K-12/airplane/icengine.html

*Internal Combustion or Reciprocating Engines.* (n.d.). U.S. Centennial of Flight Commission. Retrieved 14 March 2011 from http://www.centennialofflight.gov/essay/Dictionary/Reciprocating_Engine/DI92.htm

*Ion Engines.* (n.d.). NASA/Jet Propulsion Laboratory. Retrieved 16 March 2011 from http://dawn.jpl.nasa.gov/mission/ion_engine_interactive/index.html

Loftin, L. K., Jr. (1985). *Quest for Performance: The Evolution of Modern Aircraft: Part II: THE JET AGE: Chapter 10: Technology of the Jet Airplane.* Washington, DC: NASA Scientific and Technical Information Branch. Retrieved 16 March 2011 from http://history.nasa.gov/SP-468/ch10-3.htm

*Making Future Commercial Aircraft Quieter.* (2008). NASA/Glenn Research Center. Retrieved 16 March 2011 from http://www.nasa.gov/centers/glenn/about/fs03grc.html

Mathews, J. (2008, July 1). How Things Work: Thrust Vectoring. *Air & Space: Smithsonian.* Retrieved 21 March 2011 from http://www.airspacemag.com/flight-today/Thrust_Vectoring.html

*Pilot's Handbook of Aeronautical Knowledge.* (2008). U.S. Department of Transportation/Federal Aviation Administration. Retrieved 14 February 2011 from http://www.faa.gov/library/manuals/aviation/pilot_handbook/

*PurePower PW 1000G.* (2011). Pratt & Whitney: A United Technologies Company. Retrieved 17 March 2011 from http://www.pw.utc.com/vgn-ext-templating/v/index.jsp?vgnextoid=59ab4d845c37a110VgnVCM100000c45a529fRCRD

*Pushing the Envelope: A NASA Guide to Engines.* (2007). NASA/Glenn Research Center. Retrieved 11 March 2011 from http://er.jsc.nasa.gov/seh/ANASAGUIDETOENGINES%5B1%5D.pdf

*Safeguarding Our Atmosphere.* (2000). NASA/Glenn Research Center. Retrieved 17 March 2011 from http://www.nasa.gov/centers/glenn/about/fs10grc.html

*What Is a Rocket Engine?* (2009). NASA/Engineering & Test Directorate. Retrieved 16 March 2011 from http://sscfreedom.ssc.nasa.gov/etd/ETDFAQs_ANSWERS.asp

Woo, E. (1999, March 26). Arthur E. Raymond Dies; DC-3's Designer Changed Air Travel. *Los Angeles Times.* Retrieved 17 March 2011 from http://articles.latimes.com/1999/mar/26/local/me-21143

*Wright 1903 Engine Operation: Power Stroke.* (2008). NASA/Glenn Research Center. Retrieved 14 March 2011 from http://www.grc.nasa.gov/WWW/K-12/airplane/engpowr.html

*X-51A Makes Longest Scramjet Flight.* (2010). NASA. Retrieved 21 March 2011 from http://www.nasa.gov/topics/aeronautics/features/X-51A.html

Yetter, J. A. (January 1995). *Why Do Airlines Want and Use Thrust Reversers?* NASA/Langley Research Center. Retrieved 15 March 2011 from http://ntrs.nasa.gov/archive/nasa/casi.ntrs.nasa.gov/19950014289_1995114289.pdf

*2009 Environment Report: Alternative Energy Solutions.* (2009). Boeing. Retrieved 23 March 2011 from http://www.boeing.com/aboutus/environment/environmental_report_09/alternative-energy-solutions.html

*A380 Family.* (2011). Airbus. Retrieved 27 March 2011 from http://www.airbus.com/aircraftfamilies/passengeraircraft/a380family/

*Aviation Pioneer Richard T. Whitcomb.* (2009). NASA Langley Research Center. Retrieved 22 March 2011 from http://www.nasa.gov/topics/people/features/richard_whitcomb.html

Banke, J. (2009). *From Nothing, Something: One Layer at a Time.* NASA Aeronautics Research Mission Directorate. Retrieved 22 March 2011 from http://www.nasa.gov/topics/aeronautics/features/electron_beam.html

Banke, J. (2009). *Shhhh! Keep It Down, Please.* NASA. Retrieved 23 March 2011 from http://www.nasa.gov/topics/aeronautics/features/aircraft_noise.html

Banke, J. (2010). *The Sheer Delight of Tackling Shear Stress.* NASA. Retrieved 27 March 2011 from http://www.nasa.gov/topics/aeronautics/features/shear_stress.html

*Biofuel Demonstration.* (2011). Virgin Atlantic. Retrieved 23 March 2011 from http://www.virgin-atlantic.com/en/us/allaboutus/environment/biofuel.jsp

*Boeing 787 Dreamliner Will Provide New Solutions for Airliners, Passengers.* (n.d.). Boeing. Retrieved 27 March 2011 from http://www.boeing.com/commercial/787family/background.html

*Boeing Successfully Flies Fuel Cell-Powered Airplane.* (2008). Boeing. Retrieved 23 March 2011 from http://www.boeing.com/news/releases/2008/q2/080403a_nr.html

*Boeing Unveils Unmanned Phantom Ray Demonstrator.* (2010). Boeing. Retrieved 24 March 2011 from http://boeing.mediaroom.com/index.php?s=43&item=1202

Cole, W. (2002). *Boeing Frontiers.* The Boeing Company. Retrieved 24 March 2011 from http://www.boeing.com/news/frontiers/archive/2002/september/i_pw.html

Creech, G. (2010). *Boeing's Phantom Ray Hitches a Ride to NASA Dryden.* NASA/Dryden Public Affairs. Retrieved 24 March 2011 from http://www.nasa.gov/centers/dryden/Features/phantom_ray_arrives.html

Demerjian, D. (2008, June 24). Greener Jet Engine Could Reduce Aviation's Carbon Footprint. *Wired.* Retrieved 17 March 2011 from http://www.wired.com/cars/futuretransport/news/2008/06/ecoaviation23

Doscher, SSgt T. J. (2009). *AF Raven B Operators Maintain 'Eyes-On' for Ground Forces.* USAF. Retrieved 25 March 2011 from http://www.af.mil/news/story.asp?id=123140479

Dowdell, Maj R. (2011). *Officials Certify First Aircraft for Biofuel Usage.* USAF. Retrieved 23 March 2011 from http://www.af.mil/news/story.asp?id=123242117

*Efficient Descent Advisor (EDA).* (2011). NASA/Aviation Systems Division. Retrieved 24 March 2011 from http://www.aviationsystemsdivision.arc.nasa.gov/research/tactical/eda.shtml

*F-22 Raptor Flown on Synthetic Biofuel.* (2011). *AFNS.* Retrieved 26 April 2011 from http://www.af.mil/news/story.asp?id=123248331

Freeman, M. (2011, March 22). Air Force Receives the Final MQ-1 Predator Drone. *San Diego Union-Tribune.* Retrieved 24 March 2011 from http://www.signonsandiego.com/news/2011/mar/22/air-force-receives-final-mq-1-predator-drone/

Hockmuth, C. M. (2007). UAVs—The Next Generation. *Air Force Magazine.* Retrieved 25 March 2011 from http://www.airforce-magazine.com/MagazineArchive/Pages/2007/February%202007/0207UAV.aspx

Isherwood, M. W. (2009). Roadmap for Robotics. *Air Force Magazine.* Retrieved 25 March 2011 from http://www.airforce-magazine.com/MagazineArchive/Pages/2009/December%202009/1209roadmap.aspx

*MQ-1B Predator.* (2010). USAF. Retrieved 24 March 2011 from http://www.af.mil/information/factsheets/factsheet.asp?fsID=122

*MQ-9 Reaper*. (2010). USAF. Retrieved 24 March 2011 from http://www.af.mil/information/factsheets/factsheet.asp?fsID=6405

*NextGen 101: Giving the World New Ways to Fly*. (2011). Federal Aviation Administration. Video retrieved 24 March 2011 from http://www.faa.gov/nextgen/

*Pilot's Handbook of Aeronautical Knowledge*. (2008). U.S. Department of Transportation/Federal Aviation Administration. Retrieved 14 February 2011 from http://www.faa.gov/library/manuals/aviation/pilot_handbook/

*RQ-4 Global Hawk*. (2009). USAF. Retrieved 25 March 2011 from http://www.af.mil/information/factsheets/factsheet.asp?id=13225

Sanders, P. (2010, November 10). Boeing 787 Makes Emergency Landing. *The Wall Street Journal*. Retrieved 27 March 2011 from http://online.wsj.com/article/SB10001424052748703523604575605100601208676.html

*Science & Technology Review*. (2000 January/February). Bringing Hypersonic Flight Down to Earth. Lawrence Livermore National Laboratory. Retrieved 22 March 2011 from https://www.llnl.gov/str/Carter.html

Wallace, L. E. (2001). *The Whitcomb Area Rule: NACA Aerodynamics Research and Innovation*. NASA History Office. Retrieved 22 March 2011 from http://history.nasa.gov/SP-4219/Chapter5.html

Woods, T. (2009). *Metallic Foam Reduces Airplane Noise*. NASA/Glenn Research Center. Retrieved 23 March 2011 from http://www.nasa.gov/centers/glenn/technology/metallic_foam.html

*X-Planes 1980s and 1990s*. (n.d.). NASA. Retrieved 26 April 2011 from http://www.nasa.gov/centers/dryden/history/HistoricAircraft/X-Planes/1980/index_prt.htm

Zappei, J. (2010, November 4). Airbus A380 Fleet Grounded After Quantas Jet Engine Blowout. *The Christian Science Monitor*. Retrieved 27 March 2011 from http://www.csmonitor.com/World/Latest-News-Wires/2010/1104/Airbus-A380-fleet-grounded-after-Qantas-jet-engine-blowout

## CHAPTER 2, LESSON 1   The Atmosphere

Ambrose, S. E. (2001). *The Wild Blue: The Men and Boys Who Flew the B-24s Over Germany*. New York: Simon & Schuster.

Barker, T. (n.d.). *Jet Streams*. NOAA/National Weather Service. Retrieved 12 April 2011 from http://www.wrh.noaa.gov/mso/educate/jet1.html

*Boeing Completes Safety-of-Flight Testing of Life-Support System for F-22 Fighter Pilots*. (1997). Boeing. Retrieved 13 April 2011 from http://www.boeing.com/news/releases/1997/news.release.970522a.html

*Earth's Atmosphere*. (2010). NASA. Retrieved 7 April 2011 from http://www.nasa.gov/audience/forstudents/9-12/features/912_liftoff_atm.html

*F-22 Aircrew Life Support/Escape System*. (n.d.). USAF/Air Force HSI/Wright-Patterson Air Force Base. Retrieved 13 April 2011 from http://www.wpafb.af.mil/shared/media/document/AFD-090121-030.pdf

*F-22 Raptor*. (2009). USAF. Retrieved 13 April 2011 from http://www.af.mil/information/factsheets/factsheet.asp?id=199

*Frequently Asked Questions*. (2010). US Geological Survey. Retrieved 7 April 2011 from http://geomag.usgs.gov/faqs.php#qone

*It's a Bird! It's a Plane! … No! It's a NOAA Weather Balloon!* (n.d.). NOAA. Retrieved 18 April 2011 from http://www.noaa.gov/features/02_monitoring/balloon.html

*It's a Breeze: How Air Pressure Affects You: Feeling Pressured?* (2003). NASA. Retrieved 7 April 2011 from http://kids.earth.nasa.gov/archive/air_pressure/index.html

*Jet Stream*. (2003). Umpqua Discovery Center NOAA Weather Kiosk. Retrieved 12 April 2011 from http://www.wrh.noaa.gov/mfr/kiosk/jetstream.php

*The Jet Stream*. (2010). NOAA/National Weather Service. Retrieved 12 April 2011 from http://www.srh.noaa.gov/jetstream//global/jet.htm

*Pilot's Handbook of Aeronautical Knowledge.* (2010). U.S. Department of Transportation/Federal Aviation Administration. Retrieved 7 April 2011 from http://www.faa.gov/library/manuals/ aviation/pilot_handbook/

Wang, A., & Shinn, M. (2011, April 4). Southwest Grounds Planes After Roof Rips Open on 737. *USA Today.* Retrieved 12 April 2011 from http://travel.usatoday.com/flights/story/2011/ 04/Fuselage-hole-forces-Southwest-emergency-landing/45687532/1

*Water Cycle.* (2010). NASA. Retrieved 8 April 2011 from http://science.nasa.gov/earth-science/ oceanography/ocean-earth-system/ocean-water-cycle/

*The Water Cycle: Evaporation.* (2011). US Geological Survey. Retrieved 8 April 2011 from http://ga.water.usgs.gov/edu/watercycleevaporation.html

*The Water Cycle: Precipitation.* (2011). US Geological Survey. Retrieved 8 April 2011 from http://ga.water.usgs.gov/edu/watercycleprecipitation.html

*The Water Cycle: Sublimation.* (2011). US Geological Survey. Retrieved 8 April 2011 from http://ga.water.usgs.gov/edu/watercyclesublimation.html

*Weather for Kids: Subject: Weather Balloons.* (n.d.). NOAA. Retrieved 18 April 2011 from http://www.wrh.noaa.gov/mso/educate/balloon.html

## CHAPTER 2, LESSON 2  Weather Elements

*Air Masses.* (2010). NOAA/National Weather Service: JetStream—Online School for Weather. Retrieved 21 April 2011 from http://www.srh.weather.gov/jetstream/synoptic/airmass.htm

*Air Masses.* (2010). NOAA/National Weather Service Weather Forecast Office Corpus Christi, TX. Retrieved 19 April 2011 from http://www.srh.noaa.gov/crp/?n=education-airmasses

*Air Masses and Their Sources.* (2011). Weather Online. Retrieved 19 April 2011 from http://www .weatheronline.co.uk/reports/wxfacts/Air-masses-and-their-sources.htm

*Aviation Weather.* (n.d.). NASA Quest. Retrieved 18 April 2011 from http://quest.arc.nasa.gov/ aero/virtual/demo/weather/tutorial/tutorial1.html

*Boys' Life.* (July 1927). New York-Paris—Non-Stop. Retrieved 22 April 2011 from http://boyslife .org/wayback#issue=kG6MF1vRoZ4C&pg=24

Brain, M. (2011). *Charles Lindbergh's Transatlantic Flight.* HowStuffWorks. Retrieved 22 April 2011 from http://history.howstuffworks.com/american-history/charles-lindbergh.htm

*Evolution of the National Weather Service.* (n.d.). NOAA/National Weather Service/Eastern Region Headquarters. Retrieved 22 April 2011 from http://www.erh.noaa.gov/gyx/timeline.html

Hood, R. E. (May 1997). Lucky Lindy's Great Adventure. *Boys' Life.* Retrieved 22 April 2011 from http://boyslife.org/wayback#issue=9_4DAAAAMBAJ&pg=52

*It's a Breeze: How Air Pressure Affects You: How Air Pressure Signals Changes in the Weather.* (2005). NASA. Retrieved 20 April 2011 from http://kids.earth.nasa.gov/archive/air_pressure/weather .html

*Just What Is Lake Effect Snow?* (n.d.). NOAA/National Weather Service/Eastern Region Headquarters. Retrieved 21 April 2011 from http://www.erh.noaa.gov/buf/lakeffect/lakeintro.html

Lindbergh, C. A. (1953). *The Spirit of St. Louis.* New York: Charles Scribner's Sons.

Miller, M. (1999). *Climate of San Jose.* NOAA Technical Memorandum/NWS WR-259. Retrieved 21 April 2011 from http://www.wrh.noaa.gov/mtr/sfd_sjc_climate/sjc/SJC_CLIMATE3.php

*National Weather Service Glossary.* (2009). NOAA/National Weather Service. Retrieved 21 April 2011 from http://www.weather.gov/glossary/

*Pilot's Handbook of Aeronautical Knowledge.* (2010). U.S. Department of Transportation/Federal Aviation Administration. Retrieved 7 April 2011 from http://www.faa.gov/library/manuals/ aviation/pilot_handbook/

## CHAPTER 2, LESSON 3    Aviation Weather

*53rd Weather Reconnaissance Squadron 'Hurricane Hunters.'* (2011). USAF/403rd Wing. Retrieved 12 May 2011 from http://www.403wg.afrc.af.mil/library/factsheets/factsheet.asp?id=7483

*Clear Air Turbulence.* (2001). NOAA/National Environmental Satellite, Data, and Information Service (NESDIS). Retrieved 4 May 2011 from http://www.orbit.nesdis.noaa.gov/smcd/opdb/aviation/turb/CAT.html

*Clouds.* (2008). NOAA/National Weather Service/JetStream—Online School for Weather. Retrieved 25 April 2011 from http://oceanservice.noaa.gov/education/yos/resource/JetStream/synoptic/clouds.htm

*Convection in Thunderstorms: Building Atmospheric Giants.* (n.d.). NOAA/Earth System Research Laboratory/Physical Sciences Division. Retrieved 9 May 2011 from http://www.esrl.noaa.gov/psd/outreach/education/science/convection/Thunder.html

Fincher, L., & Read, B. (2006). *The 1943 'Surprise' Hurricane.* National Oceanic & Atmospheric Administration (NOAA), NOAA Central Library. Retrieved 12 May 2011 from http://www.history.noaa.gov/stories_tales/surprise.html

*Forecast Clear Air Turbulence Index Images.* (2006). NOAA/National Environmental Satellite, Data, and Information Service (NESDIS). Retrieved 4 May 2011 from http://www.orbit.nesdis.noaa.gov/smcd/opdb/aviation/turb/tifcsts.html

*Pilot's Handbook of Aeronautical Knowledge.* (2010). U.S. Department of Transportation/Federal Aviation Administration. Retrieved 7 April 2011 from http://www.faa.gov/library/manuals/aviation/pilot_handbook/

*Thunderstorm Hazards.* (2009). NOAA/National Weather Service Weather Forecast Office Key West, FL. Retrieved 5 May 2011 from http://www.srh.noaa.gov/key/?n=tstmhazards

*Thunderstorms and Severe Weather Spotting.* (n.d.). NOAA/National Weather Service/Eastern Region Headquarters. Retrieved 9 May 2011 from http://www.erh.noaa.gov/lwx/swep/Spotting.html

*Tropical Cyclone 101—An Introduction.* (2004). NOAA/National Weather Service/Southern Region Headquarters. Retrieved 12 May 2011 from http://www.srh.noaa.gov/tropical/awareness/tc101.htm

*What Is Lightning?* (n.d.). NOAA/National Weather Service/Flagstaff Weather Forecast Office. Retrieved 10 May 2011 from http://www.wrh.noaa.gov/fgz/science/lightnin.php?wfo=fgz

Williams, J. (2011). Airplane Turbulence Isn't as Dangerous as It Might Seem. *USA Today.* Retrieved 26 April 2011 from http://www.usatoday.com/weather/wturbwht.htm

## CHAPTER 2, LESSON 4    Weather Forecasting

*About the ROC.* (2010). NOAA's National Weather Service Radar Operations Center NEXRAD WSR-88D. Retrieved 26 May 2011 from http://www.roc.noaa.gov/WSR88D/About.aspx

*Air Force Weather History.* (2010). USAF/55th Wing Public Affairs Office. Retrieved 17 May 2011 from http://www.afweather.af.mil/library/factsheets/factsheet.asp?id=4145

The Anemometer. (2007, May 29). *Ask the Academy.* NASA. Retrieved 20 May 2011 from http://www.nasa.gov/offices/oce/appel/ask-academy/issues/ask-oce/AO_2-3_F_anemometer.html

*Automated Surface Observations.* (1999). NOAA/National Weather Service/Automated Surface Observing System. Retrieved 20 May 2011 from http://www.nws.noaa.gov/asos/obs.htm

*Automated Surface Observing Systems (ASOS).* (2010). NOAA/National Weather Service/JetStream—Online School for Weather. Retrieved 24 May 2011 from http://www.srh.noaa.gov/jetstream/remote/asos.htm

Cain, D. R. (2002). *Clear Air Mode.* NOAA National Weather Service. Retrieved 16 August 2011 from http://www.srh.noaa.gov/srh/sod/radar/radinfo/radinfo.html#clear

*Evolution of the National Weather Service.* (2010). NOAA's National Weather Service Public Affairs Office. Retrieved 17 May 2011 from http://www.weather.gov/pa/history/timeline.php

*Exactly How Does Radar Work?* (2010). NOAA National Weather Service JetStream-Online School for Weather. Retrieved 26 May 2011 from http://www.srh.noaa.gov/jetstream/doppler/how.htm

*FAI FSS—Basic Pilot Briefings.* (n.d.). Federal Aviation Administration. Retrieved 24 May 2011 from http://www.faa.gov/about/office_org/headquarters_offices/ato/service_units/systemops/fs/alaskan/alaska/fai/pfpwb/

Fenton, B. (2004, June 5). Bad Weather Nearly Brought Down D-Day. *The Telegraph.* Retrieved 26 May 2011 from http://www.telegraph.co.uk/news/uknews/1463671/Bad-weather-nearly-brought-down-D-Day.html

*GOES Project: Geostationary Operational Environmental Satellites: Project Current Status.* (2009). NASA/Goddard Space Flight Center. Retrieved 24 May 2011 from http://goespoes.gsfc.nasa.gov/goes/project/index.html

*GOES Project: Geostationary Operational Environmental Satellites: Project History.* (2008). NASA/Goddard Space Flight Center. Retrieved 24 May 2011 from http://goespoes.gsfc.nasa.gov/goes/project/history.html

*Investigative Tools.* (n.d.). The NASA Sci Files. Retrieved 20 May 2011 from http://scifiles.larc.nasa.gov/text/kids/Problem_Board/problems/stink/investtools2.html

Murnan, J., Tarp, K., & Purcell, D. (2006). *NEXRAD: Eye to the Sky.* NOAA Weather Partners Production. Video retrieved 26 May 2011 from http://www.roc.noaa.gov/WSR88D/About.aspx

*NASA and NOAA's GOES-P Satellite Successfully Launched.* (2010). NASA. Retrieved 24 May 2011 from http://www.nasa.gov/home/hqnews/2010/mar/HQ_10-056_GOES-P_launch.html

*The National Weather Service Marks Nine Decades of Aviation Weather Forecasting.* (2008). NOAA. Retrieved 17 May 2011 from http://www.noaa.gov/features/protecting_1208/weatherservice.html

*Pilot's Handbook of Aeronautical Knowledge.* (2010). U.S. Department of Transportation/Federal Aviation Administration. Retrieved 7 April 2011 from http://www.faa.gov/library/manuals/aviation/pilot_handbook/

*Psychrometer.* (n.d.). U.S. Centennial of Flight Commission. Retrieved 20 May 2011 from http://www.centennialofflight.gov/essay/Dictionary/Psychrometer/DI69.htm

*Radar, Radio Waves and Light.* (2005). NASA. Video retrieved 23 May 2011 from http://www.nasa.gov/audience/foreducators/topnav/materials/listbytype/Radar_Radio_Waves.html

Sanders, J. (2011). *Weather Team Provides Reports for Mission.* USAF. Retrieved 14 April 2011 from http://www.af.mil/news/story.asp?id=123242484

*Special Operations Weather Team.* (2010). USAF Special Operations Command Public Affairs Office. Retrieved 17 May 2011 from http://www.af.mil/information/factsheets/factsheet.asp?id=179

*Weather Instruments to Make: Sling Psychrometer.* (2010). NASA. Retrieved 20 May 2011 from http://science-edu.larc.nasa.gov/SCOOL/psychrometer.html

## CHAPTER 2, LESSON 5    The Effects of Weather on Aircraft

*Aircraft Accident Report: Air Florida, Inc., Boeing 737-222, N62AF, Collision with 14th Street Bridge Near Washington National Airport, Washington, D.C., January 13, 1982.* (1982). National Transportation Safety Board. Retrieved 1 June 2011 from http://library.erau.edu/worldwide/find/online-full-text/ntsb/aircraft-accident-reports.htm

*Aircraft Accident Report: Delta Air Lines, Inc., Lockheed L-1011-385-1, N726DA, Dallas/Fort Worth-International Airport, Texas, August 2, 1985.* (1986). National Transportation Safety Board. Retrieved 3 June 2011 from http://www.airdisaster.com/reports/ntsb/AAR86-05.pdf

*Aircraft Accident Report: In-Flight Separation of Vertical Stabilizer, American Airlines Flight 587, Airbus Industrie A300-605R, N14053, Belle Harbor, New York, November 12, 2001.* (2004). Aircraft Accident Report NTSB/AAR-04/04. Washington, DC: National Transportation Safety Board.

Barnstorff, K. (2010). *Wind Shear Accident Was Catalyst for Technology.* NASA Langley Research Center. Retrieved 3 June 2011 from http://www.nasa.gov/topics/aeronautics/features/microburst-windshear.html

Benson, J. T. (2007, February 21). Weather and Wreckage at Desert-One. *Air & Space Power Journal*. Retrieved 7 June 2011 from http://www.airpower.au.af.mil/airchronicles/cc/benson.html

Doomed Flight: Who Was Warned? (2009, March 23). *CBS News/AP*. Retrieved 8 June 2011 from http://www.cbsnews.com/stories/2004/10/26/national/main651450.shtml?source=search_story

*Dust Storm Safety*. (n.d.). NOAA/National Weather Service. Retrieved 6 June 2011 from http://www.nws.noaa.gov/om/brochures/duststrm.htm

Dust Storms, Sand Storms, and Related NOAA Activities in the Middle East. (2003, April 7). *NOAA Magazine*. US Department of Commerce/NOAA. Retrieved 6 June 2011 from http://www.magazine.noaa.gov/stories/mag86.htm

Enyedi, A. (2010). *What are PoPs?* NOAA/National Weather Service Jacksonville. Retrieved 3 June 2011 from http://www.srh.noaa.gov/jax/?n=probability_of_precipitation

*FTW88IA109*. (1991). National Transportation Safety Board. Retrieved 10 June 2011 from http://www.ntsb.gov/aviationquery/brief.aspx?ev_id=20001213X25693&key=1

Garrison, P. (2006, September 1). Flameout. *Air & Space Magazine*. Smithsonian. Retrieved 10 June 2011 from http://www.airspacemag.com/flight-today/flameout.html

Holloway, J. L. (1980). *[Iran Hostage] Rescue Mission Report August 1980: Statement of Admiral J. L. Holloway, III, USN (Ret.), Chairman, Special Operations Review Group*. The Navy Department Library. Retrieved 6 June 2011 from http://www.history.navy.mil/library/online/hollowayrpt.htm

*Hurlburt Remembers Operation Eagle Claw*. (2008). USAF/1st Special Operations Wing History Office. Retrieved 7 June 2011 from http://www2.hurlburt.af.mil/news/story.asp?id=123095779

Kamps, C. T. (2006, September 21). Operation Eagle Claw: The Iran Hostage Rescue. *Air & Space Power Journal*. Retrieved 6 June 2011 from http://www.airpower.au.af.mil/apjinternational/apj-s/2006/3tri06/kampseng.html#Kamps

Leary, W. M. (2002). *We Freeze to Please: History of NASA's Icing Research Tunnel and the Quest for Flight Safety*. NASA SP-2002-4226. NASA Office of External Relations/NASA History Office. Retrieved 1 June 2011 from http://history.nasa.gov/sp4226.pdf

*Making the Skies Safe From Windshear*. (2008). NASA Langley Research Center. Retrieved 3 June 2011 from http://www.nasa.gov/centers/langley/news/factsheets/Windshear.html

Michelle Obama's Plane Flew Less Than 3 Miles to Military Jet, Safety Officials Say. (2011, April 23). *Fox News*. Retrieved 8 June 2011 from http://www.foxnews.com/politics/2011/04/23/michelle-obamas-plane-flew-3-miles-military-jet-safety-officials-say/

NTSB Narrows Investigation of Flight 587 Crash. (2001, December 19). *CNN*. Retrieved 8 June 2011 from http://articles.cnn.com/2001-12-19/us/flight.587.crash_1_rudder-system-yaw-damper-ntsb/2?_s=PM:US

One Hundred Years of Powered Flight. (2011). *Spinoff*. NASA. Retrieved 3 June 2011 from http://www.sti.nasa.gov/tto/spinoff2003/pf_1.html

*Pilot's Handbook of Aeronautical Knowledge*. (2010). U.S. Department of Transportation/Federal Aviation Administration. Retrieved 7 April 2011 from http://www.faa.gov/library/manuals/aviation/pilot_handbook/

Shaikh, T. (2011, July 29). Air France Crash Pilots Lost Vital Speed Data, Say Investigators. *CNN*. Retrieved 17 August 2011 from http://edition.cnn.com/2011/WORLD/americas/05/27/air.france.447.crash/index.html?hpt=T1

TACA Airlines 110 'Flameout.' (1988). *MSNBC*. Retrieved 10 June 2011 from http://www.youtube.com/watch?v=IPn8G7enbF4

*VMS Flight Controls*. (2008). NASA Aviation Systems Division. Retrieved 3 June 2011 from http://www.aviationsystemsdivision.arc.nasa.gov/facilities/vms/controls.shtml#overview

*Weather*. (2010). NASA Virtual Skies. Retrieved 2 June 2011 from http://virtualskies.arc.nasa.gov/weather/4.html

*Wind Chill*. (2010). NOAA/National Weather Service/JetStream—Online School for Weather. Retrieved 2 June 2011 from http://www.srh.noaa.gov/jetstream/global/chill.htm

*Advanced Aerospace Medicine On-line: Section I—The Aerospace Environment: 1. Physical Characteristics: 1.1 The Atmosphere.* (2010). FAA/Civil Aerospace Medical Institute. Retrieved 15 June 2011 from http://www.faa.gov/other_visit/aviation_industry/designees_delegations/designee_types/ame/media/AMSACraymanFINALREV.doc

*Advanced Aerospace Medicine On-line: Section II—Aviation Operations: Aviation Physiology (Environmental Stress Factors): 1.1 Respiration and Circulation.* (2010). FAA/Civil Aerospace Medical Institute. Retrieved 16 June 2011 from http://www.faa.gov/other_visit/aviation_industry/designees_delegations/designee_types/ame/media/Section%20II.1.1%20Respiration%20and%20Circulation.doc

*Aeronautical Information Manual: Official Guide to Basic Flight Information and ATC Procedures.* (2011). Chapter 8. Medical Facts for Pilots. Federal Aviation Administration. Retrieved 17 June 2011 from http://www.faa.gov/air_traffic/publications/ATpubs/AIM/Chap8/aim0801.html

*Air Accident Brief: DCA00MA005, Sunjet Aviation, Learjet Model 35, N47BA, Aberdeen, South Dakota, October 25, 1999.* (n.d.). National Transportation Safety Board. Retrieved 14 June 2011 from http://www.ntsb.gov/Publictn/2000/aab0001.htm

Antuñano, M. J. (n.d.). *Spatial Disorientation.* FAA/Civil Aerospace Medical Institute/Aerospace Medical Education Division. Retrieved 21 June 2011 from http://www.faa.gov/pilots/safety/pilotsafetybrochures/media/SpatialD.pdf

*Atmospheric Structure.* (2010). NASA/Goddard Earth Sciences Data and Information Services Center. Retrieved 20 June 2011 from http://disc.sci.gsfc.nasa.gov/ozone/additional/science-focus/about-ozone/atmospheric_structure.shtml

*Aviation Maintenance Technician Handbook.* (2008). US Department of Transportation/Federal Aviation Administration. Retrieved 15 June 2011 from http://www.faa.gov/library/manuals/aircraft/amt_handbook/

*Aviation Weather: Atmosphere.* (n.d.). NASA/Ames Research Center. Retrieved 20 June 2011 from http://quest.arc.nasa.gov/aero/virtual/demo/weather/tutorial/tutorial2a.html

*Boyle's Law.* (2011). NASA. Retrieved 15 June 2011 from http://www.grc.nasa.gov/WWW/K-12/airplane/boyle.html

Brown, J. R., & Antuñano, M. J. (n.d.). *Altitude-Induced Decompression Sickness.* FAA/Civil Aerospace Medical Institute/Aeromedical Education Division. Retrieved 15 June 2011 from http://www.faa.gov/pilots/safety/pilotsafetybrochures/media/DCS.pdf

Cavagnaro, M. (2007). *The Good, the Bad, and the Ozone.* NASA/John F. Kennedy Space Center. Retrieved 20 June 2011 from http://www.nasa.gov/missions/earth/f-ozone.html

*Col. (Dr.) John Paul Stapp.* (n.d.). USAF. Retrieved 22 June 2011 from http://www.af.mil/information/heritage/person_print.asp?storyID=123006472

Coulter, G. R., & Vogt, G. L. (n.d.). *The Effects of Space Flight on the Human Vestibular System.* NASA. Retrieved 21 June 2011 from http://weboflife.nasa.gov/learningResources/vestibularbrief.htm

*Earth's Atmosphere.* (2010). NASA. Retrieved 20 June 2011 from http://www.nasa.gov/audience/forstudents/9-12/features/912_liftoff_atm.html

Elert, G (Ed.). (2011). *The Physics Factbook.* Retrieved 15 September 2011 from http://hypertextbook.com/facts/JianHuang.shtml

Elert, G. (2011). *The Physics Hypertextbook.* Retrieved 15 September 2011 from http://physics.info/frames/

Freudenrich, C. (2011). *How Scuba Works.* HowStuffWorks. Retrieved 18 August 2011 from http://adventure.howstuffworks.com/outdoor-activities/water-sports/scuba4.htm

*How Much Oxygen Does a Person Consume in a Day?* (2011). Discovery Communications. Retrieved 18 August 2011 from http://health.howstuffworks.com/human-body/systems/respiratory/question98.htm

*How the Heart Works: Anatomy of the Heart.* (n.d.). US Department of Health & Human Services/ National Institutes of Health/National Heart Lung and Blood Institute. Retrieved 16 June 2011 from http://www.nhlbi.nih.gov/health/dci/Diseases/hhw/hhw_anatomy.html

*How the Lungs Work: The Respiratory System.* (n.d.). US Department of Health & Human Services/ National Institutes of Health/National Heart Lung and Blood Institute. Retrieved 17 June 2011 from http://www.nhlbi.nih.gov/health/dci/Diseases/hlw/hlw_respsys.html

*Human Vestibular System in Space.* (2009). NASA. Retrieved 21 June 2011 from http://www.nasa .gov/audience/forstudents/9-12/features/F_Human_Vestibular_System_in_Space.html

*Hypoxia: The Higher You Fly … the Less Air in the Sky.* (2004). FAA/Civil Aerospace Medical Institute/ Aerospace Medical Education Division. Retrieved 15 June 2011 from http://www .faa.gov/pilots/safety/pilotsafetybrochures/media/hypoxia.pdf

*It's the Law.* (2007). NOAA. Retrieved 15 June 2011 from http://oceanexplorer.noaa.gov/ explorations/07philippines/background/edu/media/gaslaws.pdf

Jackson, S. (2011). *How Does Breathing Into a Paper Bag Help Hyperventilation?* eHow. Retrieved 17 June 2011 from http://www.ehow.com/how-does_4579147_breathing-paper-bag-help -hyperventilation.html

*Lesson 1A: Composition and Structure of the Atmosphere.* (n.d.). NASA/Goddard Space Flight Center. Retrieved 20 June 2011 from http://education.gsfc.nasa.gov/ess/units/unit2/u211a.html

Marconi, E. M. (2007). *NASA Honors Life-Saving 'Invention of the Year.'* NASA/John F. Kennedy Space Center. Retrieved 22 June 2011 from http://www.nasa.gov/missions/research/inventor_ award_0727.html

*Mixed Up in Space.* (2011). NASA. Retrieved 18 August 2011 from http://science.nasa.gov/science -news/science-at-nasa/2001/ast07aug_1/

*NBC Special Celebrates Life, Legend of Payne Stewart.* (2009). PGA Tour. Retrieved 14 June 2011 from http://www.pgatour.com/2009/tournaments/r060/09/10/payne_stewart/index.html

Pearce, J. (2009, March 26). Earl H. Wood Is Dead at 97; Helped Invent G-Suit. *The New York Times.* Retrieved 18 August 2011 from http://www.nytimes.com/2009/03/27/us/27wood. html?ref=science

*Pilot's Handbook of Aeronautical Knowledge.* (2010). U.S. Department of Transportation/Federal Aviation Administration. Retrieved 7 April 2011 from http://www.faa.gov/library/manuals/ aviation/pilot_handbook/

*Pressure Suits.* (2006). Canadian Space Agency. Retrieved 18 August 2011 from http://www.asc -csa.gc.ca/eng/astronauts/osm_aviation.asp#pressure

Spark, N. (2003). Fastest Man on Earth! *Wings and Airpower,* July 2003. Retrieved 16 September 2011 from http://www.ejectionsite.com/stapp.htm

Storey, R. A. (2010). *Trapped Gas.* FAA Airman Education Programs. Retrieved 15 June 2011 from http://www.faa.gov/pilots/training/airman_education/topics_of_interest/trapped_gas/

Wyrick, B., & Brown, J. R. (2010). *Acceleration in Aviation: G-Force.* FAA/Civil Aerospace Medical Institute/Aerospace Medical Education Division. Retrieved 21 June 2011 from http://www.faa .gov/pilots/safety/pilotsafetybrochures/media/Acceleration.pdf

## CHAPTER 3, LESSON 2   Protective Equipment and Aircrew Training

*Aboard NASA's 'Vomit Comet.'* (2008). Rochester Institute of Technology: University News. Video retrieved 1 July 2011 from http://www.youtube.com/watch?v=2V9h42yspbo

Annis, J. F., & Webb, P. (1971). *Development of a Space Activity Suit.* NASA. Retrieved 27 June 2011 from http://ntrs.nasa.gov/archive/nasa/casi.ntrs.nasa.gov/19720005428_1972005428.pdf

Benroth, N. (2010). *Not Just an Average Flight Simulator.* USAF. Retrieved 1 July 2011 from http:// www.af.mil/news/story.asp?id=123229544

*Biographical Data: Michael T. Good (Colonel, USAF, RET.).* (2011). NASA Lyndon B. Johnson Space Center. Retrieved 11 July 2011 from http://www.jsc.nasa.gov/Bios/htmlbios/good-mt.html

*A Brief History of the Pressure Suit.* (2008). NASA/Dryden Flight Research Center. Retrieved 27 June 2011 from http://www.nasa.gov/centers/dryden/research/AirSci/ER-2/pshis.html

Brown, J. J. (2010). *Egress Airmen Ensure 'Last Chance for Life.'* USAF. Retrieved 29 June 2011 from http://www.af.mil/news/story.asp?id=123213174

Brown, J. R., & Salamanca, A. (n.d.). *Oxygen Equipment: Use in General Aviation Operations.* Federal Aviation Administration/Civil Aerospace Medical Institute/Aerospace Medical Education Division. Retrieved 24 June 2011 from http://www.faa.gov/pilots/safety/pilotsafetybrochures/media/Oxygen_Equipment.pdf

*C-9B Flight Trajectory.* (2010). NASA/JSC Aircraft Operations. Retrieved 30 June 2011 from http://jsc-aircraft-ops.jsc.nasa.gov/Reduced_Gravity/trajectory.html

Congiu, S. (2010). It's a Bird. It's a Plane. It's a Student? *The Researcher News.* NASA Langley Research Center. Retrieved 1 July 2011 from http://www.nasa.gov/centers/langley/news/researchernews/rn_ZeroGravity.html

*The Effects of Space Flight on the Human Vestibular System.* (n.d.). NASA. Retrieved 30 June 2011 from http://weboflife.nasa.gov/pdf/vbrief.pdf

*Ejection Seats.* (2005). GlobalSecurity.org. Retrieved 29 June 2011 from http://www.globalsecurity.org/military/systems/aircraft/systems/eject.htm

Escutia, S. (2010). *Hollman centrifuge takes its last spin. Air Force News Service.* Retrieved 7 October 2011 from http://www.af.mil/news/story.asp?id=123229030

*First Generation X-1.* (2009). NASA/Dryden Flight Research Center. Retrieved 27 June 2011 from http://www.nasa.gov/centers/dryden/news/FactSheets/FS-085-DFRC.html

Flora, S. (2009). *AFSO 21 Makes Processes Smarter, Faster, Cheaper.* USAF. Retrieved 29 June 2011 from http://www.af.mil/news/story.asp?id=123151709

*The History of KC-135A Reduced Gravity Research Program.* (2010). NASA/JSC Aircraft Operations. Retrieved 30 June 2011 from http://jsc-aircraft-ops.jsc.nasa.gov/Reduced_Gravity/KC_135_history.html

*How Skydiving Works.* (2011). HowStuffWorks: A Discovery Company. Retrieved 28 June 2011 from http://adventure.howstuffworks.com/skydiving2.htm

*Hubble Space Telescope.* (2010). NASA. Retrieved 11 July 2011 from http://www.nasa.gov/mission_pages/hubble/story/index.html

*Hypoxia: The Higher You Fly … the Less Air in the Sky.* (2004). FAA/Civil Aerospace Medical Institute/Aerospace Medical Education Division. Retrieved 15 June 2011 from http://www.faa.gov/pilots/safety/pilotsafetybrochures/media/hypoxia.pdf

Kaufman, D. (2010). *Six altitude chambers coming to Wright-Patt.* 88th Air Base Wing Public Affairs. Retrieved 7 October 2011 from http://www.wpafb.af.mil/news/story.asp?id=123210840

May, A. M. (2010). *Mildenhall Special Operations Airmen Commemorate D-Day Landings.* Air Force Print News Today. Retrieved 28 June 2011 from http://www.af.mil/news/story_print.asp?id=123208683

McGloin, B. (2010). *Squadron Aims to Reduce Use of Air-Sickness Bags.* Air Force Print News Today. Retrieved 1 July 2011 from http://www.af.mil/news/story_print.asp?id=123228892

*NASA Physiological Training.* (n.d.). FAA. Retrieved 30 June 2011 from http://www.faa.gov/about/office_org/field_offices/cmo/coa/media/NASA%20PHYSIOLOGICAL%20TRAINING.pdf

*Parachute—History.* (2006). GlobalSecurity.org. Retrieved 28 June 2011 from http://www.globalsecurity.org/military/systems/aircraft/systems/parachute-history.htm

*Pilot's Handbook of Aeronautical Knowledge.* (2010). U.S. Department of Transportation/Federal Aviation Administration. Retrieved 7 April 2011 from http://www.faa.gov/library/manuals/aviation/pilot_handbook/

*Preflight Interview: Michael Good, Mission Specialist.* (2009). NASA. Retrieved 11 July 2011 from http://www1.nasa.gov/mission_pages/shuttle/shuttlemissions/sts125/main/interview_Good.html

*The Pull of Hypergravity.* (2003). NASA Science News. Retrieved 7 October 2011 from http://science.nasa.gov/science-news/science-at-nasa/2003/07feb_stronggravity/

*Q & A: Parachutes.* (2009). Ask the Van. Department of Physics, University of Illinois at Urbana-Champaign. Retrieved 28 June 2011 from http://van.physics.illinois.edu/qa/listing.php?id=2087

*Spacewalking History.* (n.d.). NASA Quest. Retrieved 27 June 2011 from http://quest.nasa.gov/space/teachers/suited/4space1.html

Stern, D. P. (2004). *Accelerated Frames of Reference: Inertial Forces.* NASA/Goddard Space Flight Center. Retrieved 30 June 2011 from http://www-istp.gsfc.nasa.gov/stargaze/Sframes2.htm

Stern, D. P. (2004). *Frames of Reference: The Centrifugal Force.* NASA/Goddard Space Flight Center. Retrieved 30 June 2011 from http://www-istp.gsfc.nasa.gov/stargaze/Sframes3.htm

*U-2 Dragon Lady Arrives.* (n.d.). USAF. Retrieved 28 June 2011 from http://www.af.mil/information/heritage/spotlight.asp?storyID=123021876

Wicke, R. (2008). *Air Force Trains Coast Guard Students in Altitude Chamber.* USAF. Retrieved 30 June 2011 from http://www.af.mil/news/story.asp?id=123096069

*Zero-Gravity Plane on Final Flight.* (2007). NASA. Retrieved 30 June 2011 from http://www.nasa.gov/vision/space/preparingtravel/kc135onfinal.html

## CHAPTER 4, LESSON 1  Navigational Elements

*Airway Beacons.* (n.d.). Montana Department of Transportation. Retrieved 11 October 2011 from http://www.mdt.mt.gov/aviation/beacons.shtml

Doody, D. (2011). *Basics of Space Flight, Section 1: Chapter 2. Reference Systems.* NASA/Jet Propulsion Laboratory/California Institute of Technology. Retrieved 22 July 2011 from http://www2.jpl.nasa.gov/basics/bsf2-1.php

*Earth: Prove That It Is Round.* (2011). Newton: Ask a Scientist. Department of Energy Office of Science/Argonne National Laboratory. Retrieved 21 July 2011 from http://www.newton.dep.anl.gov/askasci/gen99/gen99427.htm

*Great Circle.* (2009). US Geological Survey. Retrieved 27 July 2011 from http://earthquake.usgs.gov/learn/glossary/?term=great%20circle

*Map Projections.* (2006). US Geological Survey. Retrieved 21 July 2011 from http://egsc.usgs.gov/isb/pubs/MapProjections/projections.html

*Map Projections: From Spherical Earth to Flat Map.* (2011). US Department of the Interior. Retrieved 26 July 2011 from http://www.nationalatlas.gov/articles/mapping/a_projections.html

Mola, R. (n.d.). *Aircraft Navigation Technology.* U.S. Centennial of Flight Commission. Retrieved 21 July 2011 from http://www.centennialofflight.gov/essay/Evolution_of_Technology/navigation_tech/Tech33.htm

Mola, R. (n.d.). *The Evolution of Airway Lights and Electronic Navigation Aids.* U.S. Centennial of Flight Commission. Retrieved 20 July 2011 from http://www.centennialofflight.gov/essay/Government_Role/navigation/POL13.htm

Nalty, B. C. (Ed.). (1997). *Winged Shield, Winged Sword: A History of the United States Air Force, Volume I.* Washington, D.C.: Air Force History and Museums Program, United States Air Force.

*Navigation Services—History.* (2011). Federal Aviation Administration. Retrieved 20 July 2011 from http://www.faa.gov/about/office_org/headquarters_offices/ato/service_units/techops/navservices/history/

*Pilot's Handbook of Aeronautical Knowledge.* (2010). U.S. Department of Transportation/Federal Aviation Administration. Retrieved 7 April 2011 from http://www.faa.gov/library/manuals/aviation/pilot_handbook/

Shaw, W. (2010). *Calculations.* NASA Virtual Skies. Retrieved 25 July 2011 from http://virtualskies.arc.nasa.gov/navigation/6.html

Shaw, W. (2010). *Measurement.* NASA Virtual Skies. Retrieved 25 July 2011 from http://virtualskies.arc.nasa.gov/navigation/5.html

Snyder, J. P. (1987). *Map Projections—A Working Manual.* US Geological Survey Professional Paper 1395. Retrieved 26 July 2011 from http://onlinepubs.er.usgs.gov/djvu/PP/PP_1395.pdf

*Air Navigation: You Decide: Airspace Classes and VFR Minimums*. (2003). NASA Quest. Retrieved 10 August 2011 from http://quest.arc.nasa.gov/aero/virtual/demo/navigation/youDecide/airspace .html

*A Chart Is Not a Map*. (2011). NOAA National Ocean Service. Retrieved 8 August 2011 from http://www.nauticalcharts.noaa.gov/mcd/learn_diff_map_chart.html

*Code of Federal Regulations Sec. 91.159: Part 91 General Operating and Flight Rules: VFR Cruising Altitude or Flight Level*. (2004). Federal Aviation Administration. Retrieved 2 August 2011 from http://rgl.faa.gov/Regulatory_and_Guidance_Library/rgFAR.nsf/0/5a982143832ae11a 86256e00004f6e35!OpenDocument

*Code of Federal Regulations Sec. 91.179: Part 91 General Operating and Flight Rules: IFR Cruising Altitude of Flight Level*. (2007). Federal Aviation Administration. Retrieved 22 August 2011 from http://rgl.faa.gov/Regulatory_and_Guidance_Library/rgFAR.nsf/0/71cc6e7b8f767f748 6257384006cb2f7!OpenDocument

*Code of Federal Regulations Sec. 99.9: Part 99 Security Control of Air Traffic*. (n.d.). Federal Aviation Administration. Retrieved 1 August 2011 from http://rgl.faa.gov/Regulatory_and_Guidance_ Library/rgFAR.nsf/0/73DB8EF69BBF6171862571650050ED42?OpenDocument&Highlight =dvfr

*Compass Needle*. (1999). Newton: Ask a Scientist. Department of Energy Office of Science/Argonne National Laboratory. Retrieved 4 August 2011 from http://www.newton.dep.anl.gov/askasci/ gen99/gen99029.htm

*Differences Between Maps & Charts*. (n.d.). NOAA Office of Coast Survey. Retrieved 8 August 2011 from http://www.nauticalcharts.noaa.gov/mcd/learn_diff_map_chart.html

*Earth's Magnetism*. (n.d.). NOAA: Science on a Sphere. Retrieved 4 August 2011 from http://sos .noaa.gov/datasets/Land/earths_magnetism.html

*FAA Flight Plan*. (n.d.). Federal Aviation Administration. Retrieved 1 August 2011 from http://www.google.com/url?sa=t&source=web&cd=1&sqi=2&ved=0CBkQFjAA&url= http%3A%2F%2Fwww.faa.gov%2FdocumentLibrary%2Fmedia%2Fform%2Ffaa7233-1 .pdf&rct=j&q=faa%20flight%20plan&ei=O_U2TvTmI4Xk0QGT65XLCQ&usg= AFQjCNGiBg-C0SFfOXA4y5oBcvqZlSBJiw&cad=rja

*Federal Aviation Administration Aeronautical Information Manual: Official Guide to Basic Flight Information and ATC Procedures*. (2011). FAA. Retrieved 9 August 2011 from http://www .faa.gov/air_traffic/publications/atpubs/aim/index.htm

*Final Environmental Impact Statement (EIS): Military Training Activities at Makua Military Reservation, Hawai'i: Volume 2, Appendix D, Airspace*. (2009). Department of the Army. Retrieved 9 August 2011 from http://www.garrison.hawaii.army.mil/makuaeis/FinalDocs/Makua%20FEIS% 20Volume%202%20Appendix%20D_Airspace.pdf

Henderson, J. (2001, November 1). How Things Work: Celestial Navigation. *Air & Space*. Retrieved 3 August 2011 from http://www.airspacemag.com/flight-today/celestial.html?device=other&c=y

*The Magnetic South Pole*. (n.d.). Woods Hole Oceanographic Institution. Retrieved 4 August 2011 from http://deeptow.whoi.edu/southpole.html

McDonell, M. (1973 June). Lost Patrol. *Naval Aviation News*. Naval History & Heritage Command, Department of the Navy. Retrieved 11 August 2011 from http://www.history.navy.mil/faqs/ faq15-2.htm

*McGraw-Hill Concise Encyclopedia of Science and Technology (5th ed.)*. (2005). New York: McGraw-Hill.

*The Mystery of Flight 19—The Lost Patrol*. (n.d.). Naval Air Station Fort Lauderdale Museum. Retrieved 11 August 2011 from http://www.nasflmuseum.com/flight-19.html

*Pilot's Handbook of Aeronautical Knowledge*. (2010). U.S. Department of Transportation/Federal Aviation Administration. Retrieved 7 April 2011 from http://www.faa.gov/library/manuals/ aviation/pilot_handbook/

Shaw, W. (2010). *Aeronautical Charts*. NASA Virtual Skies. Retrieved 10 August 2011 from http://virtualskies.arc.nasa.gov/navigation/7.html

Stern, D. P. (2004). *From Stargazers to Starships: (5a) Navigation*. NASA/Goddard Space Flight Center. Retrieved 3 August 2011 from http://www-istp.gsfc.nasa.gov/stargaze/Snavigat.htm

## CHAPTER 4, LESSON 3   Dead Reckoning and Wind

*Expedition Purpose*. (2010). NOAA Ocean Explorer. Retrieved 23 August 2011 from http://oceanexplorer.noaa.gov/explorations/08auvfest/background/edu/purpose.html

*Gyroscopes*. (n.d.). U.S. Centennial of Flight Commission. Retrieved 26 August 2011 from http://www.centennialofflight.gov/essay/Dictionary/Gyroscopes/DI105.htm

Helfrick, A. D. (2010). *Principles of Avionics (6th Ed.)*. Leesburg, Virginia: Avionics Communications Inc.

*Instrument Flying Handbook*. (2009). U.S. Department of Transportation/Federal Aviation Administration. Retrieved 26 August 2011 from http://www.faa.gov/library/manuals/aviation/instrument_flying_handbook/

Lindbergh, C. A. (1953). *The Spirit of St. Louis*. New York: Charles Scribner's Sons.

*Pilot's Handbook of Aeronautical Knowledge*. (2010). U.S. Department of Transportation/Federal Aviation Administration. Retrieved 7 April 2011 from http://www.faa.gov/library/manuals/aviation/pilot_handbook/

Ruggero, E. (2003). *Combat Jump: The Young Men Who Led the Assault into Fortress Europe, July 1943*. HarperCollins. Retrieved 29 August 2011 from http://edruggero.com/combat-jump.html

Shaw, W. (2010). *Advanced Navigation*. NASA Virtual Skies. Retrieved 26 August 2011 from http://virtualskies.arc.nasa.gov/navigation/4.html

Shaw, W. (2010). *Basic Navigation*. NASA Virtual Skies. Retrieved 23 August 2011 from http://virtualskies.arc.nasa.gov/navigation/2.html

Solms, B., & Wiest, P. (1998). *Trigonometric Solutions to a Dead Reckoning Air Navigation Problem Using Vector Analysis and Advanced Organizers*. National Security Agency. Retrieved 23 August 2011 from http://www.nsa.gov/academia/_files/collected_learning/high_school/geometry/dead_reckoning.pdf

## CHAPTER 4, LESSON 4   Flight Instrumentation

*Aviation Accident Report: Loss of Thrust in Both Engines After Encountering a Flock of Birds and Subsequent Ditching on the Hudson River, US Airways Flight 1549, Airbus A320-214, N106US, Weehawken, New Jersey, January 15, 2009*. (2010). National Transportation Safety Board. Retrieved 6 September 2011 from http://www.ntsb.gov/doclib/reports/2010/AAR1003.pdf

*Horizontal Situation Indicator*. (2002). NASA. Retrieved 2 September 2011 from http://spaceflight.nasa.gov/shuttle/reference/shutref/orbiter/avionics/dds/hsi.html

*Instrument Flying Handbook*. (2009). U.S. Department of Transportation/Federal Aviation Administration. Retrieved 26 August 2011 from http://www.faa.gov/library/manuals/aviation/instrument_flying_handbook/

*Pilot's Handbook of Aeronautical Knowledge*. (2010). U.S. Department of Transportation/Federal Aviation Administration. Retrieved 7 April 2011 from http://www.faa.gov/library/manuals/aviation/pilot_handbook/

Shiner, L. (2009, February 18). A & S Interview: Sully's Tale. *Air & Space Smithsonian*. Retrieved 6 September 2011 from http://www.airspacemag.com/flight-today/Sullys-Tale.html

SRM Safety Reliability Methods. (2011). About Us. Retrieved 6 September 2011 from http://safetyreliability.com/about-us.html

Stern, J. M. (1995). Airspeed Indicator. *Microsoft Flight Simulator Handbook*. Retrieved 31 August 2011 from http://www.flightsimbooks.com/flightsimhandbook/CHAPTER_02_23_Airspeed_Indicator.php

Types of Airspeed. (n.d.). Aerospaceweb.org. Retrieved 24 October 2011 from http://www
.aerospaceweb.org/question/instruments/q0251.shtml

Welch, C. (2009). Hero of the Hudson Visits Alma Mater. *Air Force Print News Today*. USAF.
Retrieved 6 September 2011 from http://www.af.mil/news/story_print.asp?id=123144908

## CHAPTER 4, LESSON 5   Navigation Technology

*Aeronautical Information Manual: Chapter 1. Air Navigation. Section 1. Navigation Aids.* (2011).
Federal Aviation Administration. Retrieved 14 September 2011 from http://www.faa.gov/air_
traffic/publications/atpubs/aim/Chap1/aim0101.html

*Air Classics E6-B Flight Computer Instructions*. (2000). Newcastle, Washington: Aviation
Supplies & Academics, Inc. Retrieved 12 September 2011 from http://www.google.com/
url?sa=t&source=web&cd=3&sqi=2&ved=0CC8QFjAC&url=http%3A%2F%2Fsportys.
com%2Fsource%2Fimages%2F7205.pdf&rct=j&q=jeppesen%20e6b%20manual&ei=CiVu
TrbOG9DI0AHM3fmiBQ&usg=AFQjCNGGA-l9oOiG0UivCCD5iH_EOxhezA&cad=rja

Altus, S. (2009). Effective Flight Plans Can Help Airlines Economize. *Aero*. Retrieved 20 September
2011 from http://www.boeing.com/commercial/aeromagazine/articles/qtr_03_09/article_08_1
.html

*Block II Satellite Information*. (2011). United States Naval Observatory. Retrieved 19 September
2011 from ftp://tycho.usno.navy.mil/pub/gps/gpsb2.txt

*A Brief History of Satellite Navigation*. (1995). Stanford University News Service. Retrieved
19 September 2011 from http://www.stanford.edu/dept/news/pr/95/950613Arc5183.html

Bureau of Labor Statistics, US Department of Labor. (2009). Air Traffic Controllers. *Occupational
Outlook Handbook, 2010–11 Edition*. Retrieved 16 September 2011 from http://www.bls.gov/
oco/ocos108.htm

*F-22 Raptor Avionics*. (2011). GlobalSecurity. Retrieved 19 September 2011 from http://www
.globalsecurity.org/military/systems/aircraft/f-22-avionics.htm

*F-35 Joint Strike Fighter (JSF) Lightning II*. (2011). GlobalSecurity. Retrieved 19 September 2011
from http://www.globalsecurity.org/military/systems/aircraft/f-35-corps.htm

*Fact Sheet—NextGen*. (2007). Federal Aviation Administration. Retrieved 15 September 2011
from http://www.faa.gov/news/fact_sheets/news_story.cfm?newsid=8145

*Frequently Asked Questions*. (2011). National Executive Committee for Space-Based PNT.
Retrieved 16 September 2011 from http://www.pnt.gov/public/faq.shtml#difference

*GBU-15*. (2007). USAF. Retrieved 19 September 2011 from http://www.af.mil/information/
factsheets/factsheet.asp?id=105

Gordon, B. (2011). *ADS-B Broadcast Services: See What You've Been Missing*. Federal Aviation
Administration. Video retrieved 15 September 2011 from http://www.faa.gov/nextgen/
portfolio/trans_support_progs/adsb/

*GPS Privacy Jammers and RFI at Newark*. (2011 March). Federal Aviation Administration.
Retrieved 20 September 2011 from http://laas.tc.faa.gov/CoWorkerFiles/GBAS%20RFI%
202011%20Public%20Version%20Final.pdf

Hansen, R. (2006). *JDAM Continues to Be Warfighter's Weapon of Choice*. USAF. Retrieved
19 September 2011 from http://www.af.mil/news/story.asp?storyID=123017613

*Instrument Flying Handbook*. (2009). U.S. Department of Transportation/Federal Aviation
Administration. Retrieved 26 August 2011 from http://www.faa.gov/library/manuals/
aviation/instrument_flying_handbook/

*Jamming the Global Positioning System—A National Security Threat: Recent Events and Potential
Cures*. (2010, November 4). National PNT Advisory Board. Retrieved 20 September 2011
from http://www.pnt.gov/advisory/recommendations/2010-11-jammingwhitepaper.pdf

*Joint Direct Attack Munition GBU-31/32/38*. (2007). USAF. Retrieved 19 September 2011 from
http://www.af.mil/information/factsheets/factsheet.asp?id=108

Lindbergh, C. A. (1953). *The Spirit of St. Louis*. New York: Charles Scribner's Sons.

*A Look Inside the Predator.* (2011). How the Predator UAV Works. HowStuffWorks. Retrieved 19 September 2011 from http://science.howstuffworks.com/predator2.htm

*LORAN-C General Information.* (2011). US Coast Guard Navigation Center. Retrieved 15 September 2011 from http://www.navcen.uscg.gov/?pageName=loranMain

*LTJG Philip Dalton (1903–1941).* (n.d.). EAA. Retrieved 12 September 2011 from http://www.eaa.org/apps/obituaries/MemorialWall2.aspx?ID=1277

Luckett, P. D. (1983 September–October). Technology and Modern Leadership: Charles Lindbergh, A Case Study. *Air University Review.* Retrieved 29 August 2011 from http://www.airpower.au.af.mil/airchronicles/aureview/1983/sep-oct/luckett.html

*NextGen 101: Giving the World New Ways to Fly.* (2011). Federal Aviation Administration. Video retrieved 15 September 2011 from http://www.faa.gov/nextgen/

Nystrom, D. (2010). *Airmen Make Impact with First GBU-54 Combat Drop in Afghanistan.* USAF. Retrieved 19 September 2011 from http://www.af.mil/news/story.asp?id=123224886

Petosky, E. (2009). *Navigation Aids Keep F-15s on Course.* USAF. Retrieved 13 September 2011 from http://www.af.mil/news/story.asp?id=123174482

*Pilot's Handbook of Aeronautical Knowledge.* (2010). U.S. Department of Transportation/Federal Aviation Administration. Retrieved 7 April 2011 from http://www.faa.gov/library/manuals/aviation/pilot_handbook/

*Satellite Navigation: US Policy and Management.* (2011). NOAA. Retrieved 19 September 2011 from http://www.space.commerce.gov/gps/policy.shtml

*Surveillance and Broadcast Services: ADS-B Broadcast Services: See What You've Been Missing.* (2010). Federal Aviation Administration. Retrieved 15 September 2011 from http://www.faa.gov/nextgen/portfolio/trans_support_progs/adsb/broadcastservices/

# Glossary

## A

**absolute altitude**—an aircraft's vertical distance above the terrain, that is, above ground level (AGL). (p. 329)

**accelerometer**—an instrument that measures acceleration in one direction. (p. 315)

**acoustic**—having to do with sound. (p. 98)

**adiabatic cooling**—the process of cooling air through expansion. (p. 152)

**adiabatic heating**—the process of heating dry air through compression. (p. 152)

**aerodynamics**—the way objects move through air. (p. 7)

**aft**—toward the tail. (p. 45)

**agonic line**—a line where magnetic variation is zero. (p. 285)

**ailerons**—a small hinged section on the outboard portion of a wing. (p. 57)

**airfoil**—a structure—such as a wing or propeller blade—that when exposed to a flow of air generates a force. (p. 7)

**air masses**—a large body of air having fairly uniform properties of temperature and moisture. (p. 135)

**air pressure**—the force exerted by the air on a unit area of surface. (p. 12)

**airspace**—that space which lies above a country and is subject to a nation's laws. (p. 289)

**airspeed indicator**—an instrument in the cockpit that displays an aircraft's airspeed, typically in knots. (p. 323)

**air traffic control**—a management system for coordinating air traffic at airports and in the air. (p. 100)

**airway**—the path a plane follows in the sky. (p. 264)

**altimeter**—an instrument that tells the pilot how high the plane is. (p. 175)

**altimeter setting**—the station pressure—the barometric pressure at the location the reading is taken—reduced to sea level. (p. 328)

**altitude chamber**—a sealed room that reproduces a high-altitude environment. (p. 254)

**alveoli**—air sacs that are at the end of all the other passageways in the lungs. (p. 221)

**ambient pressure**—the pressure in the area immediately surrounding the aircraft. (p. 243)

**anemometer**—an instrument that measures wind speed. (p. 177)

**aneroid**—a sealed flat capsule made out of thin metal disks and emptied of all air. (p. 327)

**angle of attack**—the angle between the direction of the relative wind and the chord line of an airfoil. (p. 15)

**angular acceleration**—a simultaneous change in both speed and direction. (p. 232)

**anhedral**—a negative dihedral. (p. 40)

**anti-ice**—a system designed to prevent ice buildup on an aircraft structure. (p. 192)

**apex**—peak. (p. 273)

**arc**—a unit of measurement along a circle. (p. 267)

**arid**—hot and dry. (p. 203)

**aspect ratio**—a measure of how long and wide a wing is from tip to tip. (p. 39)

**atmosphere**—a blanket of air that surrounds Earth and consists of a mixture of gases. (p. 114)

**atmospheric pressure**—the weight of air molecules. (p. 120)

**attitude**—an aircraft's orientation, or angle, in relation to the horizon. (p. 56)

**attitude indicator**—an instrument that measures aircraft attitude in relation to the horizon. (p. 265)

**Automated Surface Observing System**—a weather reporting system that provides surface observations every minute via digitized voice broadcasts to aircraft using ground-to-air radio. (p. 178)

**automatic dependent surveillance-broadcast**—a device used in aircraft that repeatedly broadcasts a message that includes the aircraft's position and velocity. (p. 349)

**autonomous**—independent of outside control. (p. 102)

**azimuth**—a measurement of distance in degrees in the horizontal plane from a fixed point. (p. 341)

## B

**baffle**—a partition that changes the airflow direction. (p. 77)

**bank**—to roll or tilt sideways. (p. 62)

**Barany chair**—a spinning device that demonstrates the hazards of spatial disorientation. (p. 255)

**barometric pressure**—air pressure as measured by a barometer. (p. 135)

**biofuel**—a fuel made from plants. (p. 96)

**bleed air**—compressed air from the engine compressors. (p. 242)

**boundary layer**—the layer of air near a plane's surface. (p. 26)

**briefing**—the relay of information. (p. 185)

**bronchi**—big passageways at the bottom of the windpipe that get air to the lungs. (p. 221)

**bronchioles**—small passageways that branch off the bronchi inside the lungs. (p. 221)

**buffet**—vibrate. (p. 198)

## C

**cabin altitude**—cabin pressure equal to what it would be at the same altitude above sea level. (p. 128)

**calibrated airspeed**—the indicated airspeed corrected for installation error and instrument error. (p. 325)

**camber**—the curve of an airfoil. (p. 7)

**cannula**—a piece of plastic tubing that runs under the nose. (p. 246)

**capillaries**—very small blood vessels branching off arteries and veins. (p. 223)

**carbon monoxide**—a colorless, odorless, tasteless, and toxic gas. (p. 237)

**cartographers**—people who make charts and maps. (p. 272)

**ceiling**—the height above Earth's surface of the lowest layer of clouds. (p. 122)

**celestial navigation**—a method of finding your way using the stars and planets. (p. 282)

**center of gravity**—the average location of an object's weight. (p. 22)

**centrifugal effect**—the inertia that appears to fling a body away from the center of rotation. (p. 255)

**centrifuge**—a device that creates artificial gravity forces by spinning around a central point. (p. 258)

**chart projection**—the portrayal of Earth's spherical surface on a flat surface. (p. 272)

**chord line**—the straight line between the foremost and hindmost points of an airfoil viewed from the side. (p. 7)

**chronometer**—an instrument that keeps exceptionally accurate time. (p. 282)

**circulation**—the process of moving blood about the body. (p. 221)

**cirriform**—of or relating to all types of cirrus clouds. (p. 142)

**clear ice**—glossy, see-through ice formed by the relatively slow freezing of large, supercooled water droplets. (p. 197)

**cockpit voice recorder**—an instrument that records audio in the cockpit; also known as a "black box." (p. 194)

**cold front**—a boundary area that forms when a cold, dense, and stable air mass advances and replaces a warmer air mass. (p. 142)

**combustion**—the process of burning. (p. 50)

**compass rose**—a circle marked off in 360 degrees that shows direction expressed in degrees. (p. 270)

**composite**—made up of a combination of materials. (p. 89)

**compressive strength**—the ability to experience up to a certain amount of force that presses together without being crushed. (p. 104)

**compressor**—a device that compresses—or increases pressure on—air or gas. (p. 80)

**concave**—hollow and rounded like an ice cream scoop. (p. 42)

**condensation**—a change of state of water from a gas—water vapor—to a liquid. (p. 117)

**continental**—of or relating to land. (p. 135)

**contour lines**—lines on a map drawn between points of the same elevation. (p. 287)

**control column**—a device in the cockpit that a pilot uses to control pitch and roll; also known as a control yoke or a control or center stick. (p. 199)

**control stick**—a handle attached by cables, pulleys, or some other means to control surfaces for the purpose of controlling them. (p. 57)

**convection**—the atmospheric circulation that results when warm air rises and cooler air moves in to replace it. (p. 154)

**convective currents**—small areas of local circulation. (p. 154)

**convective precipitation**—weather resulting from a vertical exchange of heat and moisture. (p. 201)

**course**—the intended direction of flight in the horizontal plane measured in degrees from north. (p. 270)

**crankshaft**—a rod that converts a piston's linear up-and-down motion into circular motion. (p. 74)

**critical angle of attack**—the point at which a plane stalls. (p. 15)

**crosswinds**—winds that blow at an angle to the direction of travel. (p. 339)

**cumulonimbus clouds**—thunderstorms. (p. 123)

**cure**—to harden. (p. 104)

# D

**dead reckoning**—navigation based on computations using time, airspeed, distance, and direction. (p. 304)

**dead reckoning computer**—a device pilots use to solve equations associated with flight planning and navigation; also known as a flight computer. (p. 338)

**deadstick landing**—a type of landing made without power. (p. 193)

**decibel**—a measure of loudness. (p. 99)

**decompression**—the loss of cabin pressure. (p. 229)

**decompression sickness**—a condition resulting from exposure to low pressure that causes dissolved gases in the body to form bubbles. (p. 229)

**dehydration**—the critical loss of water from the body. (p. 238)

**deice**—to remove ice buildup from an aircraft structure, generally with a heated fluid. (p. 194)

**density altitude**—pressure altitude corrected for changes from standard temperature. (p. 329)

**deposition**—the process by which a gas changes to a solid without going through the liquid state. (p. 117)

**developable surface**—a geometric shape that doesn't stretch when flattened. (p. 276)

**deviation**—a magnetic compass error caused by local magnetic fields within aircraft. (p. 286)

**dew point**—the temperature at which air can hold no more moisture. (p. 119)

**diaphragm**—the principal muscle that helps the lungs draw in and expel air. (p. 222)

**differential pressure**—the difference in pressure between cabin pressure and atmospheric pressure outside the aircraft. (p. 243)

**dihedral angle**—the angle that a wing makes with the aircraft's horizontal. (p. 40)

**distance-measuring equipment**—an electronic navigation system that determines the number of nautical miles between an aircraft and a ground station or waypoint. (p. 343)

**ditch**—to land an aircraft on water after a loss of power or some other emergency. (p. 320)

**diuretic**—a drink or other substance that drains the body of needed water. (p. 238)

**downwash**—a downward flow of air. (p. 44)

**drag**—the pull or slowing effect of air on an aircraft. (p. 24)

**drift angle**—the angle between the heading and the track. (p. 308)

**dynamic pressure**—a type of pressure that is present when an aircraft is in motion. (p. 323)

## E

**ejection seat**—a seat that thrusts a pilot out of an aircraft in an emergency. (p. 251)

**emissions**—a discharge of gases. (p. 88)

**estimated time of arrival**—the time someone intends to arrive at a destination. (p. 295)

**Eustachian tube**—a tube leading from inside each ear to the back of the throat. (p. 227)

**evaporation**—the transformation of a liquid to a gaseous state, such as the change of water to water vapor. (p. 117)

**excess thrust**—the difference between thrust and drag. (p. 47)

**explosive decompression**—a change in cabin pressure that takes place faster than the lungs can decompress. (p. 244)

## F

**fix**—an aircraft's position in the sky over a checkpoint. (p. 305)

**flameout**—a loss of power after too much water or some other substance such as volcanic ash has extinguished or drowned an engine's heat source. (p. 192)

**flap**—a hinged device at a wing's trailing edge that produces lift. (p. 45)

**flight level**—the altitude at which an aircraft is flying. (p. 216)

**flight log**—a complete written record of a flight that includes notations on checkpoints, navigation aids, courses, and distances and elapsed time between checkpoints. (p. 296)

**flight management computer**—a computer system or software that lets aircrews electronically draw up and file flight plans. (p. 355)

**flight simulator**—a machine that imitates real-life situations and dangers that pilots may face. (p. 253)

**free fall**—descending without a device to slow the descent. (p. 231)

**fronts**—the boundary between two different air masses. (p. 135)

**full-pressure suit**—a type of suit that covers the entire body from head to toe, surrounding the wearer in a protective, contained environment similar to a pressurized cabin. (p. 248)

**fuselage**—the aircraft body. (p. 36)

## G

**general aviation**—private flights that are not commercial or military. (p. 347)

**geostationary**—to orbit at a speed and altitude that keeps satellites in the same place above the Earth at all times. (p. 181)

**g-force**—a measure of gravity's accelerative force. (p. 231)

**glide angle**—where the flight path intersects with the ground at an angle. (p. 39)

**glide slope indicator**—an instrument that vertically guides an aircraft on its final approach to a runway. (p. 331)

**gradient**—slope. (p. 64)

**great circle**—an imaginary circle on a sphere's surface whose center is the sphere's center. (p. 266)

**great circle navigation**—the shortest distance across a sphere's surface between two points on the surface. (p. 267)

**groundspeed**—an aircraft's rate of progress in relation to the ground. (p. 271)

**g-suit**—a piece of clothing that protects pilots from the effects of g-forces. (p. 233)

**gust fronts**—a rapid and sometimes drastic change in surface wind ahead of an approaching storm caused by cool air flowing out of a thunderstorm in a downdraft. (p. 163)

**gyroscope**—an instrument with a wheel spinning about a central axis that helps with stability. (p. 315)

## H

**heading**—the direction in which an aircraft nose is pointing during flight. (p. 271)

**headwinds**—wind blowing against the direction of travel. (p. 63)

**hertz**—the unit for measuring frequency. (p. 341)

**horizontal situation indicator**—a flight instrument that indicates an aircraft's position and direction in relation to the desired route. (p. 331)

**humidity**—the amount of water in the atmosphere at a given time. (p. 118)

**hyperventilation**—an abnormal increase in the volume of air breathed in and out of the lungs. (p. 226)

**hypoxia**—a state of too little oxygen in the body. (p. 224)

## I

**indicated airspeed**—the airspeed shown on the airspeed indicator. (p. 325)

**indicated altitude**—the altitude shown on the altimeter face. (p. 328)

**inertial navigation system**—a computer-based navigation system that tracks an aircraft's movements with signals produced by onboard instruments. (p. 314)

**infrared**—heat-seeking. (p. 203)

**instrument landing system**—an electronic system that provides an approach path to a specific runway. (p. 346)

**isogonic lines**—lines drawn across charts to connect points having the same magnetic variation. (p. 284)

## J

**jet propulsion**—a driving or propelling force in which a gas turbine engine provides the thrusting power. (p. 11)

**jet stream**—a strong current of air that generally sits atop the troposphere and flows from west to east. (p. 115)

## K

**kilohertz**—a frequency of 1,000 cycles per second. (p. 345)

**knot**—a nautical mile, or 1.15 statute miles (1.85 km) per hour. (p. 14)

# L

**lapse rate**—the rate at which temperature decreases with an increase in altitude. (p. 152)

**latitude**—a line north or south from Earth's equator and parallel to it. (p. 121)

**leeward**—the side facing away from the wind. (p. 146)

**legend**—a list of all symbols and their meanings on a given map. (p. 287)

**lift**—the upward force on an aircraft against weight. (p. 20)

**linear acceleration**—a change of speed in a straight line. (p. 232)

**longitude**—a measurement east or west of the prime meridian in degrees, minutes, and seconds of arc from 0 to 180 degrees. (p. 267)

**long-range navigation**—a radio navigation system that determines position by measuring the difference in time that the system receives pulses from two transmitters. (p. 348)

**lubber line**—a fixed line on a compass or other flight instrument aligned with an aircraft's longitudinal axis. (p. 331)

# M

**magnetic North Pole**—a spot on Earth where the magnetic field points vertically downward. (p. 284)

**magnetic South Pole**—a spot on Earth where the magnetic field points vertically upward. (p. 284)

**magnetos**—a unit that supplies electric current to the spark plugs. (p. 78)

**magnitude**—a measurable amount, such as a size, speed, or degree. (p. 20)

**maritime**—of or relating to the sea. (p. 135)

**mean camber line**—a reference line from leading to trailing edge drawn an equal distance from the upper and lower wing surfaces. (p. 41)

**megahertz**—a frequency of one million cycles per second. (p. 299)

**meteorologist**—a person who forecasts the weather. (p. 112)

**micro-UAV**—an aircraft that weighs as little as a few ounces or a few pounds. (p. 103)

**mixed ice**—a mixture of clear ice and rime ice formed when supercooled water droplets come in different sizes or the droplets mingle with snow or ice particles. (p. 197)

**monocoque**—a one-piece body structure that offers support and rigidity. (p. 37)

**multi-function display**—a small electronic screen at the center of the flight instrument panel that displays information such as traffic and weather. (p. 350)

# N

**nacelle**—a streamlined casing around an engine. (p. 89)

**nanotechnology**—the science and technology of building electronic circuits and devices from single atoms and molecules. (p. 104)

**nano-UAV**—a UAV so small it is invisible to the naked eye. (p. 104)

**National Airspace System**—the network of people and equipment that make up the US airspace and ensure aircraft safety. (p. 352)

**nucleus**—core. (p. 120)

## O

**occluded front**—a boundary area that forms when a fast-moving cold front catches up with a slow-moving warm front. (p. 144)

**orographic lift**—event that forces air to rise and cool. (p. 145)

**oxidizer**—a substance that includes oxygen to aid combustion. (p. 52)

## P

**parachute**—a large cloth device that slows descent. (p. 249)

**partial pressure**—the amount of pressure each gas applies individually. (p. 220)

**partial-pressure suit**—a close-fitting garment that covers most—but not all—of the body and creates pressure to support life. (p. 246)

**particulate matter**—material suspended in the air in the form of minute solid particles, such as dust, salt, or smoke particles. (p. 120)

**physiology**—the study of how the body functions. (p. 218)

**pilotage**—navigation by visual reference to landmarks or checkpoints. (p. 304)

**piston**—a metal device that moves back and forth inside a cylinder. (p. 71)

**pitch**—the up-and-down movement of the plane's nose. (p.15)

**pitch axis**—a line that starts at the center of gravity and runs from wingtip to wingtip. (p. 56)

**pitot tube**—a probe that senses temperature and relays airspeed information based on a combination of air pressures. (p. 196)

**positive pressure**—a forceful flow of oxygen that slightly overinflates the lungs. (p. 246)

**precipitation**—any or all forms of water particles that fall from the atmosphere and reach Earth's surface. (p. 119)

**pressure altitude**—the altitude indicated on the altimeter when the altimeter-setting window is adjusted to 29.92 inches of mercury. (p. 329)

**primary flight display**—a device that replaces traditional flight instruments with an electronic display of such data as airspeed, altitude, and horizon. (p. 350)

**prime meridian**—0 degrees longitude and is a line of longitude that runs through Greenwich, England. (p. 267)

**propellant**—a mix of fuel and oxidizer. (p. 52)

**propulsion**—to drive forward or to push forward. (p. 46)

**propulsion system**—an engine that generates thrust to move an aircraft forward. (p. 47)

**prototype**—a model. (p. 102)

**protractor**—a semicircular tool for measuring angles; it is also known as a plotter. (p. 311)

**psychrometer**—a weather instrument that uses two thermometers to measure relative humidity. (p. 176)

**pulmonary arteries**—blood vessels that carry blood from the heart to the organs and tissues. (p. 223)

**pulmonary veins**—blood vessels that carry blood from the organs and tissues to the heart. (p. 223)

## R

**radial**—an aircraft's position from a station. (p. 340)

**radial acceleration**—a change in direction. (p. 232)

**radio silence**—when no one is allowed to communicate by radio for security reasons. (p. 206)

**radio waves**—invisible electromagnetic waves in the portion of the electromagnetic spectrum longer than infrared light. (p. 179)

**rain shadow**—area on a mountain's leeward side with reduced precipitation because of warming air and decreasing clouds. (p. 146)

**rapid decompression**—a change in cabin pressure in which the lungs decompress faster than the cabin. (p. 244)

**reaction engine**—an engine that develops thrust by its reaction to a substance ejected from it; specifically, an engine that ejects a jet or stream of gases created by the burning of fuel within the engine. (p. 87)

**reflectivity**—the strength of a returned signal. (p. 179)

**relative bearing**—the angular difference between the aircraft heading and the direction to the station. (p. 345)

**relative humidity**—the actual amount of moisture in the air compared with the total amount of moisture the air could hold at that temperature. (p. 118)

**relative wind**—the motion of air as it relates to the aircraft within it. (p. 13)

**relief features**—symbols on a map that represent the elevation of physical features such as mountains, valleys, and cities. (p. 287)

**respiration**—breathing. (p. 221)

**rhumb line**—a line that runs across meridians at the same angle. (p. 276)

**rime ice**—brittle and frostlike ice formed by the instantaneous freezing of small, supercooled water droplets. (p. 197)

**rocket sled**—a rocket-powered vehicle that rides at ground level on tracks and is used for a variety of speed, acceleration, and deceleration tests. (p. 234)

**roll**—the up-and-down motion of an aircraft's wings. (p. 56)

**roll axis**—a line that begins at the center of gravity, is perpendicular to the yaw and pitch axes, and runs from nose to tail. (p. 56)

**rotate**—to raise an aircraft's nose wheel off the runway. (p. 211)

## S

**saturated**—as full of moisture as something can get. (p. 119)

**secant**—intersecting. (p. 273)

**situational awareness**—a pilot's understanding of where his or her aircraft is in relation other aircraft, weather, location, and airspace regulations. (p. 351)

**slat**—a movable, hinged part that pivots down to generate more force. (p. 45)

**slide rule**—a device with two moving parts used to make calculations. (p. 339)

**smart bomb**—a precision weapon guided by GPS, INS, and/or laser technology to its target. (p. 353)

**spark plugs**—parts that ignite the fuel-air mixture. (p. 74)

**spatial disorientation**—the lack of knowing an aircraft's position, attitude, and movement. (p. 235)

**spoiler**—a small, flat plate that attaches to the tops of the wings with hinges; it increases drag. (p. 45)

**squall line**—a narrow band of active thunderstorms. (p. 144)

**stall**—a rapid reduction in lifting force caused by exceeding the critical angle of attack. (p. 15)

**static line**—a cord attached to the aircraft and pack that automatically deploys the parachute as a jumper exits the aircraft. (p. 250)

**static pressure**—a type of pressure that is always present whether an aircraft is moving or at rest. (p. 323)

**stationary front**—a boundary area that forms between two air masses of relatively equal strength. (p. 144)

**stratiform**—of or pertaining to all types of stratus clouds. (p. 141)

**sublimation**—the process by which a solid changes to a gas without going through the liquid state. (p. 117)

**symmetrical airfoil**—an airfoil in which the upper surface and the lower surface have the same shape. (p. 41)

## T

**tactical air navigation**—an electronic navigation system used by military aircraft that provides distance and direction. (p. 344)

**tailwinds**—wind blowing from behind. (p. 63)

**tangent**—touching. (p. 273)

**taxi**—to move slowly on the ground before takeoff and after landing. (p. 63)

**temperature inversion**—an increase in temperature with altitude. (p. 158)

**tensile strength**—the ability to experience up to a certain amount of tension or stretching without tearing. (p. 104)

**terminal velocity**—the point at which drag on an object equals the force of gravity and the object stops accelerating. (p. 231)

**terrain**—the landscape, anything from mountains to coastlines and lakeshores. (p. 145)

**theory**—a set of principles scientists use to explain something they have observed. (p. 6)

**thermal**—the property of being warm or hot. It also refers to the rising current of warm air that causes turbulence. (p. 154)

**thrust**—a forward force that moves an aircraft through the air. (p. 24)

**tissue**—the cells and any substance found between cells that make up muscles, skin, or other body parts. (p. 222)

**TOGA**—a command that crews use to abort a landing; it stands for "takeoff/go around." (p. 202)

**torque**—a twisting or rotating force. (p. 38)

**trachea**—the windpipe, which moves air drawn in through the nose or mouth into the lungs. (p. 221)

**track**—the actual path taken over the ground in flight. (p. 271)

**transatlantic**—across the Atlantic Ocean. (p. 132)

**transcontinental**—coast-to-coast. (p. 132)

**transonic**—aircraft speeds nearing the speed of sound. (p. 92)

**trim**—balance. (p. 24)

**trough**—a long stretch of low pressure. (p. 162)

**true airspeed**—the calibrated airspeed corrected for altitude and temperature. (p. 325)

**true altitude**—an aircraft's vertical distance above sea level. (p. 329)

**true north**—a meridian line on a chart that runs to the North Pole. (p. 270)

**tunnel vision**—a condition in which the edges of your sight gray out to a point where you only have a narrow field of vision straight ahead. (p. 225)

## V

**variation**—the difference between true direction and magnetic direction. (p. 284)

**vector**—an aircraft's direction and speed. (p. 271)

**velocity**—speed and direction. (p. 20)

**viscosity**—stickiness, or the extent to which air molecules resist flowing past a solid or other object. (p. 26)

**visual flight rules**—also referred to simply as VFR, these rules are in play when the weather is clear enough that a pilot can operate an aircraft by sight as opposed to relying on aircraft instruments. (p. 141)

**vortex**—a kind of whirlpool of wind or water with a vacuum at its center. (p. 26)

## W

**warm front**—a boundary area that forms when a warm air mass advances and replaces a colder air mass. (p. 140)

**waypoints**—geographical locations in latitude and longitude along a route to mark a route and follow an aircraft's advance along that desired route. (p. 315)

**weight**—the downward force that directly opposes lift. (p. 22)

**windblast**—the effect of air friction on a pilot who ejects from a high-speed aircraft. (p. 251)

**wind chill**—the cooling effect of wind on surfaces. (p. 197)

**wind correction angle**—the angle between an aircraft's desired track and the heading needed to keep the aircraft flying over its desired track. (p. 309)

**wind shear**—a sudden, drastic shift in wind speed, direction, or both that may take place in the horizontal or vertical plane. (p. 123)

**wind shifts**—abrupt changes in the direction of wind by 45 degrees or more in less than 15 minutes with winds of at least 10 knots. (p. 139)

**wind triangle**—a method for dealing with the effect of wind on flight. (p. 310)

**wind tunnels**—controlled spaces for testing airflow over a wing, aircraft, or other object. (p. 8)

**windward**—the side facing the wind. (p. 145)

**winglet**—a plate attached to a wingtip or a wingtip bent upward. (p. 41)

## Y

**yaw**—the side-to-side motion of an aircraft's nose. (p. 56)

**yaw axis**—a line that starts at the center of gravity, runs perpendicular to the wings, and is directed toward the aircraft's lower surface. (p. 56)

# Index

## L

Lake effect snow, 146
Lambert conformal conic, 274
Land breeze, 147, 155
Landing
    deadstick, 193
    descent and, 65–66
    instrument landing system
        (ILS), 346
Landing threshold, 346
Lapse rate, 152
Latent heat of evaporation, 117
Latitude, 121, 267–269
Law of inertia, 10
Layers of Earth's atmosphere,
    114–116, 126–130
Leading edge of an airfoil, 7
Leading edge cuff, 61
Leading edge flaps, 61
Leans, 236
Leeward side, 146
Legend for map, 287
Lenticularus clouds, 126
Lift, 20–22, 198
Lightning, 165–166
Lindbergh, Charles, 132–134, 308
Linear acceleration, 232
Localizer, 346
Longitude, 267–269
Long-range navigation
    (LORAN), 348
Lost, procedures to perform
    when, 298–300
Lost Patrol story, 280–281
Low altitude charts
    (IFR enroute), 292
Low clouds, 122
Lowe, Richard, 54–55
Low-level wind shear alert
    system, 158–159, 201
Low-pressure systems, 137–139
Low-speed flight, 26
Lubber line, 331

## M

Mach speed, 83
Magnetic compass, 283–284
Magnetic directions, 284
Magnetic field, 114
Magnetic North Pole, 284
Magnetic South Pole, 284
Magnetos, 78–79
Magnitude, 20

Map elements, 287–288
Map scale, 287
Maritime air mass, 135
Mature stage of thunderstorm,
    161
Mean camber line, 41–42
Mechanical system of
    reciprocating engines, 74–77
Mechanical turbulence, 156–157
Mercator projections, 273
Meridians, 267
Mesopause, 116
Mesosphere, 116
METARs (aviation routine
    weather reports), 186
Meteorologist, 112
MFD (multi-function display),
    350
Microburst, 200–203
Micro-UAV, 103–104
Middle clouds, 123
Middle marker, 346
Military operation airspace, 291
"Miracle on the Hudson,"
    320–322
Mixed ice, 197
Modification, 137
Monocoque shell, 37
Montgolfier brothers, 6
Morse code, 264
Motion. *See* Aircraft motion
    and control
Motion sickness, 237
Movable slats, 60
Multi-function display (MFD),
    350
Multi-tube-type noise
    suppressor, 87
Murphy's law, 234

## N

Nacelle, 89
Nanotechnology, 104
Nano-UAV, 104
National Airspace System
    (NAS), 352
National security areas, 291
National Transportation Safety
    Board, 195
Navigational aids, 280–301.
    *See also* Dead reckoning
    charts, 288–292
    clock and compass functions,
    282–287

Navigational aids, *continued*
    flight planning, 292–295
    lost, procedures to perform
        when, 298–300
    Lost Patrol story, 280–281
    map elements, 287–288
    preflight plan, drafting of,
        296–298
Navigational elements, 262–278
    chart projections, 272–276
    Earth's size and shape
        relationship to, 266–267
    history of, 263–266
    latitude and longitude
        correlation to flight position,
        267–269
    navigation direction
        determination, 270–271
    projections, problems
        associated with, 276–277
    World War I incident, 262
Navigation direction
    determination, 270–271
Navigation technology, 336–357.
    *See also* Flight instrumentation
    computer flight-planning
        tools, 354–356
    current developments in,
        348–352
    dead reckoning computers,
        338–340
    GPS and inertial navigation,
        353–354
    GPS jammers, 336
    plotter, 298, 311, 337–338
    radio aids, 340–348
NDB (nondirectional radio
    beacon), 345
Negative dihedral, 40
Net torque, 63
Newton, Sir Isaac, 10
Newton's law of gravity, 23
Newton's laws of motion,
    10–11
NEXRAD, 183–184
NextGen (Next Generation),
    348–349
Nimbostratus clouds, 122
Nimbus clouds, 126
Noise-reduction innovations,
    98–99
Noise suppressors, 86–87
Nondirectional radio beacon
    (NDB), 345
Nucleus, 120